MW01273214

Aquatic ecosystems can be adversely affected by human activities such as intensification of agricultural activity, or erosion and sedimentation due to irrigation projects, or groundwater pollution and eutrophication, and so on. Interfaces, or ecotones, between terrestrial and aquatic ecosystems have an essential role in the movement of water and materials throughout the landscape. Ecotones are zones where ecological processes are more intense and resources more diversified. They are also zones which react quickly to human influences and changes of environmental variables.

This volume is derived from an international conference of the Ecotone project, established under the dual responsibility of the UNESCO International Hydrological Programme and the Man and Biosphere Programme, and summarizes the results of the subnetwork activities devoted to the study of groundwater/surface water interactions. The contributors were carefully selected on their international scientific reputation to represent the multidisciplinary viewpoints of hydrologists, biologists and ecologists. Topics covered include interrelationships between surface water and groundwater in riparian forests, wetlands, areas surrounding lakes and alluvial flood plains.

This book defines strategies for the integration of data obtained by different disciplines in order to provide a scientific basis for the sound ecological management of water resources leading to sustainable development of the environment. It addresses areas of active research in hydrology and biology, and is therefore aimed towards researchers, water resource project managers, and policy makers.

Groundwater/Surface Water Ecotones: Biological and Hydrological Interactions and Management Options

INTERNATIONAL HYDROLOGY SERIES

The **International Hydrological Programme** (IHP) was established by the United Nations Educational, Scientific and Cultural Organisation (UNESCO) in 1975 as the successor to the International Hydrological Decade. The long-term goal of the IHP is to advance our understanding of processes occurring in the water cycle and to integrate this knowledge into water resources management. The IHP is the only UN science and educational programme in the field of water resources, and one of its outputs has been a steady stream of technical and information documents aimed at water specialists and decision-makers.

The **International Hydrology Series** has been developed by the IHP in collaboration with Cambridge University Press as a major collection of research monographs, synthesis volumes and graduate texts on the subject of water. Authoritative and international in scope, the various books within the Series all contribute to the aims of the IHP in improving scientific and technical knowledge of fresh water processes, in providing research know-how and in stimulating the responsible management of water resources.

INTERNATIONAL HYDROLOGY SERIES

Groundwater/Surface Water Ecotones: Biological and Hydrological Interactions and Management Options

Edited by
Janine Gibert *University of Lyon*
Jacques Mathieu *University of Lyon*
Fred Fournier *UNESCO, Division of Water Sciences*

CAMBRIDGE
UNIVERSITY PRESS

PUBLISHED BY THE PRESS SYNDICATE OF THE UNIVERSITY OF CAMBRIDGE
The Pitt Building, Trumpington Street, Cambridge CB2 1RP, United Kingdom

CAMBRIDGE UNIVERSITY PRESS
The Edinburgh Building, Cambridge CB2 2RU, United Kingdom
40 West 20th Street, New York, NY 10011-4211, USA
10 Stamford Road, Oakleigh, Melbourne 3166, Australia

First published 1997

Printed in the United Kingdom at the University Press, Cambridge

Typeset in 9½/13pt Times

A catalogue record for this book is available from the British Library

Library of Congress Cataloguing in Publication data

Groundwater/surface water ecotones : biological and hydrological
  interactions and managment options / [edited by] Janine Gibert,
  Jacques Mathieu, Fred Fournier.
      p.   cm. – (International hydrology series)
    ISBN 0-521-57254-1 (hbk.)
    1. Groundwater ecology.   2. Hydrogeology.   3. Groundwater.
  I. Gibert, Janine.   II. Mathieu, Jacques.   III. Fournier, Frédéric.
  IV. Series.
  QH541.5.G76G77   1997
  574.5′2632–dc20   96–18933 CIP

ISBN 0 521 57254 1 hardback

Contents

List of authors

A. BOTHAR
Hungarian Danube Research Station, Hungarian Academy of Sciences, H-2131 Göd, Hungary

C. BRADLEY
School of Geography, Birmingham University, Edgbaston, Birmingham B15 2TT, England

A.G. BROWN
Department of Geography, Leicester University, Leicester LE1 7RH, England

R. CARBIENER
Laboratoire de Botanique et Ecologie végétale, CEREG URA CNRS 95, Institut de Botanique, 28 rue Goethe, F-67083 Strasbourg cedex, France

B. CAUSSADE
Institut de Mécanique des Fluides de Toulouse, URA CNRS 0005, Allée du Professeur Camille Soula, 31400 Toulouse, France

D.L. CORRELL
Smithsonian Environmental Reasearch Center, P.O. Box 28, Edgewater, Maryland 21037, USA

D.L. DANIELOPOL
Limnological Institute, Austrian Academy of Sciences, A-5310 Mondsee, Austria

T.V. DIKARIOVA
Water Problems Institute, Russian Academy of Sciences, 10 Novaya Basmannaya str., P.O. Box 524, Moscow 107078, Russia

M.-J. DOLE-OLIVIER
Université Lyon 1, URA CNRS 1974, Ecologie des Eaux Douces et des Grands Fleuves, HydroBiologie et Ecologie Souterraines, 43 Bd du 11/11/1918, 69622 Villeurbanne cedex, France

J.E. DREHER
Breitenfurterstrasse 458, A-1236 Vienna, Austria

I. EGLIN
Laboratoire de Botanique et Ecologie végétale, CEREG URA 95 CNRS, Institut de Botanique, 28 rue Goethe, F-67083 Strasbourg cedex, France

B.K. ELLIS
Flathead Lake Biological Station, University of Montana, Polson, Montana 59860, USA

F. FOURNIER
UNESCO, Division des Sciences Ecologiques, 7 Place de Fontenoy, 75700 Paris, France

O. FRÄNZLE
Projektzentrum Ökosystemforschung, Christian-Albrechts-Universität zu Kiel, Schauen-burger str. 112, 24118 Kiel, Germany

J. GIBERT
Université Lyon 1, URA CNRS 1974, Ecologie des Eaux Douces et des Grands Fleuves, HydroBiologie et Ecologie Souterraines, 43 Bd du 11/11/1918, 69622 Villeurbanne cedex, France

T.A. GLUSHKO
Water Problems Institute, Russian Academy of Sciences, 10 Novaya Basmannaya str., P.O. Box 524, Moscow 107078, Russia

C.C. HAKENKAMP
Department of Zoology, University of Maryland at College Park, 1200 Zoology-Psychology Building, College Park, MA 20742-4415, USA

T.W. HOBMA
Vrije Universiteit Amsterdam, Faculty of Earth Sciences, De Boelelaan 1085, 1081 HV Amsterdam, The Netherlands

G. JACKS
Land and Water Resources, Royal Institute of Technology, S-100 44 Stockholm, Sweden

A. JOELSSON
Halland County Board, S-310 86 Halmstad, Sweden

U. JOHANSSON
Tönnersjöheden Experimental Forest, Swedish University of Agricultural Sciences, S-310 38 Simlängsdalen, Sweden

T.E. JORDAN
Smithsonian Environmental Reasearch Center, P.O. Box 28, Edgewater, Maryland 21037, USA

W. KLUGE
Projektzentrum Ökosystemforschung, Christian-Albrechts-Universität zu Kiel, Schauen-burger str. 112, 24118 Kiel, Germany

M. LEICHTFRIED
Biological Station Lunz, Institute of Limnology, Austrian Academy of Sciences, A-3293 Lunz am See, Austria

S. LIU
Department of Hydrology, Institute of Geography, Chinese Academy of Sciences, Beijing 100101, People's Republic of China

G. MAGNIEZ
Laboratoire de Biologie Animale et Générale, Université de Bourgogne, 6 Bd Gabriel, 21000 Dijon, France

F. MALARD
Université Lyon 1, URA CNRS 1974, Ecologie des Eaux Douces et des Grands Fleuves, HydroBiologie et Ecologie Souterraines, 43 Bd du 11/11/1918, 69622 Villeurbanne cedex, France

L. MARIDET
CEMAGREF, Division BEA, Hydroécologie Quantitative, BP 220, 69009 Lyon, France

P. MARMONIER
Université de Savoie, GRETI, 73376 Le Bourget du Lac, France

J. MATHIEU
Université Lyon 1, URA CNRS 1974, Ecologie des Eaux Douces et des Grands Fleuves, HydroBiologie et Ecologie Souterraines, 43 Bd du 11/11/1918, 69622 Villeurbanne cedex, France

A.M.J. MEIJERINK
International Institute for Aerospace Surveyard Earth Sciences (ITC), Enschede, 7500 AA, The Netherlands

F. MÖSZLACHER
Institute of Zoology, University of Vienna, Althanstrasse 14, A-1090 Vienna, Austria

R.J. NAIMAN
Center for Streamside Studies, AR-10, University of Washington, Seattle, Washington 98195, USA

A.C. NORRSTRÖM
Land and Water Resources, Royal Institute of Technology, S-100 44 Stockholm, Sweden

J. NOTENBOOM
RIVM, Laboratory of Ecotoxicology, P.O. Box 1, Bilthoven 3720 BA, The Netherlands

N.M. NOVIKOVA
Water Problems Institute, Russian Academy of Sciences, 10 Novaya Basmannaya str., P.O. Box 524, Moscow 107078, Russia

J. OLAH
Fisheries Research Institute, H - 5541 Szarvas, Hungary

J. PASTOR
Natural Resources Research Institute, University of Minnesota, Duluth, Minnesota 55811, USA

M. PHILIPPE
CEMAGREF, Division BEA, Hydroécologie Quantitative, BP 220, 69009 Lyon, France

G.C. POOLE
Center for Streamside Studies, AR-10, University of Washington, Seattle, Washington 98195, USA

P. POSPISIL
Institute of Zoology, University of Vienna, Althanstrasse 14, A-1090 Vienna, Austria

M. PUSCH
Institut für Gewässerökologie und Binnenfischerei, Müeggelseedamm 260, D-12587 Berlin, Germany

B. RATH
Hungarian Danube Research Station, Hungarian Academy of Sciences, H-2131 Göd, Hungary

F. ROBACH
Laboratoire de Botanique et Ecologie végétale, CEREG URA 95 CNRS, Institut de Botanique, 28 rue Goethe, F-67083 Strasbourg cedex, France

U. ROECK
Laboratoire de Botanique et Ecologie végétale, CEREG URA 95 CNRS, Institut de Botanique, 28 rue Goethe, F-67083 Strasbourg cedex, France

R. ROUCH
Centre de Recherches Souterraines et Edaphiques, Laboratoire Souterrain, CNRS, F-09200 Moulis, France

M. RULIK
Department of Ecology, Faculty of Natural Sciences,
Palacky University, Svobody 26, 771 46 Olomouc, Czech
Republic

J.-M. SANCHEZ-PEREZ
Laboratoire de Botanique et Ecologie végétale, CEREG
URA 95 CNRS, Institut de Botanique, 28 rue Goethe, F-
67083 Strasbourg cedex, France

P.E. SCHMID
Biological Station Lunz, Institute of Limnology, Austrian
Academy of Sciences, A-3293 Lunz am See, Austria

J.M. SCHMID-ARAYA
Biological Station Lunz, Institute of Limnology, Austrian
Academy of Sciences, A-3293 Lunz am See, Austria

J. SIMONS
Environmental Protection Agency, 401 M Street SW,
Washington DC 20460, USA

J. A. STANFORD
Flathead Lake Biological Station, University of Montana,
Polson, Montana 59860, USA

A. SZÖLLÖSI-NAGY
UNESCO, Division of Water Sciences, 7 Place de
Fontenoy, 75700 Paris, France

C. THIRRIOT
Institut de Mécanique des Fluides de Toulouse, URA 0005
CNRS, Allée du Professeur Camille Soula, 31400 Toulouse,
France

P. TORREITER
Institute of Zoology, University of Vienna, Althanstrasse
14, A-1090 Vienna, Austria

M. TRÉMOLIÈRES
Laboratoire de Botanique et Ecologie végétale, CEREG
URA 95 CNRS, Institut de Botanique, 28 rue Goethe, F-
67083 Strasbourg cedex, France

M.J. TURQUIN
Université Lyon 1, URA CNRS 1974, Ecologie des Eaux
Douces et des Grands Fleuves, HydroBiologie et Ecologie
Souterraines, 43 Bd du 11/11/1918, 69622 Villeurbanne
cedex, France

M. H. VALETT
Department of Biology, University of New Mexico,
Albuquerque, New Mexico 97131, USA

V. VANEK
VVB VIAK Consulting Engineers, Geijersgaten 8, S-216 18
Malmö, Sweden
Université Lyon 1, URA CNRS 1974, Ecologie des Eaux
Douces et des Grands Fleuves, HydroBiologie et Ecologie
Souterraines, 43 Bd du 11/11/1918, 69622 Villeurbanne
cedex, France

Ph. VERVIER
CERR - CNRS, 29 rue Jeanne Marvig, 31055 Toulouse,
France

N.J. VOELZ
Department of Biology, St. Cloud State University, St.
Cloud, Minnesota 56301, USA

J.V. WARD
Department of Biology, Colorado State University, Fort
Collins, Colorado 805 23, USA

J.-G.WASSON
CEMAGREF, Division BEA, Hydroécologie Quantitative,
BP 220, 69009 Lyon, France

D.E. WELLER
Smithsonian Environmental Reasearch Center, P.O. Box 28,
Edgewater, Maryland 21037, USA

W. VAN WIJNGAARDEN
International Institute for Aerospace Surveyard Earth
Sciences (ITC), Enschede, 7500 AA, The Netherlands

R.C. WISSMAR
School of Fisheries, Wh-10, and Center for Streamside
Studies, AR-10, University of Washington, Seattle,
Washington 98195, USA

I.N. ZABOLOTSKY
Water Problems Institute, Russian Academy of Sciences, 10
Novaya Basmannaya str., P.O. Box 524, Moscow 107078,
Russia

V.S. ZALETAEV
Water Problems Institute, Russian Academy of Sciences, 10
Novaya Basmannaya str., P.O. Box 524, Moscow 107078,
Russia

Preface

When the UNESCO research programme on Man and the Biosphere was launched in 1971, aquatic ecosystems immediately became the subject of a special project due to the important changes that can be produced in them by human activities. In the beginning, attention was focused on the consequences of intensification of agriculture : erosion and sedimentation, groundwater pollution, eutrophication, and other adverse events.

After 15 years, the first phase of the project was concluded with a final meeting devoted to the use of scientific information in understanding the impact of land use on aquatic ecosystems. It highlighted the major role of interfaces (ecotones) between terrestrial and aquatic ecosystems in the regulation of biogeochemical cycles and in the structuring of landscape mosaics. Viewing aquatic systems as a collection of resource patches separated by ecotones allows to determine the relative importance of upstream - downstream linkages, lateral linkages and vertical linkages to be examined, and provides a conceptual framework for the understanding of factors regulating the exchange of energy and materials between identifiable resource patches.

Therefore, it was recommended by this meeting that UNESCO's future work on ecosystems should emphasize the in-depth study of ecotones, their management and restoration. Thus the Ecotone project was established under the double responsability of the International Hydrological Programme and the Man and Biosphere Programme with a general objective to determine the management options for the conservation and restoration of the land/inland water ecotones through increased understanding of ecological processes. A global network of cooperative research sites has been developed to study such ecotones like riparian forests, wetlands, areas surrounding lakes, alluvial plains, flood plains, as well as two specific fields of major interest : fish populations and land/inland-water ecotones, and interrelationships between surface water and groundwater.

At the end of a first phase of the Ecotone project, time arrived to synthesize the results of activities. The present volume is born from the efforts made by the surface water / groundwater ecotones network and materialized by an International Conference. We should like to express our thanks here to the leaders of the network as well as to the University of LYON 1, France, and to the Professor Janine GIBERT for the organization of the International Conference during which all the participants in the network have had the opportunity to present and compare the results of their studies and to propose future prospects.

Fred FOURNIER   Andras SZOLLOSI-NAGY

I

Introduction

1 The groundwater/surface water ecotone perspective: state of the art

J. GIBERT*, F. FOURNIER** & J. MATHIEU* EDITORS

* *Université Lyon 1, URA CNRS 1974, Ecologie des Eaux Douces et des Grands Fleuves, Hydrobiologie et Ecologie Souterraines, 43 Bd du 11 novembre 1918, 69622 Villeurbanne cedex, France*

** *UNESCO, Division des Sciences Ecologiques, 7 Place de Fontenoy, 75700 Paris, France*

ABSTRACT   This paper considers the increased interest in groundwater/surface water ecotones. That clearly appeared within the framework of the UNESCO/MAB and IHP projects on the role of land/inland water ecotones in landscape management and restoration. Groundwater/Surface water ecotones are transition zones, the limits between very contrasted systems. At different space and time scales they provide, favour, filter or stop exchanges and they can also modify interactions between ecosystems. This paper outlines the content and the structure of the book.

BACKGROUND

January 1991 marks the official start of the joint project between the Man and the Biosphere Programme (MAB) and the International Hydrological Programme (IHP) of UNESCO on the role of land/inland water ecotones in landscape management and restoration. The launching of this project was due to the increasing awareness of the important role played by ecotones between terrestrial and aquatic systems in the landscape and in natural resource management (Di Castri *et al.*, 1988; Naiman & Décamps, 1990; Holland & Risser, 1991; Hansen & Di Castri, 1992). They play an essential role as controls for the movement of water and materials throughout the landscape. They are zones where ecological processes are more intense and resources more diversified. The interface favours species dispersal. They are also zones which react quickly to human influences and changes in environmental variables.

Considering the key role of land/inland water ecotones, UNESCO/MAB and IHP launched a collaborative research project with the aim of determining the management options for the conservation and restoration of land/inland-water ecotones through increased understanding of ecological processes. Within this framework, a sub-network was devoted to groundwater/surface water ecotones.

In our understanding of the structure and function of freshwater ecosystems it is now increasingly realized that the surface and subsurface aquatic ecosystems interact in various ways. A wide range of contacts exist between the two environments. Within this diversity in mind four major characteristics of these ecotones emerge: 1) their elasticity in space and time, 2) their permeability to fluxes, 3) their frequent intermediate biodiversity and 4) their connectivity (Gibert *et al.*, 1990). These basic characteristics and the questions raised in previous studies justify the continuing research into groundwater/surface water ecotones, and led the UNESCO /MAB and IHP Ecotones Project to organize a special workshop on this subject in Lyon in July 1993.

INTRODUCTION

In the water cycle, the contact zones between surface ecosystems and groundwater ecosystems are often ignored, even kept out. Their importance seems negligible in comparison with the two adjacent worlds: the surface world and the underground world. In addition their study increases the complexity of the approaches to look at the water problems in the environment as the level of knowledge is quite different between surface water and groundwater. Groundwater is the hidden and less known phase of the continental waters.

However, an increasing number of scientists are paying more and more attention to the waterbodies interfaces as well as to land/water interfaces as they are convinced that they play a determining role in the ecosystem evolution and that their extensive study may have in the future a great practical

3

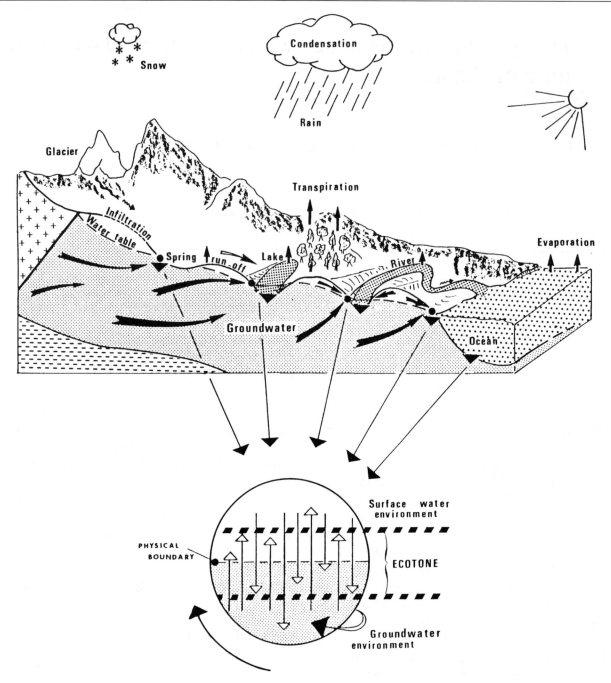

Fig. 1 Groundwater/surface water ecotones in the mountain and plain landscapes (from Gibert, 1991).

and theoretical value, for the following reasons. They are sites where intense hydraulic exchanges exist and where the bio-geochemical activity is higher than in the adjacent systems that influence the quality of the water flowing through the interface. The human impacts on these contact zones are increasing and their effects are often adverse. Their sensitivity and their role in global change are being increasingly questioned. The interfaces are very exciting study subjects because of their complexity. In fact the groundwater/surface water ecotone concept is more complex than previously thought.

The groundwater/surface water ecotones are existing in many different types of environment and in all countries (Fig. 1). Their starting point is the rainfalls which are distributed over the land in two ways: by infiltration when the water percolates through the soil to become groundwater and by surface runoff when the water is collected by the hydrological network. The surface runoff has close links with the under-flow and the groundwater either along the river sides or in the bottom of rivers or lakes (Fig. 2). The main characteristics of these interfaces are their great variety of elasticity, permeability, biodiversity and connectivity (Gibert et al., 1990).

Having recognized that the ecotones have fundamental characteristics and before evaluating their main functions, it

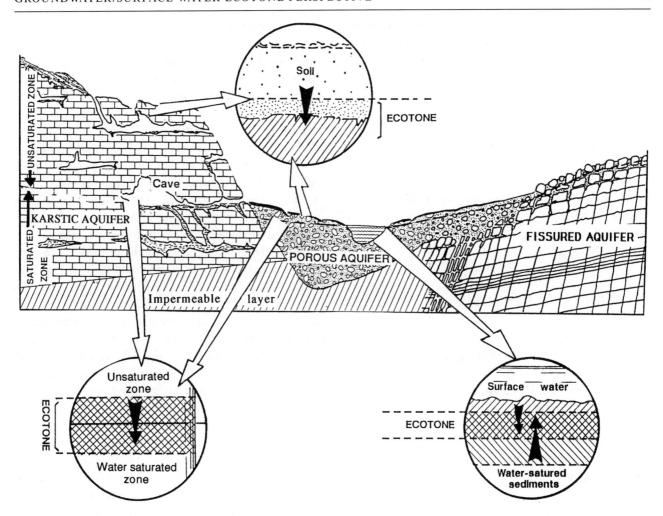

Fig. 2 Three types of groundwater/surface water ecotones (vertical and horizontal types): soil-karstic aquifer interface (epikarst), spring in karstic and porous aquifer and stream underflow (adapted from Gibert *et al.*, 1990).

is necessary to underline the problem of the spatial boundaries of these ecotones. When a hydrologist or hydrogeologist delimits the water compartments, he considers surface water flow, water/sediment interface and groundwater. It would never occur to a hydrogeologist to delimit his systems from a level which is, for example, one meter below the surface because inputs and outputs take place at this interface. For a groundwater ecologist, inputs of water, organic matter and energy into the groundwater system come from the interface to deep water. A stream ecologist, to define his system, incorporates a part of the sediment of the surface ecosystem, then a part of the alluvial aquifer. Many biotic and abiotic tracers can be used in attempts to delineate the transition zone between surface and groundwater ecosystems but they provide many different responses (Gibert *et al.*, 1990). Hence considerable spatial and temporal variations may exist and it is impossible to find a unanimous definition for the exact limits of the interface between surface water and ground-

water because of the different points of view of researchers. Consequently the solution is to think in terms of a dynamic transition zone (see Vervier *et al.*, 1992; Stanford & Ward, 1993, Gibert *et al.*, 1994). In any case, these problems of spatial delimitation must not hold up progress towards a better understanding of the processes within this zone. In this book, the focus will be placed more on the internal and external processes occuring in these ecotones than on their limits.

OBJECTIVES OF THE BOOK

Crucial questions concern the structure, function/dysfunction and evolution of groundwater/surface water ecotones as well as their use in management of terrestrial and aquatic environments, for example, adaptability, buffer capacities to resist and level out changes, cybernetic control and the emergence of spatial and temporal patterning, management, etc. At present, we are not able to glimpse answers to questions such as: how does the possibility of self-organization arise within these ecotones? What are the controls and the pressures exerted by the adjacent ecosystems on the ecotones and what is their essential nature?

The present book is a presentation and an evaluation of research directions on interactions between surface water and groundwater environments within a multidisciplinary framework, and a statement of the points of view of hydrologists, biologists and ecologists. Emphasis has been placed on the development of comparative studies of structure, function and succession of the interfaces between surface water and groundwater as well as of hydrological and biological mechanisms and their interactions. The aim of this book is also to define strategies for the integration of results and data obtained by the different disciplines in order to provide a scientific basis for the sound ecological management of water resources leading to sustainable development of the environment.

CONTENT AND STRUCTURE OF THE BOOK

The content of this book is wide-ranging and includes all the main mechanisms and processes – viewed at different space and time scales – that take place in the groundwater/surface water ecotones. It provides an up-to-date evaluation of groundwater/surface water ecotones. It is composed of 28 papers grouped into three sections excluding this introduction. Each section begins with a paper that provides an overview of the given topic.

Section II: Function of groundwater/surface water interfaces

Section II is concerned with the structure, succession and properties of these ecotones. It provides the scientific basis of the book through an examination of the physical and biological processes related to the hydrological cycle.

As an introduction to this section Danielopol *et al.* recall the concept of a global aquifer/river system viewed by hydrogeologists and show that the structure of surface/groundwater organism assemblages depends on the boundary conditions of the groundwater system and on the initiating conditions offered by the ecological tolerances of each species. Ecotonal animal assemblages are in a permanent dynamic state influenced largely by the hydrologic regime and associated factors such as oxygen. Their studies contribute largely to general ecology debates.

The following six chapters (chap. 3 Schmid; chap. 4. Schmid-Araya; chap. 5 Ward & Voelz; chap. 6 Malard *et al.*; chap. 7 Pusch; chap. 8 Bothar *et al.*) focus on biodiversity and on the spatial and temporal dynamics of interstitial micro-meio-mesofaunal assemblages and macrophyte organization of communities. Communities are organized in patches and there is a high probability for a community assembly, such as

insect larvae (chironomids), to be governed by random patch formation; this reduces the probability of strong competition among species. Highly fluctuating patterns in time of faunal densities may be promoted by a fluctuating and unpredictable environment. A marked faunal gradient is observed on a scale of meters across the groundwater/surface water ecotone and a differential permeability effect can be expressed on the invertebrate population structure. Community respiration in the ecotone seems to be much higher than previously estimated and it is of the order of epibenthic community respiration. At the scale of the landscape, connectivity between side arms and the main channel has been demonstrated with macrophytes and crustaceans.

The spatial and temporal patterns of matter transport, retention, uptake and transformation in the ecotone are dealt with in five chapters (chap. 9 Maridet *et al.*, chap. 10 Leichtfried; chap. 11 Rulik; chap. 12 Trémolières *et al.*). The ecotone, a temporary or permanent sink of organic matter from the drainage basin, is determined by such diverse and interactive external factors as riparian vegetation, channel and bank morphology, flood regime of the surface water, with seasonal variations of retention and storage. Moreover the field ecotone produces additional nutrient fluxes, for example DOC increases, resulting in a rise in microbial nitrate and oxygen consumption. The distribution of bacteria indicates a very high metabolic rate. The soil/vegetation system and macrophytes act as a self-purification filter for the surface and groundwater. The importance of the deposition function in the ecotone has been emphasized in this process of self-purification.

The common theme of the following five chapters (chap. 13, Bradley & Brown; chap. 14 Miejerink & van Wijngaarden; chap.15 Dreher *et al.*; chap. 16 Fränzle & Kluge; chap. 17 Liu) is the interactions between hydrology and ecology and the importance of hydrologic dynamics at the landscape scale whatever the area studied, whether a floodplain peat bog in Central England, a Danubian wetland, a tropical semi-arid wetland in Kenya, the German Bornhöved Lakes area or fields in Northern China. Changes in flow rates and gradients in groundwater/surface water ecotones were examined within small stream and large riverine ecosystems, lakes and adjacent soil systems. Typology characteristics, water transport and chemical reactions were used to present approaches for identifying transition zones that couple different groundwater basins within catchments. Some models are proposed for solute transport as well as for hydraulics or heat flow. Remote sensing and sequential aerial photographs offer a unique way of studying the dynamics of the ecotone.

The last paper of this session (chap. 18 Poole *et al.*) is a critical analysis of ground penetrating RADAR for exploration and mapping ecologically significant ecotonal structures.

Section III: Malfunction of groundwater/surface water interfaces: causes and methods of evaluation

Malfunctioning occurs as a result of hydraulic stress, loads of particulate and dissolved matter, and excessive leaching of fertilizers or other substances from the catchment areas. As an introduction to this section, Vanek (chap. 19) emphasizes the heterogeneity of groundwater/surface water ecotones caused by several thousand years of human impact. The heterogeneity is measurable and predictable. Many ecotones are no longer able to assimilate or buffer the resulting inputs. Studies carried out across different space scales have revealed that two of the key factors for protection of both surface and underground systems are water exchanges and the biological properties of the interface. If we want to use the landscape in a sustainable way, ecotone-oriented measures must be combined with modified land use of the whole catchments.

In the drainage situation, from groundwater to surface streams (chap. 20, Correll *et al.*) there is a failure of the agricultural riparian forest to protect surface water from groundwater nitrate contamination. So we can observe an increasing impact of negative aspects on water quality compared with positive aspects (such as denitrification, degradation and self-purification). To examine the trophic relations in river and hyporheic habitats an innovative *in situ* procedure has been developed (chap. 21 Wissmar *et al.*) and it indicates that stable isotopes (such as ∂^{15} N) may be powerful tools for determining nitrogen source and processing information in the ecotone.

An analytical model is proposed to evaluate the fate of phytosanitary products in the soil via groundwater (chap. 22 Thirriot & Caussade); it takes into account sorption when there is washing of in-solution products.

The last chapter of this section (chap. 23 Glushko) considers soils of the northeastern coast of the Caspian Sea as the zone of sea water/groundwater interaction. Changes in the sea water level may either cause the accumulation or dissolution of soil minerals (particularly salt). Groundwater contributes to a decrease in salt content in the upper horizons of the soil.

Section IV: Management and restoration of groundwater/surface water interfaces

Anthropic pressure has important consequences for water resource management. Fluvial deposits, urban runoff, agricultural waste, industrial effluents, and civil engineering schemes such as hydropower development on the river all participate in the deterioration of the quality of waters. As an introduction to this section Zaletaev (chap. 24) discusses the problems of the management of ecotones in irrigation regions. Changes in the hydrologic and hydrogeological regimes of a territory lead to the intensification of succes-

sions of biotic complexes and the formation of numerous interlinked irrigation ecotones.

The last four chapters of the management and restoration section (chap. 25 Hobma; chap. 26 Jacks *et al.*; chap. 27 Dikariova; chap. 28 Novikova & Zabolosky) attempt to link ecohydrology and hydrochemistry in the study of these ecotones. With anthropization in temperate zones we observe a drastic reduction of contact zones (straightening and regulation of rivers, drainage of lakes . . .) and in arid zones an increase in the number of these frontiers. The effects of excessive groundwater exploitation in the recharge area on ecohydrological relations in the seepage zone are difficult to determine. The strategies for and benefits of ecotone protection for management of groundwater quality have been assessed. For example the use of planting on mounds has preserved a bog as a nitrogen sink and made rapid reforestation possible. Controlled surface runoff increases the downward infiltration that recharges the existing shallow aquifers, decreases flooding hazards, prevents soil erosion and assists the growth of plant cover. The monitoring of riparian ecosystems is a necessary condition for management of water bodies. Management of ecotone systems in irrigational regions in arid, semiarid and subhumid zones is possible on the basis of cooperation and special hydro-land-improvement measures.

Chapter 29 synthesizes the information given in this book and points to prospects for the future, in a concluding section.

CONCLUSION

This volume does not claim to describe all the dynamics, management and restoration of groundwater/surface water ecotones. It should be considered only as a point of departure for future studies. Recent data allow us to understand present and past functioning and the interactions with adjacent ecosystems.

The book shows managers and water agencies how to integrate the concept of the groundwater/surface water ecotone into new laws, new statutes for conservation and sustainable exploitation of streams, lakes and aquifers.

All in all, it is hoped that this book will have a very large audience. It will be useful not only for ecologists who study landscape boundaries but also for hydrologists, modellers and other scientists who are interested in the quality and quantity of waterflows and managers who may find in its pages both case studies and answers to their questions.

ACKNOWLEDGEMENTS

We wish to express our appreciation to all participants of the Groundwater/Surface Water Ecotones Conference held in

Lyon who contributed significantly to the success of this meeting. All of the papers in this volume were evaluated by two anonymous reviewers for each paper.

Financial support was provided by UNESCO (Man and the Biosphere International Hydrological Programmes) and by scientific organizations like the CNRS; the Ministry for the Environment; political organizations like the City of Villeurbanne, le Grand Lyon, le Conseil Général et Envirhonalpes, le Conseil Régional and all the organizations concerned with water management like the Agence de l'Eau, Rhône Méditerranée Corse and the Compagnie Nationale du Rhône.

Particular thanks are due to Robert R.J. Naiman for initiating the programme on the role of land/inland water ecotones for management and landscape and for his active participation in the organization of the conference.

REFERENCES

di Castri, F., Hansen, A.J. & Holland, M.M (1988). *A new look at ecotones.* Emerging Internationnal Projects on Landscape Boundaries. Biology International. Special issue 17. UNESCO MAB, 163p.

Gibert, J. (1991). Groundwater systems and their boundaries: conceptual frame work and prospects in groundwater ecology. *Verh. Internat. Verein Limnol.*, **24**, 1605–1608.

Gibert, J., Dole-Olivier, M.J., Marmonier, P. & Vervier, P. (1990). Groundwater ecotones. In *Ecology and Management of Aquatic-Terresrial Ecotones*, ed. R.J. Naiman & H. Décamps, pp. 199–225. Paris – London, Man and the Biosphere Series Unesco, & Parthenon Publishing Carnforth.

Gibert, J., Stanford, J. A., Dole-Olivier, M. J. & Ward J. V. (1994). Basic attributes of groundwater ecosystems and prospects for research. In *Groundwater Ecology*, ed. J. Gibert, D. L. Danielopol & J. A. Stanford, pp. 7–40, San Diego: Academic Press.

Hansen, A.J.& di Castri, F. (ed.) (1992). *Landscape boundaries: consequences for biotic diversity and ecological flows.* New York: Springer-Verlag.

Holland, M.M. & Risser P.G. (1991). Introduction. In *Ecotones the role of landsape boundaries in the Management and Restoration of Changing Environments*, ed. M.M. Holland, P.G. Risser & R.J. Naiman, pp. 1–8. New York-London: Chapman and Hall.

Naiman, R.J. & Décamps, H. (ed.) (1990). *The ecology and management of aquatic-terrestrial ecotones.* Paris-Carnforth Hall (Man and the Biosphere Series): UNESCO – The Parthenon Publishing Group.

Stanford, J.A. & Ward, J.V. (1993). An ecosystem perspective of alluvial rivers: connectivity and the hyporheic corridor. *J. N. Am. Benthol. Soc.*, **12**, 1, 48–60.

Vervier, P., Gibert, J., Marmonier, P. & Dole-Olivier, M.J. (1992). A perspective on the permeability of the surface freshwater-groundwater ecotone. *J. North Am. Benthol. Soc.*, **11**, 1, 93–102.

II

Function of groundwater/surface water interfaces

2 Ecotonal animal assemblages; their interest for groundwater studies

D.L. DANIELOPOL*, R. ROUCH**, P. POSPISIL***,
P. TORREITER*** & F. MÖSZLACHER***

* *Limnological Institute, Austrian Academy of Sciences, A-5310 Mondsee, Austria*

** *Centre de Recherches Souterraines et Edaphiques, Laboratoire Souterrain, CNRS, F-09200 Moulis, France*

****Institute of Zoology, University of Vienna, Althanstrasse 14, A-1090 Vienna, Austria*

ABSTRACT   In order to better understand the structure of the surface-groundwater organismal assemblages (SGOA), the concept of global aquifer/river system (GARS) of Castany is recalled. The pattern of SGOAs emerges within different space and time scales. Examples of SGOAs and of the processes which form them are presented using case studies from the alluvial sediment fauna of the rivers Rhone, the Danube at Vienna, and the Lachein brook, in southern France. The structure of these assemblages depends on the boundary conditions of the groundwater system, mainly the hydrologic dynamics, and on the initiating conditions offered by the ecological tolerances of each species which contributes in the assemblages. The possible role of the macroorganisms in the functioning of the groundwater ecosystem is discussed. Their role appears less important than was thought earlier. Finally we emphasize the necessity of protecting the diversity of ecotonal assemblages. Scientific, cultural and practical arguments are presented.

INTRODUCTION

The review deals with studies on surface-groundwater organismic assemblages (SGOA) in unconsolidated, porous media, mainly alluvial sediments. These animal assemblages are formed by hypogean dwelling organisms (Hy) and epigean ones (Epi) which live permanently or temporarily in such subsurface habitats. At a first approximation, the SGOA definition agrees with the concept of ecotone, *sensu* Odum (1971, p.157): 'a transition between two or more diverse communities'. The study of such ecotonal animal assemblages represents one of the core aspects of groundwater ecology (GW).

Following the discovery by Stanko Karaman and Pierre-Alfred Chappuis, more than 50 years ago, that the alluvial sediments closely located to surface running waters harbour an exceptionally abundant and diverse interstitial fauna, several generations of biologists concentrated their efforts to describe this type of faunistical assemblage. Many students tried to understand the origin, the evolution and/or the function of these interstitial organisms (historical review in Danielopol, 1982, Gibert, 1992, Valett *et al.*, 1993).

The ecological investigations of the pioneers of the groundwater ecology of alluvial sediments, like E. Angelier, T. Orghidan, J. Schwoerbel, S. Husmann, pointed out the fact that Epi animals do not penetrate, in most cases, deep into the substrate, where Hy animals dominate. This situation was recently rediscussed by Bretschko (1992), and presented as a paradox. With a broader (holistic) view of the GW systems and their biota an answer to Bretschko's paradox is possible. In the present study we adopt a systemic approach founded on the hydrological concept of global aquifer/river system (GARS) developed by Castany (1982; 1985) and use it as an ecological scene (or landscape) within which we shall place the SGOA. As such, our approach will be different from those who treat the SGOA as inhabitants of the 'hyporheic habitat', a transitional zone between surface and 'true' groundwaters (discussion in Orghidan, 1959; Williams, 1984; White, 1993).

For present day GW ecologists as for their former colleagues the study of SGOA has *inter alia* the following interests: (1) to understand what kind of pattern such communities display, (2) to evaluate what role the fauna plays in the functioning of a GW system, (3) to better understand the diversity of GW organisms. We have discussed similar topics in earlier papers (Danielopol, 1989; Rouch,

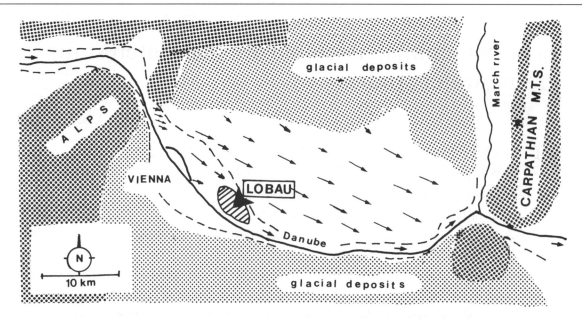

Fig. 1 Location of the Marchfeld aquifer in the Vienna Basin with the research area (Lobau).

1986; Rouch & Danielopol, 1987; Danielopol *et al.*, 1991). Here we should like to review recent research undertaken mainly in Europe. Finally we emphasize the necessity to protect the diversity of SGOA. Scientific, cultural and practical arguments will be presented.

THE GLOBAL AQUIFER/RIVER SYSTEM

Castany (1982; 1985) used the term of global aquifer/river system (GARS) for the unconfined aquifers in alluvial sediments, which interact connected to streams or rivers. Such a system has three units, which are hydrodynamically, hydrochemically and hydrobiologically, in a permanent and close way, (Castany, 1985, p.2): (1) the reservoir which has a heterogenous structure, because it is formed by alluvial sediments (often including also the permeable underlying rocks), (2) the groundwater, flowing through the reservoir and (3) the water of the surface running water system. The reservoir and the groundwater constitute the aquifer. These three units produce a global system which integrates aspects of both ground and surface waters. The hydraulic relationships between the aquifer and the river are induced by the relative position of the water table and the water level of the river. With this approach the physical boundary between the surface water and the groundwater environment lies immediately below the bottom of the surface water body.

The Marchfeld, an alluvial aquifer of the Danube in Lower Austria (Fig. 1), responds to Castany's concept. It is connected to the river, which, globally, drains it; locally,

depending on the piezometric head, it is recharged by the river (see Danielopol, 1983; Danielopol *et al.*, 1991; Dreher *et al.*, 1985). The physical boundaries of this large-scale aquifer (km scale) are the Danube and the March-Thaya river beds, the impervious floor is made up of Pannonian (marine) sediments; other boundaries are the hill-slopes on the north-western part which close, on the left-hand side of the Danube, the Vienna basin. However the global functioning of the aquifer is locally modified, especially within the wetland zone, along the Danube, in the dead arms of the river, which are still hydrologically connected to the main channel. The hydrologic regime of these surface waters models, at a lower scale the behaviour of the aquifer, like those of the Eberschüttwasser, which are currently being investigated by our research group (see this volume, Dreher *et al.*).

With the global aquifer/river system approach it is easier to describe the SGOAs and their formation process. Surface dwelling organisms penetrate actively or passively into the aquifer and disperse further, depending on their own ecological tolerances as well as on the ecological conditions of the subsurface environment. Some of the Epi animals remain closely located below the surface sediment layer because of their biological requirements (Bretschko, 1992); others can take advantage of the interstitial topography and subsurface water flow and move deeply into the aquifer (remember also the distribution of the plecopters within the aquifer of the Flathead River, in Stanford & Ward, 1988; 1993). Hypogean dwelling animals are able to colonize both the deeper layers of the alluvium as well as superficial layers very closely located to the surface waters (e.g. the ostracods in the wetland of the Danube; Danielopol, 1991). In this way, we obtain through the overlapping of Epi and Hy organisms, groundwater ecotonal assemblages (SGOAs). Both the Epi and the Hy ones are components of the global aquifer/river system. Within the latter the GW dynamics in connection with the surface waters models, in a significant way, the changes in the behaviour of the other components of the system, like the SGOA (see below). Due to this framework it is no longer necessary to separate in a precise way a hyporheic zone from the groundwater one, as in the models proposed by running water ecologists (Triska *et al.*, 1989; Williams, 1989; White, 1993 *inter alia*) or those of the early GW biologists (e.g. Orghidan, 1959). Depending on the relationships between surface and groundwater and depending on how fast and strongly the water changes in the subsurface 'black box', we shall have in the global aquifer/river system areas of various size harbouring, for different periods of time, ecotonal assemblages. Interestingly enough this global approach, in some ways foreseen by Schwoerbel (1967), is now effectively used by a new generation of ecologists in Europe (Gibert *et al.*, 1990) and North America (Bencala, 1993; Harvey & Bencala, 1993; Dahm *et al.*, 1993; Stanford & Ward, 1993).

The interest of adopting this global approach lies in its capacity to better integrate particular ecological situations, whatever the spatial and temporal scale used. Indeed, the multitude of unique SGOAs appear as aspects of a general pattern. With this approach GW ecologists have to define at which dimension of spatial and temporal scale the SGOA type patterns are apparent. Moreover, GW ecologists should try to understand patterns in terms of the processes that produce them. These are the topics of the next section.

SURFACE-GROUNDWATER ORGANISMIC ASSEMBLAGES; PATTERN AND PROCESS FORMATION

In the concluding remarks to the workshop 'Perspectives on the hyporheic zone' Hakenkamp *et al.* (1993, p. 97) noted: 'each hyporheic system is unique, but cross-system comparisons may be possible if we begin to examine the nature and extent of hyporheic differences and stream orders'. It is with this hierarchical approach in mind that we shall present three examples with which we demonstrate how patterns of SGOA emerge at various spatial scales, from the regional one (km-scale) to the local habitat scale (m-scale)

The organization of interstitial animal assemblages of the Rhone river (Creuzé des Châtelliers, 1991) at a 100-km scale (i.e. comparative data from different regions) reflects in an exemplary way the type of relationships existing between the river and the alluvial aquifer.

Within the Brégnier-Cordon area, where the aquifer is poorly developed, the superficial sediments (the so-called 'hyporheal') are mainly recharged by the Rhone River.

Hence, the Hy animals are scarcely represented or completely absent. Moreover, the interstitial space is often reduced by the accumulation of fine sediment eroded from the surrounding area. The SGOA, with low diversity, is dominated by a few Epi species, with low abundances; the densities and diversities fluctuate seasonally.

The opposite occurs within the Miribel-Jonage sector, where a large aquifer is drained into the Rhone. Here, the Hy animal fraction of the SGOA is important. Because fine-grained sediments are periodically flushed out from the interstitial space, a SGOA with well diversified and relatively abundant Epi animals is also found. Unlike the Brégnier-Cordon site, the permanent discharge of GW determines a relative stability in the environment of surficial sediments. One notes a very attenuated seasonality of the animal population dynamics.

Within the Donzère-Mondragon zone, the aquifer is poorly developed; the impervious surrounding rocks lie in proximity to the riverbed. This situation determines a reduced area through which the aquifer/river exchanges occur. The subsurface sediments are mainly fed by surface running water which infiltrates rapidly through the very coarse sediment of the riverbed. The animal assemblages found in this environment are dominated by Epi organisms. Their densities fluctuate following the seasonality rhythm of the surface water. Low abundances are found during winter and higher densities and diversities during summer. In particular during this latter season there is an important intrusion of surface (benthic) dwelling cladocers and ostracods.

The pattern of SGOA formation is visible also at intermediate (100 m) scales. Here we shall describe the animal assemblages that we observed at the margin of the Marchfeld aquifer, in the Danube wetland at Vienna. Previously Danielopol (1983; 1989; 1991) described the SGOA of a 30 m^2 area within the surficial alluvial sediments (from the water/sediment interface down to 2 m deep) of the Eberschüttwasser (Fig. 2, site A) and one of us (Pospisil, in prep.) investigated the GW fauna of the surrounding area (Fig. 2, site B).

Within site C (an area of about 900 m^2), located close to the Eberschüttwasser (Fig. 2, 4) we studied the SGOA structure during 1991–1993. The description of the sampling site is presented in Danielopol *et al.* (1992); for the hydrology of this GW area, see Dreher *et al.* (this volume). The site like site B belongs to the Marchfeld aquifer, but it has a particular situation. The Eberschüttwasser infiltrates into site C, crosses the alluvial sediments and exfiltrates, mainly, in the next dead arm, the Mittelwasser (see arrows on Fig. 2, 4). As the hydrologic regime of the Eberschüttwasser follows those of the Danube it models also the behaviour of site C. For instance, several times a year the Eberschüttwasser floods parts or all of site C (Dreher *et al.*, this volume). Within this

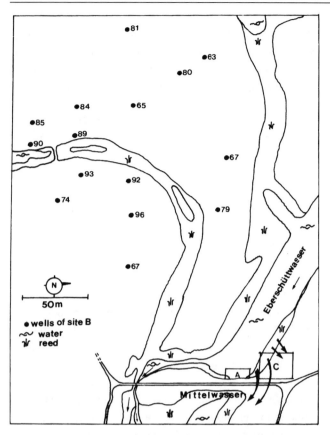

Fig. 2 Location of sampling sites A, B, C in the Lobau; black arrows – waterflow direction through the sediments at site C.

Table 1. *Species list of Ostracoda Lobau – Site C*

a. *Cryptocandona kieferi* (Hy)
b. *Mixtacandona laisi vindobonensis* (Hy)
c. *Kovalevskiella sp.* (Hy)
d. *Pseudocandona albicans* (Epi)
e. *Pseudocandona lobipes* (Epi)
f. *Fabaeformiscandona wegelini* (Hy)(Ec)
g. *Candonopsis kingsleii* (Epi)

Notes:
Hy – Hypogean dwelling species
Epi – Epigean dwelling species
Ec – Ecotonal dwelling species

reorganization of the SGOA followed (see the spring 1993 data). Once again the species richness and the ostracod densities increased (Fig. 5). Generally speaking, we see a gradual decrease in the diversity, starting from site D3, located in the area submitted to strong surface water infiltration (Fig. 4) towards D17, which is fed mainly by GW from the inland zone of the Marchfeld aquifer. While in the former case the habitat is well oxygenated (Fig. 3) during a rather long period of time, in the latter case (D17) the area remains, for the same period of time, hypoxic (i.e. the oxygen concentration is less than 0.5 mg l^{-1}).

The variability of the ostracod samples, both quantitative and qualitative, in space and time, suggests that these crustaceans move steadily through the interstitial space; they form aggregates which, sometimes, as in the area of well T3, persist for longer periods of time while at other sites they have an ephemeral existence. The spatial distribution of each species over the whole area of site C is determined not only by the dynamics of the GW which brings or regulates the necessary energy resources (like the organic matter and the oxygen) of the ostracods, but also by the ecological tolerances of each ostracod species to the environment with which it is confronted. For instance the Hy species *C. kieferi* has a wide tolerance to hypoxia. This is documented by field and laboratory observations. Species which are known to live in well oxygenated environments and/or ones with high amounts of food supply, like *Cypridopsis vidua* (Roca & Danielopol, 1991), do not penetrate deep into the aquifer. Such species are to be found at site A at 0.5 m deep but not far into the aquifer (e.g. sites B or C).

The interstitial space at site C is far from being saturated with meio or macroorganisms. Most of the samples from the surrounding piezometers (Möszlacher, unpubl.) show a very poor fauna, many of the samples are bare. A decrease in the abundance of the dominant species, *C. kieferi*, is not correlated with significant increases in the densities of other species (Fig. 6). These are arguments against the existence of a biologically structured community at this site.

area oxygen (Fig. 3) and temperature fluctuate markedly during the year. The GW system is hypoxic during the summer and autumn, while during winter and spring becomes better oxygenated (Fig. 3).

The SGOA formed by the ostracod taxocoene at site C, becomes apparent only if we consider the whole area (in this case about 900 m^2). There are seven ostracod species (Table 1), four hypogeans, the others are epigeans that also live in the Eberschüttwasser. The most common and abundant species is a Hy species (*Cryptocandona kieferi*). The spatial distributions of the other Hy and Epi species overlap in an irregular way. During spring (April, 1992) when the GW system is better oxygenated and the water temperature is low, due to the important infiltration of surface water from the Eberschüttwasser, Epi species like *Pseudocandona albicans* and *Ps. lobipes* (Fig. 4) colonize the margin of this area (see wells D3–D5 in Fig. 4) and the subsurface zone with the highest infiltration capacity and through flow (i.e. around the axis D3 – T3, e.g. wells D3, D9, T3 in Fig. 4). During summer and autumn when the surface water slows down its influx into the aquifer at this site (Dreher *et al.*, this volume), the Epi species disappear and the ostracod taxocoene becomes gradually less diverse (Fig. 5). With the floods and the reincrease of the infiltration regime into the alluvial sediments a

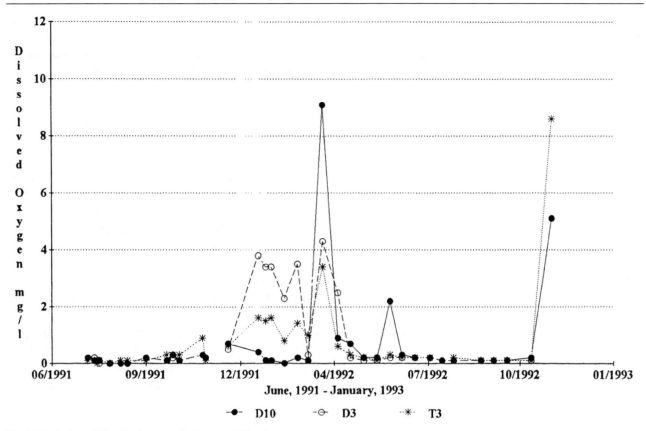

Fig. 3 Evolution of dissolved oxygen in three multi-level monitoring wells (location on Fig. 4).

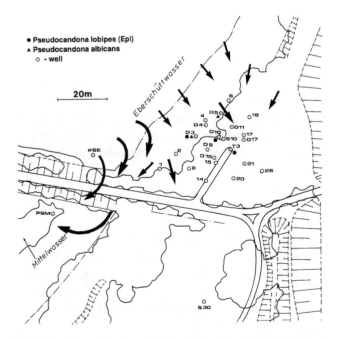

Fig. 4 The Eberschüttwasser GW-area (site C) with distribution of 2 Epi species; arrows – waterflow direction through the sediments.

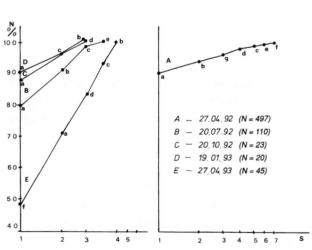

Fig. 5 Seasonal evolution of the ostracod assemblage at site C (cumulated data from the D1 – D3 wells, see Fig. 4 for location) expressed as 'diversity (k) curves'; a-f ostracod taxa (see Table 1); A-E – sampling dates; N – Total ostracod individuals for one sampling date; N% same expressed as cumulated percentage.

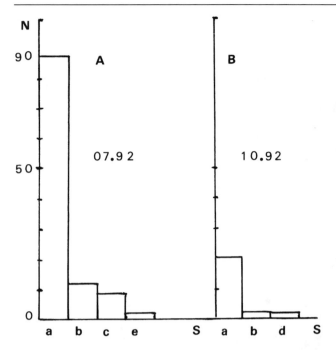

Fig. 6 Ostracod densities at site C (cumulated data from the same wells as in Fig. 5); A, B sampling dates; N – number of individuals; S – species (a – e ostracod taxa, see Table 1).

The analysis of ecotonal assemblages can be studied also at metre or decimetre scales; this is the case for the Lachein stream (Ariege, France). Here the crustaceans were investigated within an area of ≤75 m at 0.6 m deep in the streambed sediment, using 130 sampling units for a year, 20 every two months (Rouch, 1988). Within this area (Fig. 7) four zones can be defined, the flowing surface water, the main channel (sectors I,G,E,C), two gravel bars (sectors I,H,F,A), and a zone of dead (stagnant) waters (sector B,D). These four zones differ through their porosity, grain size, oxygen concentration (Rouch, 1988) as well as through other chemical characteristics (Rouch *et al.*, 1989).

A rich crustacean fauna (ostracods, copepods, syncarids, isopods, amphipods) lives within these sediments. Fifty species were identified, of which 22 are exclusively Hy dwelling animals (stygobites). Most of this latter group are stenotopic species (Rouch, 1988). Some of them are restricted to the hypoxic sector, located in the dead-water zone (e.g. the syncarids), others colonize the well oxygenated areas (e.g. the isopod Microparasellidae).

The best represented group, i.e. the most abundant and species-rich at this study site is the copepod Harpacticoida. There are 21 species (9 are exclusively dwelling Hy species) which represent 60 per cent of the total number of individuals caught. The harpacticoid SGOA of the Lachein site is stable within a year's seasonal cycle, i.e. no significant difference can be observed between the different series of samples. Neither the global densities nor the

rank abundances of the dominant species change. However this assemblage displays a strong spatial heterogeneity. Four groups can be distinguished (Fig. 7). Their distribution correlates with the four sectors of the streambed defined above (Rouch, 1991).

The first group is dominated by the Hy species *Elaphoidella bouilloni*, and is spread within the hypoxic area; the mean density is 7 ind./l and the Shannon diversity index $H'=2.087$. The second group is characterized by the Epi species *Limocamptus echinatus* and *Attheyella crassa* and occur along the main channel (mean density 10 ind./l, $H'=2.253$). Within the gravel bars two groups can be distinguished, one dominated by the Hy species *Ceuthonectes gallicus* (mean density 28 ind./l, $H'=1.870$) and another at the periphery of these bars, dominated by the Epi species *Bryocamptus minutus* (mean density 23 ind./l, $H'=1.714$). The very strong analogies which exist between the spatial distribution of the oxygen concentration and of the various chemical facies on the one hand, and the harpacticoid assemblages on the other hand, suggest that a key factor determines these spatial distribution patterns.

The study of the piezometric chart of the hydraulic conductivity (Fig. 8) for this site (Rouch, 1992) shows that a strong drainage zone exists which crosses the site diagonally below the gravel bars (here the permeability is about 10^{-3} m s^{-1}). It is within the area with higher permeability that the oxygen concentrations are the highest, the carbon dioxide and the hydrogen carbonates have the lowest concentrations and the harpacticoid populations are the most abundant. It appears that on the Lachein, the surface water flow modulated the morphology of the streambed, implicitly the sediment porosity and grain size. But the subsurface water flow generates the spatial heterogeneity of the substrate due to its different velocity patterns.

In contrast to the Lobau site where the surface water which infiltrates underground determines the seasonal changes of the ostracod SGOA, the Lachein site presents a relative stability of its abiotic and biotic parameters. In this case the spatial heterogeneity we mentioned is due, mainly, to the subsurface water flow (differences of hydraulic conductivity).

With regard to the biology of the Lachein site it should be mentioned that the Hy dwelling harpacticoids build the most abundant populations close to the surface of the stream bottom, as in the Lobau area studied by Danielopol (1991). It is at the sites with the highest densities (the gravel bars) that the diversity is the lowest. It is very plausible that due to interspecific competition between harpacticoid species, the diversity of their assemblage is maintained at lower values (Rouch, 1991).

The data presented here suggest the following conclusions.

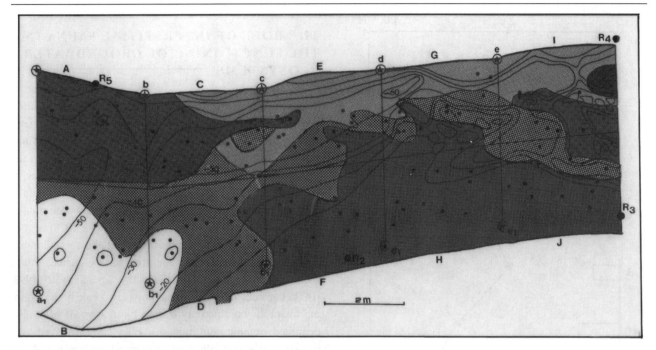

Fig. 7 Distribution chart of 4 harpacticoid groups produced with a discriminant factorial analysis (see text); colours: white – group 1, light gray – group 2, dark gray – group 3, stippled gray – group 4.

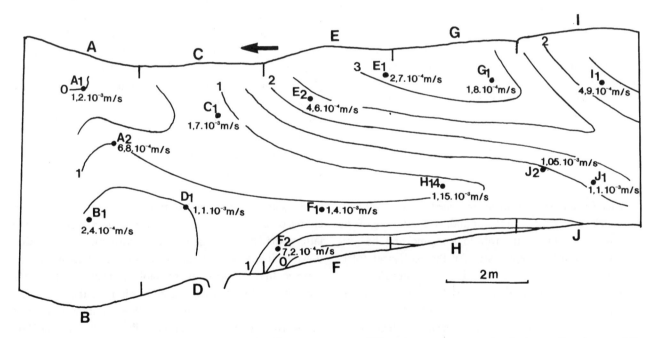

Fig. 8 Piezometric chart of the Lachein area and hydraulic conductivity (izopiezes at 0.5 cm intervals).

1. The water dynamics in both surface and subsurface space are the key factor for the global aquifer/river system (see also Marmonier & Dole, 1986).

2. Viewed at a regional scale, along large rivers or drainage basins, the SGOA appear like beads on a string. Along the river (the string), SGOA formation depends on the different processes occuring between the river and the aquifer (the beads), e.g. the Rhone case (Creuzé des Châtelliers, 1991).

3. At the local scale of a global aquifer/river system we can have different types of SGOA. For instance those located in zones which have a more stable hydrological regime have SGOAs with higher diversity (e.g. the Lachein case). Assemblages that are submitted to a complex and variable hydrologic impact, display lower

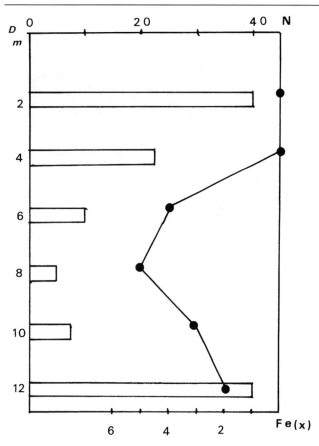

Fig. 9 Distribution of densities of the Hy macrocrustaceans (*Proasellus slavus* and *Niphargus sp.*) and their faeces in the well T3 (cumulated data for 7 series of samples at each depth; *D* – depth in metres below the surface; bars – total number of individuals (*N*); *Fe(x)* – number of samples with high amount of macrocrustacean faeces.

THE ROLE OF INTERSTITIAL FAUNA IN THE FUNCTIONING OF GROUNDWATER ECOSYSTEMS

The corollary of the discovery of rich and abundant interstitial organisms in the surficial layers of riverine sediments was to question the possible role of this fauna for the functioning of the ecosystem to which they belong. This is not only an academic question but it also has practical value, as it is plausible that meio- and macroorganisms contribute in the mineralization of the organic matter which accumulates within the interstitial space. Husmann (1974; 1978) argued that such a process occurs in both natural alluvial sediments, and in the slow sand filters operating for the purification of the drinking water plants in Germany. Sinclair *et al.* (1993, p.471) consider that 'Protozoa may play a role in maintaining hydraulic conductivity during biotreatement of readily degraded organic contaminant'. Eder (1980) arrived at a similar conclusion, following experiments with nematodes in slow filtration sand columns.

Danielopol (1983; 1989) showed that Hy dwelling isopods, amphipods and oligochaetes which live within the riverine sediments of the Danube modify the microenvironment by their biological activity, either by feeding on fine sediment or by their movement through such sediment. Danielopol (*op.cit.*) suggested that such activities could improve the hydraulic conductivity of the sediment. Recent experiments (P. Torreiter, in prep.) with slow filtration columns where the isopod *Asellus aquaticus* was added, showed contradictory results. When isopods were added at the top of the sand column and allowed to feed on conditioned litter, one could see in one of the experiments an increase of the waterflow through the column, suggesting the positive role of the crustaceans on the porosity. In other subsequent experiments, no obvious improvement of filtration was obtained. At low discharge rates and high density of asellids the respiration of the latter exhausted oxygen in their environment and slowed down their activity. Subsequently the whole column became hypoxic. In this precise case the biological activity of isopods did not effectively change large portions of the interstitial environment in such a way as to allow the improvement of the hydraulic conductivity of the sediments, as postulated by Danielopol (1983; 1989).

One of us (P.P.) examined in the field (samples from Lobau, site C, well T3) the possible correlation between the accumulation of macrocrustacean faeces (those of the isopod *Proasellus slavus* and the amphipod *Niphargus* sp. are easily identifiable) and the abundance of living animals in the interstitial space. Fig. 9 shows that no positive correlation exists between these two parameters. Nor is there any correlation between the abundance of these macrocrustaceans and those

diversity and a more variable pattern (e.g. the Lobau site C case).

4. The SGOAs (and implicitly the spatial distribution of each Epi and Hy species) are moulded by the interaction which occurs between the boundary conditions of the aquifer system (e.g. the hydrologic regime) and the initiating conditions represented by the ecological tolerances of the various species of the community. From this point of view the SGOA does not appear as a simple transition between two or more communities. It corresponds to the dynamic relationships which form between species with different origins, Epi and Hy species, as well as from their interactions with the local conditions of the river/aquifer system.

5. In those cases where the abiotic conditions in the aquifer allow the development of abundant organismic populations, the SGOA could be moulded also by biotic factors, like interspecific competition and/or predation. For the documentation of these aspects careful investigations are needed in the future.

Table 2. *Correlation of macro – meiofauna densities (Wells – T3 & D3, at 27/04/1992)*

Macrofauna/Ostracoda	Macrofauna/Cyclopoida
$n=7$ $r=0.47$	$n=7$ $r=0.32$
$r^2=0.22$	$r^2=0.1$

Notes:
Macrofauna=Isopoda (*Proasellus slavus*)
Amphipoda (*Niphargus sp.*)
Meiofauna=Ostracoda, Cyclopoida
n=number of samples
r – coefficient of correlation
r^2 – coefficient of determination

of the interstitial meiofauna like ostracods or cyclopoids (Table 2). One could expect that through the biological activity of macrocrustaceans, the microenvironment would become more favorable for the development of meiofauna too, as in a deep-sea benthic fauna case recently discussed by Thistle *et al.* (1993).

In conclusion, the question remains open: to what extent and how effectively can the hypogean meio- and macro-organisms transform their environment, in order to modify effectively the functioning of other parts of their ecosystem? More careful experiments and field observations are obviously needed.

WHY PROTECT THE DIVERSITY OF ECOTONAL ASSEMBLAGES

Several generations of biologists described a large number of new taxa, mainly hypogean-dwelling animals that live in porous unconfined aquifers, sometimes very closely located to contact with surface waters. *Stygofauna Mundi* edited by Botosaneanu (1986) gives an impressive list of such animals. The 'hyporheic habitat' became famous not only for its diverse animal assemblages, but also for its abundant populations of meio- and macroorganisms. They should therefore be protected for scientific, cultural and practical reasons. Indeed, there are still plenty of unknown questions concerning the origin and evolution of these GW animals. We do not know, for instance, what the genetical structure of interstitial populations within a metapopulation is. Do populations of species like *Fabaeformiscandona wegelini*, which live along the Danube, Sava or Rhone rivers, in various isolated habitats, display the same genetical structure and ecological requirements?

Understanding why some SGOAs have a higher species richness than the Epi-dwelling assemblages while in other cases the diversity is much lower (the 'intermediary diversity'

of Gibert *et al.*, 1990) needs causal explanations. Marmonier *et al.*, (1993) suggested that one can approximate the potential diversity of a GW habitat. This is an interesting prospect that needs further investigation. For instance it is obvious that the hypoxic environment in the Lobau prevents colonization of many surface dwelling animals and there is a reduced number of species which will contribute in this SGOA.

Rouch & Danielopol (1987) showed that GW dwelling species originate from Epi organisms with wide ecological tolerances, which expand and diversify during favorable, stable, periods. This view is at variance with the classic idea that GW animals stem from specialized organisms which took refuge in the subterranean environment under the constraints of abiotic or biotic factors. The possibility of investigating present day surficial alluvial sediments in tropical and/or subtropical zones makes it possible to test such a hypothesis. For instance the Sycamore Creek in the Sonoran Desert, Arizona (Boulton *et al.*, 1992a; 1992b; Valett *et al.*, 1992) has a high diversity of interstitial animals, both Epi and Hy organisms. No evidence could be found for a retreat of surface animals into underground habitats during dry periods, when desiccation of the stream occurs. Therefore the origin of GW animals should not be looked for in 'the refuge-under-constraints' model but in the alternative hypothesis proposed by Rouch & Danielopol (*op.cit.*).

Answering such questions and contributing in scientific debates dealing with GW organisms, represents a cultural activity which challenges our intellectual creativity. Therefore the protection of the diversity of GW organisms and their habitats, represents a necessity for which we need not just effective measures, but also educational programs (Pospisil & Danielopol, 1990).

ACKNOWLEDGEMENTS

We are much indebted to friends and colleagues who helped during the field work and/or discussed with us aspects of this publication. A. Mangin, M. Bakalowicz, A. Descouens (Moulis), J. Dreher (Vienna), G. Bretschko, P, Schmid (Lunz), M. Creuzé des Châtelliers, J. Gibert (Lyon), P. Marmonier (Chambéry). H. Bennion and I. Gradl (Mondsee) helped to edit the manuscript. Financial support for the Austrian team is provided by the Fonds zur Förderung der wissensschaftlichen Forschungì (Project P-7881 atributed to DLD). DLD's visits on various occasions to the Laboratoire Souterrain in Moulis and the Department of Subterranean Ecology of the University of Lyon were possible through the CNRS – Austrian Academy of Sciences cultural exchange programs.

REFERENCES

Bencala, K.E. (1993). A perspective on stream-catchment connection. *J. N. Am. Benthol. Soc.*, **12**, 44–47.

Botosaneanu, L. (1986). *Stygofauna Mundi. A faunistic distributional and ecological synthesis of the world fauna inhabiting subterranean waters.* Leiden: E.J. Brill.

Boulton, A.J., Peterson, C.G., Grimm, N.B. & Fisher, S.G. (1992 a). Stability of an aquatic macroinvertebrate community in a multiyear hydrologic disturbance regime. *Ecology*, **73**, 2192–2207.

Boulton, A.J., Valett, H.M. & Fisher, S.G. (1992 b). Spatial distribution and taxonomic composition of the hyporheos of several Sonoran Desert streams. *Arch. Hydrobiol.*, **125**, 37–61.

Bretschko, G. (1992). Differentiation between epigeic and hypogeic fauna in gravel streams. *Regulated Rivers*, **7**, 17–22.

Castany, G. (1982). *Principes et méthodes de l'hydrologie.* Paris: Dunod.

Castany, G. (1985). Liaisons hydrauliques entre les aquifères et les cours d'eau. *Stygologia*, **1**, 1–26.

Creuzé des Châtelliers, M., (1991). *Dynamique de répartition des biocénoses interstitielles du Rhône en relation avec des caractéristiques géomorphologiques.* Doctoral Thesis, Univ. Claude Bernard, Nr. 33–91, Lyon, 1–160.

Dahm, C.N., Valett, H.M., Morrice, J.A., Wroblicky, G.J. & Campana, M.A. (1993). Nutrient dynamics and hydrology of hyporheic zone of montane catchments. *Bull. NABS*, **10**, 105.

Danielopol, D.L. (1982). Phreatobiology reconsidered. *Pol. Arch. Hydrobiol.*, **29**, 375–386.

Danielopol, D.L. (1983). *Der Einfluss organischer Verschmutzung auf das Grundwasser-Ökosystem der Donau im Raum Wien und Niederösterreich.* Forschungsberichte BMGU, Wien 5, 5–159.

Danielopol, D.L. (1989). Groundwater fauna associated with riverine aquifers. *J. N. Am. Benthol. Soc.*, **8**, 18–35.

Danielopol, D.L. (1991). Spatial distribution and dispersal of interstitial Crustacea in alluvial sediments of a backwater of the Danube at Vienna. *Stygologia*, **6**, 97–110.

Danielopol, D.L., Dreher, J., Gunatilaka, A., Kaiser, M., Niederreiter, R., Pospisil, P, Creuzé des Châtelliers, M. & Richter, A. (1992). *Ecology of organisms living in a hypoxic groundwater environment at Vienna (Austria); Methodological questions and preliminary results.* Proceedings 1st Internat. Conference on Ground Water Ecology ed. by J. A. Stanford & J.J. Simons, 79–90, AWRA, Bethesda, Maryland.

Danielopol, D.L., Pospisil, P. & Dreher, J. (1991). *Ecological basic research with potential application for the groundwater management. Hydrological Basis of Ecologically sound Management of Soil and Groundwater.* Proceedings of the Vienna Symposium, August 1991. IAHS Publ., 202, 215–228.

Dreher, J., Pospisil, P. & Danielopol, D.L. (this volume). The role of hydrology in defining a groundwater ecosystem.

Dreher, J., Pramberger, F. & Rezabek, H. (1985). Faktorenanalyse, eine neue Möglichkeit zur Ermittlung hydrogragraphisch ähnlicher Bereiche in einem Grunwassergebiet. *Mitteilungsblatt hydrogr. Dienst. Österr.*, **54**, 1–12.

Eder, R. (1980). *Beitrage zur Kenntnis der interstitiellen nematodenfauna am Beispiel eines Schotterkörpers der Donau bei Fischamend.* Doctoral Thesis, Univ. Vienna, 1–142.

Gibert, J. (1992). *Groundwater ecology from the perspective of environmental sustainability.* Proceedings 1st Internat. Conference on Ground Water Ecology, ed. by J. A. Stanford & J. J. Simons, 3–13, AWRA, Bethesda, Maryland.

Gibert, J., Dole-Olivier, M.J., Marmonier, P. & Vervier, Ph. (1990). Surface water – groundwater ecotones. In *The ecology and management of aquatic – terrestrial ecotones*, ed. by R.J. Naiman & H. Decamps, pp. 199–225, Man & Biosphere Ser., 4, Paris: Parthenon Publ. Group.

Hakenkamp, C.C., Valett, H.M. & Boulton, A.J. (1993). Perspectives on the hyporheic zone: integrating hydrology and biology. Concluding remarks. *J. N. Am. Benthol. Soc.*, **12**, 94–99.

Harvey, J.W. & Bencala, K.E. (1993). The effect of streambed topography on surface-subsurface water exchange in mountain catchments. *Water Resour. Res.*, **29**, 89–98.

Husmann, S. (1974). Die ôkologische Bedeutung der Mehrzellerfauna bei der natürlichen und künstlichen Sandfiltration. Wiss. *Berrichte Untersuchungen u. Planungen Stadtwerke Wiesbaden AG.*, **2**, 173–183.

Husmann, S. (1978). Die Bedeutung der Grundwasserfauna für biologische Reinigungsvorgänge im Interstitial von Lockergesteinen. *"gwf" wasser/abwasser*, **119**, 293–302.

Marmonier, P. & Dole, M.J. (1986). Les Amphipodes des sédiments d'un bras court-circuité du Rhône: logique de répartition et réaction aux crues. *Sciences de l'eau*, **5**, 461–486.

Marmonier, P., Vervier, Ph., Gibert, J. & Dole-Olivier, M.J. (1993). Biodiversity in ground waters: a research field in progress. *TREE*, **8**, 392–395.

Odum, E. P. (1971). *Fundamentals of ecology* (3rd ed). W.B. Saunders co., Philadelphia, Pennsylvania.

Orghidan, T. (1959). Ein neuer Lebensraum des unterirdischen Wassers, der hyporheische Biotop. *Arch. Hydrobiol.*, **55**, 392–414.

Pospisil, P. & Danielopol, D.L. (1990). Vorschläge für den Schutz der Grundwasserfauna im geplanten Nationalpark "Donauauen" östlich von Wien, österreich. *Stygologia*, **5**, 75–85.

Roca, J.R. & Danielopol, D.L. (1991). Exploration of interstitial habitats by the phytophilous ostracod *Cypridopsis vidua* (O. F. Müller): experimental evidence. *Annls Limnol.*, **27**, 243–252.

Rouch, R. (1986). Sur l'écologie des eaux souterraines dans le karst. *Stygologia*, 2, 352–398.

Rouch, R. (1988). Sur la répartition spatiale des Crustacés dans le sous-écoulement d'un ruisseau des Pyrénées. *Annls Limnol.*, **24**, 213–234.

Rouch, R. (1991). Structure du peuplement des Harpacticides dans le milieu hyporhéique d'un ruisseau des Pyrénées. *Annls Limnol.*, **27**, 227–241.

Rouch, R. (1992). Caractéristiques et conditions hydrodynamiques des écoulements dans les sédiments d'un ruisseau des Pyrénées. Implications écologiques. *Stygologia*, **7**, 13–25.

Rouch, R. & Danielopol D.L. (1987). L'origine de la faune aquatique souterraine, entre le paradigme du refuge et le modèle de la colonisation active. *Stygologia*, **4**, 345–372.

Rouch, R., Bakalowicz, M., Mangin, A. & D'Hulst D., (1989). Sur les caractéristiques chimiques du sous-écoulement d'un ruisseau des Pyrénées. *Annls Limnol.*, **25**, 3–16.

Schwoerbel, J. (1967). Die Stromnahe phreatische Fauna der Donau (hyporeische Fauna). In *Limnologie der Donau*, ed. R. Liepolt, pp. 284–294, Stuttgart: Schweizerbart.

Sinclair, J.L., Kampbell, D.H., Cook, M.L., & Wilson, J.T. (1993). Protozoa in subsurface sediments from sites contaminated with aviation gasoline or jet fuel. *Appl. Environ. Microbiol.*, **59**, 467–472.

Stanford, J.A. & Ward, J.V. (1988). The hyporheic habitat of river ecosystems. *Nature*, **335**, 64–66.

Stanford, J.A. & Ward, J.V. (1993). An ecosystem perspective of alluvial rivers: conectivity and the hyporheic corridor. *J. N. Am. Benthol. Soc.*, **12**, 48–60.

Thistle, D. Hilbig, B. & Eckman, J. (1993). Are polychaetes sources of habitat heterogeneity for harpacticoid copepods in the deep sea? *Deep-Sea Res.*, **40**, 151–157.

Triska, F.J., Kenedy, V.C., Avanzino, R.J., Zellweger, G. W. & Bencala, K.E. (1989). Retention and transport of nutrients in a third-order stream in northwestern California: hyporheic processes. *Ecology*, **70**, 1893–1905.

Valett, H.M., Fischer, S.G., Grimm, N.B., Stanley, E.H. & Boulton, A.J. (1992). *Hyporheic-surface water exchange. Implications for the structure and functioning of desert stream ecosystems.* Proceedings 1st. Internat. Conference on Ground Water Ecology, ed. J.A. Stanford & J.J. Simons, pp. 395–405, AWRA, Bethesda, Maryland.

Valett, H.M., Hakenkamp, C.C. & Boulton, A.J. (1993). Perspectives on the hyporheic zone: integrating hydrology and biology. Introduction. *J. N. Am. Benthol. Soc.*, **12**, 40–43.

White, D.S. (1993). Perspectives on defining and delineating hyporheic zones. *J. N. Am. Benthol. Soc.*, **12**, 61–69.

Williams, D.D. (1984). The hyporheic zone as a habitat for aquatic insects and associated arthropods. In *The ecology of aquatic insects*, ed V.H. Resh & D.M. Rosenberg, pp. 430–451, New York: Praeger Publs..

Williams, D.D. (1989). Towards a biological and chemical definition of the hyporheic zone in two Canadian rivers. *Freshwater Biol.*, **22**, 189–208.

3 Stochasticity in resource utilization by a larval Chironomidae (Diptera) community in the bed sediments of a gravel stream

P. E. SCHMID

Biological Station Lunz, Institute of Limnology, Austrian Academy of Sciences, A-3293 Lunz am See, Austria

ABSTRACT The larval chironomid community of the bed sediment surface and the hyporheic interstitial was examined in a gravel stream between September 1984 through August 1985 and between March and June 1993.

Spatial and temporal species turnover between horizontally adjacent sampling sites fluctuated distinctly in all sediment depth layers. The species composition showed a significantly lower spatial turnover in the upper 10 cm of the bed sediments than in deeper layers ($P<0.05$). Moreover, species abundance patterns of a five-species assemblage implied random assortment (*sensu* Tokeshi, 1990) and indicated a high probability for a species assembly to be dictated by environmental stochasticity. Neutral models were developed to evaluate the significance of observed overlap values in spatial distribution amongst abundant chironomid species (*sensu* Schmid, 1993). The spatial organization of a larval chironomid assemblage in the gravel stream Oberer Seebach seemed to be governed by coexistence due to random colonization processes, which reduce the probability of strong competitive interactions. Larval species colonization into open microhabitat patches of interstitial space (47.8 cm^3) was rapid with representatives of a species assemblage arriving within less then 24 hours of the start of the colonization experiment. Moreover, possible random movement across and between sediment depth layers of larval chironomid species may promote a rapid colonization in this gravel stream.

INTRODUCTION

The recent emphasis on a non-equilibrium view of communities (Strong *et al.*, 1984), as opposed to an equilibrium one, has drawn attention to the importance of environmental stochasticity, habitat heterogeneity and patchiness in running water ecosystems (Tokeshi, 1994; Schmid, 1993). Moreover, whatever view one holds on the importance of competition and other biotic interactions such as predation in structuring stream communities, assemblages consist of different kinds of species, distributed in patches of variable spatial extension (Schmid, 1993). While the relative abundance of species may be understood as a basic measure with which the biotic and abiotic effects on a community are estimated (May, 1975; Ugland & Gray, 1982; Tokeshi, 1990, 1993; Schmid, 1992a), the analysis of species abundance patterns has only been progressed beyond mainly statistical models recently by Tokeshi (1990).

In the present study, I test data on a larval chironomid assemblage inhabiting the bed sediments of a gravel stream against niche-oriented models of species abundance following Tokeshi (1990) and some newly developed models. An examination of the spatial resource utilization using neutral model approaches (Schmid, 1993) should reveal information about the spatial aspect of coexistence within the chironomid assemblage at the sediment surface and within the hyporheic interstitial of a gravel stream.

MATERIAL AND METHODS

The data for this study come from investigations on a chironomid community within a 100 m stretch of the second order limestone stream Oberer Seebach in Lower Austria (research area 'Ritrodat': 47°51′N, 5°04′E, 615 m a.s.l.). The study site is described in detail in Schmid (1992c,1993) and a brief description of the sampling process is given here. From September 1984 through August 1985 stratified random sampling of the bed sediments was performed at roughly monthly intervals using the freeze-coring method with prior electropositioning (Klemens, 1991). On each sampling occasion 10 freeze-cores were taken to sediment depth of 70 cm, obtaining 70 sampling units (7 depth layers). In the laboratory larvae were sorted and identified to species level. For species abundance calculations data thus gathered were pooled over all sediment depth layers for each sampling occasion, consisting of up to 6000 individuals per sample.

Furthermore, colonization experiments were conducted between March and June 1993 using a modified cage-pipe method. This cage-pipe sampler allows an instantaneous, undisturbed sampling at different sediment depth layers as well as any horizontal direction (Panek, 1991). The sampler consists of a plastic tube with rings of holes permanently installed in the stream bed. The topmost ring of holes is leveled with the topmost 5 cm of the bed sediments. A steel tube is inserted into the plastic tube and has corresponding rings of holes. The entrance holes into the pipe can be closed by pulling up the steel tube. Four small cages, made out of brass lattice with a solid bottom and lid, are inserted one above the other into the steel tube. Each cylindrical cage consists of one opening (630 mm^2) surrounded by a 80 μm mesh net. Five cage-pipes with four depth layers each (0–10 cm, 10–20cm, 20–30cm and 30–40cm) were exposed in a riffle zone of the gravel stream. Depending on the depth layer where the cages are exposed, the cages were filled with substrate from different depth layers of the brook. This substrate was rinsed and kept for 24 hours in filtered water at room temperature to remove all meio- and macrofauna, before it was exposed in the cages. The mean substrate volume of these cages is 85 cm^3 (SE±5.7) with a mean interstitial space of 47.8 cm^3 (SE±9.3). The cage-openings were exposed for these colonization experiments in the riffle-area (upstream) for 15, 24, 48, 96, 118, 264, 504 and 600 hours. After exposure the cages were removed and all individuals sorted live and identified to species and instar level.

Data analysis

In order to analyse the spatial and temporal variation in the total species composition a turnover index was used following Schmid (1993):

$$t = |1 - \sum^n \min [p_{ij}, p_{ik}]| \tag{1}$$

where p_i denotes the proportional abundance of species i in each of two spatially or temporally adjacent sample units (j, k) and n is the total number of species found in both sample units.

SPECIES ABUNDANCE MODELS

Two basic approaches are followed within niche-oriented model conceptions: (a) species assemble randomly within resource space with little or no interaction between them, and (b) species apportion randomly a multidimensional niche space in terms of sequential breakage processes which implies that species invasions in an ecological community are sequential, spatio-temporal processes.

Random Assemblage models
Random Assortment model (*cf.* Tokeshi, 1990; RA): in this model each species is assumed to carve out its own niche independently of other species in a temporally variable basis. Because of possibly continual change in resource space, species are unlikely to fill up the niche most of the time (Tokeshi, 1990). This model involves a random collection of species niches of any fraction length.
Complete Random Assemblage model (Schmid, this paper; CRA): this model stipulates the hypothesis that inter- and intraspecific competition between species plays a minor role in community organization and that the resource use of species is governed by environmental stochasticity. Here, the community is characterized by random colonization and dispersal processes, each species redistributing before space could become a limiting factor within a species assemblage. In this model each colonizing species takes independently from other species an arbitrary portion of the resource space available. The total resource space available is rarely completely filled by species because species continuously redistribute.

Random Resource-apportionment models
Random Fraction [1] model (*cf.* Tokeshi, 1990; RF1): under this model assumption species which colonize resource space randomly select one of the existing species in order to get a random portion of its niche.
Random Fraction [2] model (Schmid, this paper; RF2): this model is analogous to a colonization process where a species gets a random resource fraction left by its temporal predecessor.
Dominance Decay model (*cf.* Tokeshi, 1990; DD): in this model species invade the resource space only of the most dominant species in a species assemblage.
Dominance Preemption model (*cf.* Tokeshi, 1990; DP): this model is the inverse of the Dominance Decay model and it stresses that dominance of a species is considered absolute in that its abundance exeeds that of all the lower ranking species combined.
MacArthur Fraction model (*cf.* Tokeshi, 1990; MF): this model assumes that a new species invades the resource space of a more abundant species with a higher probability than a less abundant one.

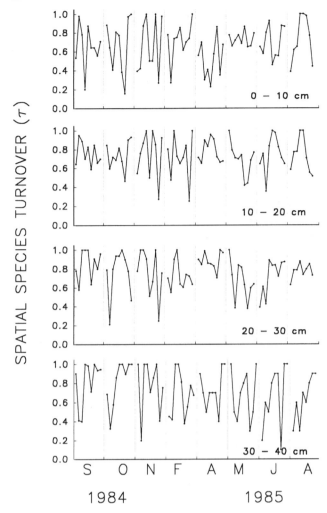

Fig. 1 Species turnover values (τ) between spatially adjacent core samples in each of four sediment depth layers in the stream Oberer Seebach during the observation period 1984/1985. Each dot represents the calculated spatial species turnover index for two adjacent core positions of a particular month.

For each of these models, 10 000 simulations were performed to create 10 000 five-species assemblages. To judge if the observed relative abundance of the chironomid species assembly is in conformity with one of the random models, 95% confidence limits (CL) were calculated for the mean abundances of each species from the simulated assemblages (cf. Tokeshi, 1990). If for all species the observed species abundance values fell within the corresponding 95% confidence limits from one of the models, the observed pattern did not significantly depart from the model's prediction ($P<0.05$).

Furthermore, these random models are compared to the deterministic Geometric Series model (May, 1975). An agreement with the Geometric Series (GS) should show reasonable constancy in the calculated k-values and a low value of discrepancy. The k-value in the GS as a niche-apportionment model defines the fraction of the total niche occupied by the first invading species.

NEUTRAL RESOURCE UTILIZATION MODELS

Neutral models of species assemblages may provide a way of analysing if a certain observed pattern in data is or is not likely to have arisen by chance. Thus, observed ecological patterns can be evaluated against appropriate randomization models which retain most of the characteristics of real communities, to aid in the interpretation of observed patterns (Caswell, 1976; Lawlor, 1980; Tokeshi & Townsend, 1987; Schmid, 1992b). Neutral models were developed by Monte-Carlo simulations for evaluating the significance of the observed overlap in spatial distribution in the most abundant five-species chironomid assembly (Schmid, 1993). Segregation or significantly small overlap in distribution was tested by comparing the observed overlap of pair-wise associations of larval species with the probability distribution of the values obtained from randomization models (Lawlor, 1980; Tokeshi & Townsend, 1987; Schmid, 1992b, 1993). To test the random dispersion/colonization hypothesis two random occurrence models were developed in which different aspects of randomness were incorporated. The first model assumes that 'patches' of larval individuals are randomly distributed over the habitat units (=sample units) and the second model stipulates the hypothesis that all larval individuals are randomly distributed over habitat units of the bed sediments (after Schmid, 1993).

RESULTS

Spatial and temporal species turnover

Eighty chironomid species and species groups were found within the bed sediments of the study site during the sampling period 1984 and 1985. The vertical distribution pattern from 0 to 70 cm revealed that maximum larval densities were recorded in the top 10 cm of the bed sediments, whereas 93.2% of all larval chironomids occurred between 0 and 40 cm. In view of this depth distribution, further analyses of the chironomid community concentrated on these upper 40 cm of the sediments. The spatial species turnover showed significant variation between the four sediment depth layers (ANOVA: $F_{3,308}=2.83$; $P<0.05$; Fig.1) and between horizontally adjacent core sites (ANOVA: $F_{77,234}=2.03$; $P<0.001$; Fig.1). Species turnover increased distinctly from the surface layer to the deeper sediment layers. This increased replacement of species by other species in deeper depth layers may be due to increasing variability in water through-flow rates and consequently higher sedimentation processes of particulate matter. Moreover, temporal species turnover significantly increased from September/October to November 1984/February 1985 and showed distinct fluctuations in the following months (ANOVA: $F_{6,7}=4.79$; $P<0.05$). On

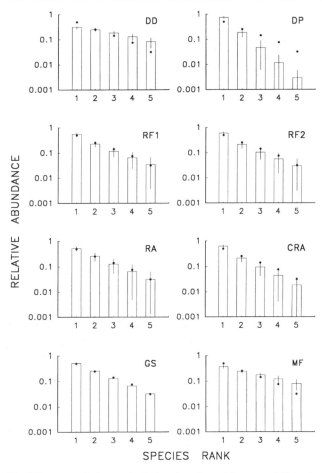

Fig. 2 Patterns of observed relative abundance (shown as full dots) with patterns derived from eight different models (shown as histograms). Vertical lines are expected 95% CL for simulated species-abundance models: DD: Dominance Decay, DP: Dominance Preemption, RF1: Random Fraction, RF2: Random Fraction [2], RA: Random Assortment, CRA: Complete Random Assemblage, GS: Geometric Series and MF: MacArthur Fraction.

the other hand, no significant differences in spatial species turnover were found between the two transects sampled (ANOVA: $F_{1,12}=0.47$; $P>0.50$) during the sampling period.

Species abundance patterns

Further analyses concentrated on a five-species assemblage of the most abundant detritivorous larval species *Corynoneura lobata* (Edwards), *Tvetenia calvescens* Edwards, *Synorthocladius semivirens* (Kieffer), *Micropsectra atrofasciata* Kieffer and *Rheotanytarsus nigricauda* Fittkau, which accounted for 54.2% of the total chironomid community during the observation period (Schmid, 1993). These five larval species were present in most stages of larval development throughout the year. Fig. 2 shows the observed relative abundance of species, together with eight model patterns. The observed pattern of rank abundance for the five-species assemblage departed significantly from the Dominance

Preemption model (DP) towards a more even situation, whereas an inverse trend was observed from the Dominance Decay (DD) and the MacArthur Fraction model (MF). Two of the resource apportionment models, the Random Fraction (RF1) and Random Fraction (RF2) model and the two random assemblage models, the Random Assortment (RA) and the Complete Random Assemblage (CRA) model significantly fitted with all observed relative abundance values ($P<0.05$). However, the Random Assortment model showed the most significant fit to the observed species-abundance pattern of the larval chironomid assemblage in the gravel stream (Fig.2).

The deterministic Geometric Series (GS) model with a mean k of 0.513 resembled the observed relative abundance values, although two of the five-species ranks (rank 3,4; Fig.2) lay outside the mean model prediction. Moreover, if the Geometric Series model gives an appropriate mechanism for species-abundance pattern, the resource-preemption constant k should be rather constant in value. The results of separately fitted monthly data sets to the Geometric Series show that this is not the case. The parameter k fluctuated distinctly between 0.396 and 0.634 during the observation period. Moreover, the mean discrepancy calculated as the sum of squared Euclidian distance between observation and model expectation ($D=0.641;\pm SE\ 0.015$) indicated a distinct departure of the observation from the model prediction.

Spatial stochasticity

Fig. 3 shows the probability for random colonization/dispersion processes based on comparisons of observed species pair-wise overlap values in spatial distribution with those expected from a patch model and an individual model. In the five-species assembly observed pattern of spatial overlap within the same depth layer resemble those generated by random placement of individuals and patches of individuals among habitat units in the same depth. Among 61.5% of all species-pairs the aggregation pattern was not strong enough to be significantly different from random colonization processes by individuals. Furthermore, 97.8% of all species-pairs indicated no significant departure from random patch formation processes. Thus, groups of individuals are formed independently by different species, leading to a decline in spatial overlap between species. Moreover, the fluctuations in aggregation patterns of larval individuals showed no significant differences between depth layers (ANOVA: $F_{3,269}=1.90$; $P>0.10$), whereas significant temporal variations in aggregation were evident (ANOVA: $F_{7,265}=2.08$; $P<0.05$). On the other hand, patches of larval individuals evidenced a significant increase for random colonization processes with increasing sediment depth (ANOVA: $F_{3,269}=2.95$; $P<0.05$) without distinct fluctuations in time (ANOVA: $F_{7,265}=2.02$; $P>0.05$).

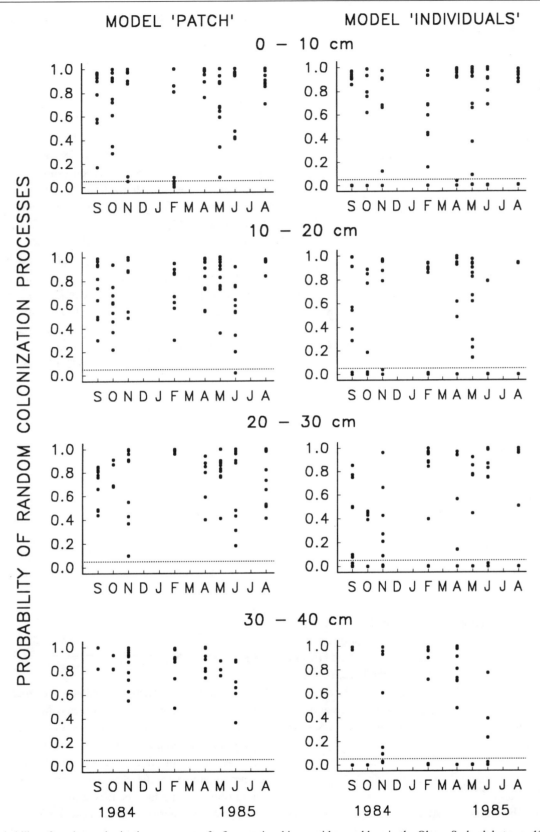

Fig. 3 Probability of random colonization processes of a five-species chironomid assemblage in the Oberer Seebach between 1984 and 1985. Full dots give the probability of random colonization processes (based on species pair-wise spatial overlaps) derived from two neutral models: patch-model and individual model. Horizontal, broken line gives level above which aggregation processes did not depart significantly from randomness ($P>0.05$).

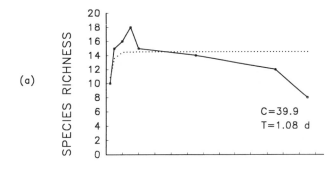

(a)

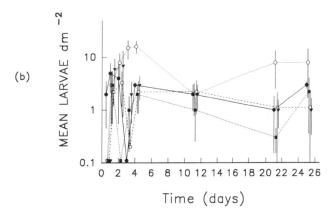

(b)

Fig. 4 Relationship between exposure time (days) and (a) species richness; superimposed dotted line represents the colonization curve (modified after Tokeshi & Townsend, 1987): $N(d) = M[1 - e^{((-C^*d)/M)}]$, where M is the mean species richness and C is a constant; T is the expected time to reach 95% of the species richness; and (b) mean number of individuals (±95%CL) per dm^2 (logarithmic scale) of *Corynoneura lobata* (solid dots), *Micropsectra atrofasciata* (open dots), *Synorthocladius semivirens* (solid square), *Tvetenia calvescens* (open square) and *Rheotanytarsus nigricauda* (solid triangle) during the colonization period between March and June 1993.

The spatial segregation between individuals of the five-species assemblage was quantified by comparison of observed spatial overlap values with those generated by randomizations of individuals of species (Tokeshi & Townsend, 1987; Schmid, 1992b, 1993). Spatial segregation oscillated around a mean value of 0.519 (±SE 0.033), which indicates that species show a tendency for colonization processes dictated by chance.

Colonization experiments

Fig. 4 shows the temporal pattern of species colonization during the observation period between March and June 1993. It is evident that the highest species number was achieved within a colonization period of less than 48 hours (Fig. 4a). In order to examine the rate at which new species immigrate into open microhabitat spaces until the larval

community approaches the mean species richness, an asymptotic colonization model was fitted to the data (Fig. 4a). The expected time of the larval chironomid community to reach 95% of the mean species richness was 1.08 days during the observation period (Fig. 4a). These results indicate fast colonization processes of larval species in the bed sediments of this gravel stream during the observation period.

However, comparing the colonization pattern of the five-species assemblage analysed, the results show similar species-specific responses in colonization capacity (Fig. 4b). The fastest colonizer among these larval species was *Corynoneura lobata*, which achieved the mean maximum density within a period of 24 hours. The tube-building species *Micropsectra atrofasciata* and *Rheotanytarsus nigricauda* reached higher abundances compared to all species, although the mean period to approach the maximum density was 2 and 5 days, respectively. On day 4 of the observation a distinct decline in densities of all larval species of the assemblage occurred except *R. nigricauda*. This density reductions were due to a spate following snow melt in April 1993 (Fig. 4b). Despite the decline in number of individuals, the number of species colonizing open microhabitat patches was not affected by the high water (Fig. 4a). Moreover, comparing the colonization pattern of all larvae evidenced a significant decline of colonizing individuals with increasing sediment depth (ANOVA: $F_{3,3096} = 17.42$; $P < 0.001$) and significant variation between horizontally adjacent cage-pipe positions (ANOVA: $F_{4,3095} = 7.30$; $P < 0.001$).

Thus, marked variations in the colonization rates in the downstream direction were observed for all larvae and instar stages of the five-species assemblage, suggesting random movements across and between adjacent microhabitat patches in a riffle area of the gravel brook Oberer Seebach.

DISCUSSION

Factors such as disturbance, habitat heterogeneity and patchiness may be important characteristics in structuring running water communities (Reice, 1974; Schmid, 1993). However, recent emphasis on the examination of concepts such as equilibrium versus non-equilibrium based hypotheses, accompanied by the necessity of critical appraisals of relationships among factors, may lead to an understanding of the given complexity of species coexistence in communities (Ugland & Gray, 1982; Tokeshi & Townsend, 1987; Tokeshi, 1993, 1994; Schmid,1992b; 1993).

In running water ecosystems aggregation of food resources such as particulate organic matter (Leichtfried, 1986) and specimens might be buffered by disturbance-mediated processes such as discharge fluctuations (Schmid-Araya, 1993; Schmid, 1992a,b, 1993), and variable dispersal processes (Tokeshi & Pinder, 1986; Tokeshi & Townsend, 1987). The degree of aggregation of larval chironomids varies substantially in time and in space as observed on leaves of two submerged macrophytes (Tokeshi & Pinder, 1986), at the sediment surface of the River Danube (Schmid, 1992b) and in the bed sediments of the stream Oberer Seebach (Schmid, 1993). A general approach to explain variable patchy distribution patterns may be given when considering stochasticity in dispersal, microhabitat colonization and resource availability in running water ecosystems.

Furthermore, species-abundance patterns are assumed to reflect structures in ecological communities (May, 1975), both driven by biotic and abiotic influences on the community. Sugihara (1980) and Tokeshi (1990) have introduced the hypothesis of resource apportionment in species-abundance patterns. If a community consists of species with similar resource use and tendencies to interact within the same resource space, the process of resource-apportioning among species can be linked to a unit stick being fractioned into pieces according to some deterministic or/and stochastic division rules. The resource apportioning models analysed in this study are based on sequential breakage processes of the resource spectrum (Pielou, 1975; Sugihara, 1980) incorporating randomness in the process of resource apportionment (Tokeshi, 1990, 1993). Moreover, random assemblage models envisage random colonization and dispersal processes by species more or less indepedent from other species within the community. The significant fit of the observed species-abundance pattern with the Random Fraction models and the random assemblage models, indicate that species may both randomly colonize the bed sediments of the gravel stream and temporally select a random fraction of the available habitat resource. The greater significance at the fit of the observed five-species assemblage data to the Random Assortment model, is evidence of a tendency of the chironomid assemblage to colonize stochastically any given habitat unit without competition as strong structering force in this particular species assembly. Similar results were found by Tokeshi (1990) for an epiphytic chironomid assemblage, where both the Random Fraction model and Random Assortment model fitted significantly to the observed relative abundances of six species in terms of numbers. The same author also compared species-abundance pattern in terms of biomass with most of those models and found that none of the resource-apportionment models achieved a successful fit, whereas the Random Assortment model fitted significantly to the biomass data. These results may relate generally to a stochastically, dynamic non-equilibrium situation where species interactions are of minor importance.

Although the Geometric Series model appeared to resemble the observed relative abundance values, the parameter k varied distinctly during the observation period. A similar, temporal variation of the niche preemption constant was observed for an ostracod assemblage in the bed sediments of the Oberer Seebach (Marmonier, 1984) which also indicated for the ostracode assemblage a distinct departure from the model prediction.

Moreover, the mean spatial segregation among the five-species assemblage did not depart from the expectations of a random colonization process, implying that the amount of overlap in spatial distribution is governed by stochastic dispersal processes of individuals between habitat units of the gravel stream. These dispersal and colonization processes may be driven by the fluctuations in discharge, which continously re-structure the chironomid community in the Oberer Seebach (Schmid, 1993). Furthermore, results based on colonization experiments in the period between March and June 1993 evidenced that within 26 hours, 95% of the mean species richness is reached in open microhabitat patches of the bed sediments. This rapid species re-colonization period is comparable to results obtained in a small river in eastern England (Tokeshi & Townsend, 1987). However, the high variation in larval densities between spatially adjacent colonization sites reflects a high probability for random movements of species in both horizontal and vertical directions of the bed sediments.

In conclusion, the larval chironomid community in the gravel stream Oberer Seebach may be governed by environmental stochasticity, reducing the possibility for interspecific interactions. Randomly formed patches and variable degrees of aggregation between and among species may support a species-rich community at the bed sediment surface and within the hyporheic interstitial of a gravel stream. Random movements across and between sediment depth layers promote a rapid colonization of open microhabitat patches in this stream. At the same time, larval species seemed to colonize indepedently from other species microhabitat patches and may redistribute before space could become a limiting factor within the species assemblage.

ACKNOWLEDGEMENTS

Parts of this study were conducted while PES was in receipt of a grant from the Austrian Fond zur Förderung der wissenschaftlichen Forschung (FWF-Projekt: P 8007). I am deeply indebted to Dr Jenny Schmid-Araya and Dr Mutsunori Tokeshi for inspiring discussions concerning aspects of this study. Furthermore, I thank two referees for their interesting comments on the manuscript and Mr Rob Foote for checking the English.

REFERENCES

Caswell, H. (1976). Community structure: a neutral model analysis. *Ecological Monographs*, **46**, 327–354.

Klemens, W. E. (1991). Quantitative sampling of bed sediments (Ritrodat-Lunz study area, Austria). *Verh. Internat. Verein. Limnol.*, **24**, 1926–1926.

Lawlor, L. R. (1980). Structure and stability in natural and randomly constructed competitive communities. *Amer. Nat.*, **116**, 394–408.

Leichtfried, M. (1986). *Räumliche und zeitliche Verteilung der partikulären organischen Substanz (POM-Particulate organic matter) in einem Gebirgsbach als Energiebasis der Biozönose.* Ph.D Thesis, University of Vienna, pp.360.

Marmonier, P. (1984). Vertical distribution and temporal evolution of the ostracod assemblage of the Seebach sediments. *Jahresbericht der Biologischen Station Lunz*, **7**, 49–82.

May, R.M. (1975). Patterns of species abundance and diversity. In *Ecology and Evolution of Communities*, ed. M. L.Cody & J.M.Diamond, pp.81–120, Cambridge, Mass., Belknap/Harvard University Press.

Panek, K. L. J. 1991. *Dispersionsdynamik des Zoobenthos in the Bettsedimenten eines Gebirgsbach.* PhD thesis. University of Vienna, pp. 190.

Pielou, E. C. (1975). *Ecological Diversity*. New York, Wiley-Interscience.

Reice, S. R. (1974). Environmental patchiness and the breakdown of leaf litter in a woodland stream. *Ecology*, **55**, 1271–1282.

Schmid, P. E. (1992a). Community structure of larval Chironomidae (Diptera) in a backwater area of the River Danube. *Freshwater Biology*, **27**, 151–168.

Schmid, P. E. (1992b). Population dynamics and resource utilization by larval Chironomidae (Diptera) in a backwater area of the River Danube. *Freshwater Biology*, **28**, 111–127.

Schmid, P. E. (1992c). Habitat preferences as patch selection of larval and emerging chironomids (Diptera) in a gravel brook. Netherlands *Journal of Aquatic Ecology*, **26**, 419–429.

Schmid, P. E. (1993). Random patch dynamics of larval Chironomidae (Diptera) in the bed sediments of a gravel stream. *Freshwater Biology*, **30**, 239–255.

Schmid-Araya, J. M. (1993). Spatial distribution and population dynamics of a benthic rotifer, *Embata laticeps* (Murray) (Rotifera, Bedlloidea) in the bed sediments of a gravel brook. *Freshwater Biology*, **30**, 395–408.

Strong, D. R., Simberloff, D., Abele, L. G. & Thistle, A. B. (1984). *Ecological communities. Conceptual Issues and the Evidence.* Princeton University Press, pp.613.

Sugihara, G. (1980). Minimal community structure: an explanation of species abundance patterns. *Amer. Nat.*, **116**, 770–787.

Tokeshi, M. (1990). Niche apportionment or random assortment: Species abundance patterns revisited. *J. An. Ecol.*, **59**, 1129–1146.

Tokeshi, M. (1993). Species abundance patterns and community structure. *Adv. Ecol. Res.*, **24**, 111–186.

Tokeshi, M. (1994). Community ecology and patchy freshwater habitats. In *Aquatic Ecology: Scale, Pattern & Process*, ed. P. S. Giller *et al.*, pp.63–90. Blackwell Scientific Publications.

Tokeshi, M & Pinder, L.C.V. (1986). Dispersion of epiphytic chironomid larvae and the probability of random colonization. *Internationale Revue der gesamten Hydrobiologie*, **71**, 613–620.

Tokeshi, M. & Townsend, C.R. (1987). Random patch formation and weak competition: coexistence in an epiphytic chironomid community. *J. An. Ecol.*, **56**, 833–845.

Ugland, K. I. & Gray, J. S. (1982). Lognormal distribution and the concept of community equilibrium. *Oikos*, **39**, 171–178.

4   Temporal and spatial dynamics of meiofaunal assemblages in the hyporheic interstitial of a gravel stream

J.M. SCHMID-ARAYA

Biological Station Lunz, Institute of Limnology, Austrian Academy of Sciences, A-3293 Lunz am See, Austria

ABSTRACT The meiofaunal assemblages within the interstitial hyporheic (0–40 cm) of a gravel stream were studied at fortnightly sampling intervals between October 1991 and October 1992. Four taxa constituted 79.7% of the community: Nematoda, Rotifera, Cyclopoida and Gastrotricha. Significant seasonal variation occurred in most meiofaunal groups, and were characterized by either one abundance peak (gastrotrichs and harpacticoids) or two abundance maxima (microturbellarians, nematodes, rotifers and cyclopoids). These taxa showed significant differences in their mean densities between sediment depths. Most meiofaunal groups exhibited highest densities at sediment depths between 20 and 40 cm, whereas the depth distribution of Rotifera was more variable among the sites without a distinct depth maxima.

The effect of variables such as sediment depth, water depth, temperature, variation of groundwater levels and discharge was tested upon the mean abundances of each taxa. Except for Rotifera, a combined effect of some of these variables was detected in most taxa. Sediment depth was positively related to meiofauna densities suggesting that densities increased with increasing depth.

INTRODUCTION

In freshwater pelagic ecosystems the role of micro-meiozooplankton (i.e. Protozoa and Rotifera) as important groups in the food chain is well known (Pace & Orcutt, 1981). In streams there is an increasing awareness that these micro-meiofaunal groups may constitute a potential link between the heterotrophic production and the meio- and macro-invertebrate predators. Recently, Schmid (1994) demonstrated that rotifers are important prey items for early instars of predatory chironomids in a gravel stream.

On the other hand, the early recognition of a distinct hyporheic fauna by Orghidan (1959) and Schwoerbel (1961), later revisited by many authors (i.e. Stanford & Ward, 1988, Gibert *et al.*, 1990, Marmonier & Creuzé des Châtelliers, 1991, Boulton *et al.*, 1992, among many others) emphasizes the increasing interest towards the understanding of the hyporheic community. However, most of the studies deal with invertebrate size classes which do not include meiofaunal taxa such as Rotifera, Gastrotricha, and Microturbellaria. The

reasons might be related to the fact that mesh nets were too large to retain some of these taxa (Palmer, 1990), or simply the fixation methods used. Ruttner-Kolisko (1971) already mentioned that these three taxa can not be efficiently preserved.

Thus, in the literature some of these meiofaunal groups have been considered as poor representative groups (i.e. Boulton *et al.*, 1992), or they have been omitted from taxonomic lists. In contrast, Palmer (1990) has shown that some of the meiofaunal groups (i.e. Rotifera, Nematoda, Copepoda) can achieve high densities in sandy streams.

Compared to macroinvertebrates such as most insect larvae, meiofaunal groups (Microturbellaria, Gastrotricha, Rotifera, Nematoda, Oligochaeta, Tardigrada, and Microcrustacea) are particularly suited by their size and shape to interstitial life, and are important permanent inhabitants of the stream surface and/ or the hyporheic zone.

The present study assesses the spatio-temporal dynamics of meiofaunal taxa inhabiting the hyporheic interstitial of a gravel stream. The possible effect of some hydrophysical variables on their distribution pattern is also analysed.

(a)

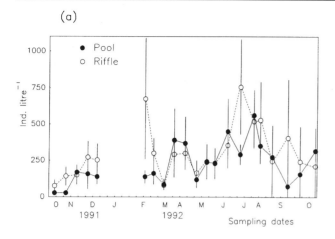

(b)

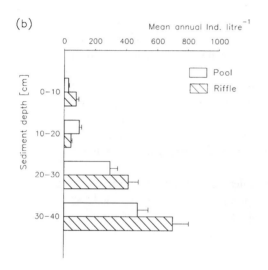

Fig. 1 The total meiofauna densities in the hyporheic interstitial of the gravel stream Oberer Seebach between October 1991 and October 1992. (a) Mean densities of all meiofaunal taxa in pool and riffle sites during the year (vertical lines:±1SE). (b) The annual mean depth distribution of densities in pool and riffle sites (horizontal lines:±1SE).

MATERIAL AND METHODS

Sampling was carried out in the experimental research area (Ritrodat) of the Oberer Seebach in Lower Austria (47°51′N, 5°04′E; 615 m a.s.l.) a calcareous and summer cold gravel stream (for further description, see Bretschko, 1991).

For sampling the meiofauna, quantitative samples were collected using stand-pipe traps (Bretschko & Klemens, 1986). These traps were installed at two locations within the research area characterized as riffle and pool sites (Schmid-Araya, 1993b). Replicated samples were collected every fortnight at each of four sediment depths (0–10, 10–20, 20–30 and 30–40 cm) between October 1991 through October 1992. The traps are made of plastic tubes with a circle of catching

holes (diameter 8 mm) near the bottom of each tube. The catching holes can be closed between sampling dates using a smaller piece of tube with a foam rubber collar. While sampling, this piece of tube is removed, so that invertebrates can enter via the catching holes. In this study, one liter of sediment-water sample was collected after keeping the traps open for 24 h.

These samples were kept cool, and transported to the laboratory, then were homogenized and divided into non-sieved and three sieved fractions (>250 μm, >100 μm, and >30 μm), and stored at 4 °C. Identification and enumeration of microfauna (see Schmid-Araya, 1994b) and meiofauna were carried out on each fraction of the live material as recommended for interstitial fauna by Ruttner-Kolisko (1971).

Water temperature, water level and groundwater level are continuously recorded in the experimental area, and at each sampling occasion measurements on water depth at the stand-pipe traps were also made.

Data analysis

Parametric statistical tests were used, and when appropiate densities were log $(x+1)$ transformed. Significance tests of differences between mean densities of each taxa among sampling dates and sediment depth layers were done with one-way analysis of variance (ANOVA). To examine the relationship between hydrophysical variables and meiofaunal densities, stepwise multiple regression was used. The significance of each variable was judged by the t value associated with each regression coefficient, and ß coefficients were used as indicators of relative importance of the various hydrophysical variables (Zar, 1984).

RESULTS

Temporal and depth distribution

Total meiofauna densities in the hyporheic interstitial of the Oberer Seebach fluctuated significantly throughout the year (ANOVA $P<0.001$; Fig. 1a). Total density reached a maximum value of 3475 ind.l^{-1} at the end of winter 1991/92, and 2225 ind.l^{-1} at the beginning of summer 1992. Throughout the year, meiofaunal taxa were found in all depth layers, but higher mean annual densities occurred at 30–40 cm sediment depth in both pool and riffle sites (Fig. 1b).

The most abundant meiofauna taxa were nematodes (31.4%), rotifers (26.9%), cyclopoid copepods (11.6%), and gastrotrichs (9.8%). All other meiofaunal groups accounted for the remaining 20.3%.

Significant seasonal variations occurred in most meiofaunal groups (ANOVA, $P<0.05$; Fig. 2), with the exception of Cladocera, Tardigrada and cyclopoid copepods (ANOVA,

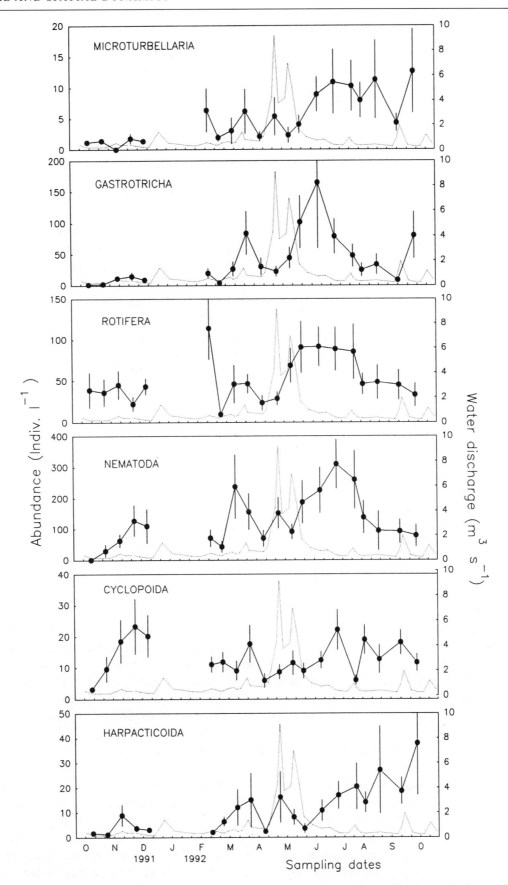

Fig. 2 Temporal mean densities of each representative meiofaunal taxon in the hyporheic interstitial (0–40 cm) of a gravel stream (vertical lines: ±1SE). Dotted lines represent the surface discharge for the same period of time.

P>0.05). The first two taxa were not sufficiently abundant to show meaningful density differences. During the year, the peak abundances of each taxa did not coincide. Taxa that exhibited one abundance peak were gastrotrichs with a maximal peak in spring-summer 1992, and harpacticoid copepods in autumn 1992 (Fig. 2). The other meiofaunal taxa had two abundance peaks during the year. Microturbellarians were more abundant in summer and autumn 1992. Both, interstitial rotifers and nematodes had higher densities in winter 1991/92 (mid- and late, respectively), and during the summer of 1992 (Fig. 2).

The meiofaunal taxa shown in Figs. 3 and 4, were found at all depth layers of the hyporheic interstitial. These taxa showed significant differences in their mean densities between sediment depths (ANOVA, *P*<0.05). The exception was the cladocerans, probably due to their low densities. Microturbellarians and gastrotrichs exhibited highest densities at sediment depths between 20 and 30 cm, and between 30 and 40 cm in pool and riffle sites. In contrast, the depth distribution of Rotifera was more variable among the sites without a distinct depth maxima (Fig. 3). Maximal abundances for nematodes, cyclopoid and harpacticoid copepods were found at depth layers between 20 and 40 cm in the riffle and pool areas (Fig. 4).

The effects of environmental variables

Table 1 illustrates the results of multiple regression analysis performed to investigate the relationship between meiofaunal densities and the variables sediment depth, water temperature, water depth, discharge and variations of the groundwater levels.

In the hyporheic interstitial the combination of sediment depth and water temperature was significantly positively related to densities of microturbellarians and oligochaetes (Table 1). In both taxa, sediment depth had a higher influence (ß coefficients), which indicates that abundances increased with increasing depth. Abundances of gastrotrichs and nematodes were significantly related to a combination of sediment depth, water depth, water temperature and discharge. In both groups, sediment depth layer and water levels showed a higher and positive relation (ß coefficients, Table 1). In Rotifera only the variable sediment depth had a significant and positive relation to their abundances (Table 1). The same variable combined with discharge was related to cyclopoid densities. A significant combined effect of sediment depth, variation on groundwater level, water temperature and discharge was found upon the densities of harpacticoid copepods. In taxa which were significantly related to the variable water discharge (Gastrotricha, Nematoda, Cyclopoida and Harpacticoida), the effect was negative implying that at increasing discharge values densities may decrease.

DISCUSSION

The variety of sampling methods employed in other studies contrast to the methodology used here and makes it difficult to compare results. Nevertheless, the use of stand-pipe traps is a suitable method for sampling in gravel streams, and may have the advantage of estimating animal activity and abundances if exposure time is kept constant throughout the sampling period (Waringer, 1987). The mean densities of meiofauna taxa found in the hyporheic interstitial of the Oberer Seebach fall within the reported abundance ranges of some taxa at the sediment surface and per volume of sediment in other types of streams (Williams & Hynes, 1974; Godbout & Hynes, 1982; Whitman & Clark, 1984). However, the rotifer mean densities are lower in the gravel stream Oberer Seebach when compared to total densities reported by Palmer (1990) in a sandy sediment stream. The same author evidenced that pronounced seasonal distribution occurred as also found in this study. Nevertheless, for comparable groups such as rotifers and nematodes the seasons at which peak maxima were found, slightly differed between these two stream types. A similar seasonal variation was observed when comparing Palmer's total copepods with harpacticoids in this study. Moreover, Kowarc (1991) using freeze-core techniques found that harpacticoids in the Oberer Seebach had two peaks, one in spring and one in late autumn, and their vertical distribution is species-specific.

On the other hand, a significant effect of variables such as temperature, mean current velocity and water discharge upon the densities of some meiofaunal taxa has been shown by Palmer (1990). The results of this study coincide with the significant effect of water temperature and water discharge, but differed for each meiofaunal taxa with the previous study. In contrast, total number of rotifers in the Oberer Seebach were not significantly related to these variables. Furthermore, because of the lack of information about the seasonality and the relations to hydrophysical variables in other streams, the results found here for microturbellarians and gastrotrichs await further research to confirm these findings. This study evidences that: a) sediment depth had a high and positive influence upon the meiofaunal taxa implying that densities increase with increasing depth, and b) the negative influence of discharge upon some taxa suggests that densities decrease with increasing discharge values. Changes within the hyporheic interstitial communities after spates by migrating downwards have been demonstrated in other studies (i.e. Williams & Hynes, 1974; Dole-Olivier & Marmonier, 1992). Similarly, Schmid-Araya (1996) demonstrated that discharge had a significant effect on the dispersion patterns of some rotifer species, and that increasing values of water levels and discharge coincided with the

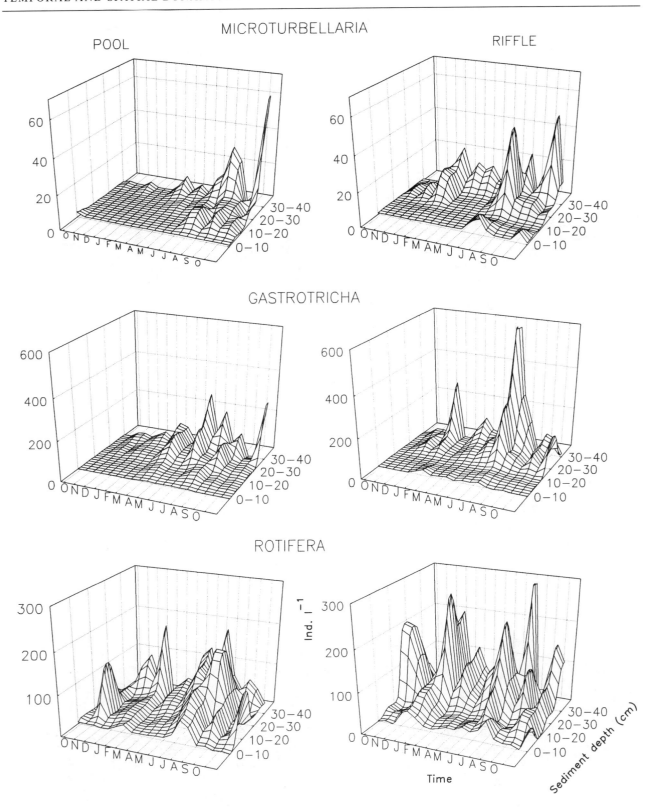

Fig. 3 Vertical distribution of microturbellarians, gastrotrichs and rotifers in the hyporheic interstitial of the Oberer Seebach.

increases on densities of rotifer interstitial populations. In the gravel stream Oberer Seebach higher numbers of bacteria (Kasimir, 1996), and meio-microfauna (Schmid-Araya, 1994a,b) have been recorded in the hyporheic interstitial than at the surface of the stream. This vertical distribution coincides with previous results showing an increase of organic

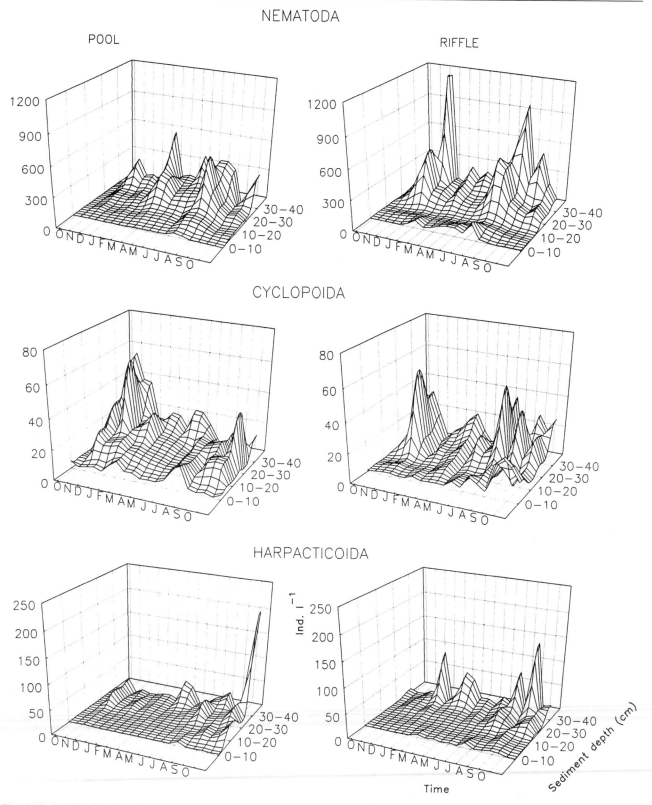

Fig. 4 Vertical distribution of nematodes, cyclopoids and harpacticoids in the hyporheic interstitial of the Oberer Seebach.

matter with depth (Leichtfried, 1991). Similar depth distributions have been reported for Ostracoda in the same gravel stream (Marmonier, 1984).

In general, studies dealing mainly with macrofauna taxa and some meiofaunal groups (i.e. Nematoda, Copepoda) have reported two different vertical distribution patterns: a)

Table 1. *The results of a multiple regression for the prediction of abundances of meiofaunal taxa inhabiting within the hyporheic interstitial of a gravel stream. Coef. is the regression coefficient associated with each independent variable; the t-value and associated probability (P) indicate of each variable in the prediction. ß-coef. is the standarized partial regression coefficient. (SE: standard error)*

Variable	Coef.	±SE	ß-coef.	t-value	Coef.-P
1. MICROTURBELLARIA	(F: 39.858; df: 2332; Model $P<0.001$)				
Constant	−0.447	0.204		−2.196	0.030
Sediment depth	0.325	0.041	0.397	8.004	<0.001
Temperature	0.111	0.022	0.241	4.864	<0.001
2. GASTROTRICHA	(F: 50.967; df: 5329; Model $P<0.001$)				
Constant	−2.064	0.369		−5.598	<0.001
Sediment depth	0.935	0.064	0.616	14.630	<0.001
Water depth	0.024	0.004	0.311	5.347	<0.001
Water temperature	0.150	0.036	0.176	4.134	<0.001
Discharge	−0.161	0.043	−0.210	−3.764	<0.001
3. ROTIFERA	(F: 95.014; df: 1333; Model $P<0.001$)				
Constant	1.316	0.182		7.235	<0.001
Sediment depth	0.649	0.067	0.471	9.748	<0.001
4. NEMATODA	(F: 80.400; df: 4330; Model $P<0.001$)				
Constant	−1.964	0.419		4.677	<0.001
Sediment depth	1.227	0.072	0.666	16.867	<0.001
Water depth	0.033	0.005	0.347	6.468	<0.001
Water temperature	0.145	0.041	0.140	3.515	<0.001
Discharge	−0.182	0.048	−0.200	−3.753	<0.001
5. OLIGOCHAETA	(F: 22.316; df: 2332; Model $P<0.001$)				
Constant	0.001	0.203		0.007	0.994
Sediment depth	0.246	0.041	0.315	6.075	<0.001
Water temperature	0.079	0.022	0.180	3.470	<0.001
8. CYCLOPOIDA	(F: 34.600; df: 2332; Model $P<0.001$)				
Constant	1.023	0.145		7.043	<0.001
Sediment depth	0.412	0.051	0.397	7.955	<0.001
Discharge	−0.063	0.026	−0.120	−2.440	0.020
9. HARPACTICOIDA	(F: 33.394; df: 4330; Model $P<0.001$)				
Constant	−1.310	0.357		−3.672	<0.001
Sediment depth	0.529	0.049	0.497	10.600	<0.001
Variation groundwater level	0.018	0.008	0.143	2.232	0.030
Water temperature	0.158	0.028	0.265	5.560	<0.001
Discharge	−0.088	0.033	−0.170	−2.638	0.008

greater abundance occurs at the upper layers of the hyporheic interstitial (10–20 cm below surface (i.e. Williams & Hynes, 1974, Bretschko, 1992), and b) highest densities at the surface with decreasing abundances with increasing depth (Pugsley & Hynes, 1986, McElravy & Resh, 1991, among others).

Palmer (1990), reported meiofaunal densities at all depth sampled and occasionally down to 60 cm; however some dominant taxa achieved the highest abundances at the top 10 cm of the sediment. In contrast, in this gravel stream highest densities of meiofauna were found at deeper sediment depths (e.g. 20–30 and 30–40). The well-oxygenated hyporheic interstitial (Bretschko, 1991), and a heterogeneous grain size dis-

tribution of the Oberer Seebach might explain the high meiofaunal abundances at deeper sediment depths.

The relative importance of meiofaunal groups has not been yet fully incorporated within the functioning of a stream ecosystem. But in marine habitats, meiofauna have been shown to enhance the availability of detritus to macroconsumers (Tenore *et al.*, 1977). It is plausible that during the shredding process by large macroinvertebrates, finer particles which are not efficiently grazed are released, and in turn can be further processed by micro-meiofaunal taxa. An example in this gravel stream is the epizoic bdelloid rotifer *Embata laticeps* which settles on different body parts of larger macrofauna taxa and filters the particles which are released when the host feeds (Schmid-Araya, 1993a).

In addition, meiofaunal taxa such as rotifers, gastrotrichs, and copepoda, often scrape sediment particles, and faecal pellets of larger macrofauna searching for the attached bacteria (Schmid-Araya, personal observation). Lately, Findlay & Arsuffi (1989) found that meiofauna taxa contribute significantly to the total secondary production, and may be an important part in the food web. In the Oberer Seebach, Schmid (1994) has found up to 30 rotifer species, and other diverse meiofaunal groups (e.g. Tardigrada, Harpacticoida) in the guts of first and second instar of predatory larval Tanypodinae (Chironomidae).

In conclusion, this study evidences that at least two abundant taxa such as Rotifera and Gastrotricha achieved high densities throughout the year in this gravel stream, and these taxa may have a relevant role within the meiofauna in such a fluctuating stream. Similarly, the distinct seasonal abundance of each meiofaunal group and the combined effect of several hydrophysical variables which varies among taxa, emphasize the complex dynamics of meiofauna groups in the hyporheic interstitial of a gravel stream.

ACKNOWLEDGEMENTS

This study is part of a project financed by the Austrian Fonds zur Förderung der wissenschaftlichen Forschung (FWF P8035–BIO). The author thanks Dr P.E. Schmid for valuable help while sampling and valuable discussions on the topic. I gratefully acknowledge the logistic help of Dr M. Leichtfried. I also specially thank to Frau E. Kronsteiner for helpful assistance.

REFERENCES

Boulton, A.J., Valett, H.M. & Fischer, S.G. (1992). Spatial distribution and taxonomic composition of the hyporheos of several Sonoran Desert streams. *Arch. Hydrobiol.*, **125**, 37–61.

Bretschko, G. (1991). The limnology of a low order alpine gravel stream (Ritrodat-Lunz study area, Austria). *Verh. Internat. Verein. Limnol.*, **24**, 1908–1912.

Bretschko, G. (1992). Differentiation between epigeic and hypogeic fauna in gravel streams. *Regulated Rivers*, **7**, 17–22.

Bretschko, G. & Klemens, W. (1986). Quantitative methods and aspects in the study of the interstitial fauna of running waters. *Stygologia*, **2**, 279–316.

Dole-Olivier, M.J. & Marmonier, P. (1992) Effects of spates on the vertical distribution of the interstitial community. *Hydrobiologia*, **230**, 49–61.

Findlay, S.E.G. & Arsuffi, T.L. (1989). Microbial growth and detritus transformations during decomposition of leaf litter in a stream. *Freshwater Biology*, **21**, 261–269.

Gibert, J., Dole-Olivier, M.J., Marmonier, P. & Vervier, P. (1990). Surface water/groundwater ecotones. In *Ecology and management of aquatic-terrestrial ecotones*, ed. R.J. Naiman & H. Décamps, pp. 199–225,

Man and the Biosphere Series, Volume 4. UNESCO, Paris and the Parthenon Publishing Group, Carnforth, England.

Godbout, L. & Hynes, H.B.N. (1982). The three dimensional distribution of the fauna in a single riffle in a stream in Ontario. *Hydrobiologia*, **97**, 87–96.

Kasimir, G. (1996). *Microbial biomass and activities in a second order mountain brook*. 5th International Worshop on the Measurement of Microbial Activities in the Carbon Cycle in Aquatic Environments, University of Copenhagen, Denmark.

Kowarc, V. (1991). Distribution of harpacticoids in a second order mountain stream (Ritrodat-Lunz study area, Austria). *Verh. Internat. Verein. Limnol.*, **24**, 1930–1933.

Leichtfried, M. (1991). POM in a gravel stream (Ritrodat-Lunz Study Area, Austria, Europe). Verh. Internat. *Verh. Internat. Verein. Limnol.*, **24**, 1921–1925.

Marmonier, P. (1984). Vertical distribution and temporal evolution of the Ostracod assemblage of the Seebach sediments (Lunz, Austria). *Jahresbericht Biologische Station Lunz*, **7**, 49–82.

Marmonier, P. & Creuzé des Châtelliers, M. (1991). Effects of spates on interstitial assemblages of the Rhône River. Importance of spatial heterogeneity. *Hydrobiologia*, **210**, 243–251.

McElravy, E.P. & Resh, V.H. (1991). Distribution and seasonal occurrence of the hyporheic fauna in a northern Californian stream. *Hydrobiologia*, **220**, 233–246.

Orghidan, T. (1959). Ein neuer Lebensraum des unterirdischen Wassers: Der hyporheische Biotop. *Arch. Hydrobiol.*, **55**, 392–414.

Pace, M.L. & Orcutt, J.D. Jr. (1981). The relative importance of protozoans, rotifers and crustaceans in a freshwater zooplankton community. *Limnol. Oceanogr.*, **26**, 822–830.

Palmer, M.A. (1990). Understanding the movement dynamics of a stream-dwelling meiofauna community using marine analogs. *Stygologia*, **5**, 67–74.

Pugsley, C.W. & Hynes, H.B.N. (1986). A freeze-core technique to quantify the three dimensional distribution of fauna and substrate in stony streambeds. *Can. J. Fish. Aqua. Sci.*, **40**, 637–643.

Ruttner-Kolisko, A. (1971). *Interstitial Fauna. A Manual on Methods for the Assessment of Secondary Productivity in Fresh Waters*, ed. W.T. Edmondson & G.G. Winberg, pp. 122–124. IBP Handbook no. 17, Oxford, Edingburgh: Blackwell Scientific Publications.

Schmid, P.E. (1994). Is prey selectivity by predatory Chironomidae a random process? *Verh. Internat. Verein. Limnol.*, **25**, 1656–1660.

Schmid-Araya, J.M. (1993a). Benthic Rotifera inhabiting the bed sediments of a mountain gravel stream. *Jahresbericht Biologische Station Lunz*, **14**, 75–101.

Schmid-Araya, J.M. (1993b). Spatial distribution and population dynamics of a benthic rotifer, *Embata laticeps* (Murray) (Rotifera, Bdelloidea) in the bed sediments of a gravel brook. *Freshwater Biology*, **30**, 395–408.

Schmid-Araya, J.M. (1994a). Spatial and temporal distribution of micro-meiofaunal groups in an alpine gravel stream. *Verh. Internat. Verein. Limnol.*, **25**, 1649–1655.

Schmid-Araya, J.M. (1994b). The temporal and spatial distribution of benthic microfauna in the bed sediments of a gravel stream. *Limnol. Oceanogr.*, (in press).

Schmid-Araya, J.M. (1995). Disturbance and population dynamics of rotifers in bed sediments. *Hydrobiologia*, **313/314**, 279–301.

Schwoerbel, J. (1961). Über die Lebensbedingungen und die Besiedlung des hyporheischen Lebensraum. *Arch. Hydrobiol./ Suppl.*, **25**, 182–214.

Stanford, J.A. & Ward, J.V. (1988). The hyporheic habitat of river ecosystems. *Nature*, **335**, 64–65.

Tenore, K.R., Tietjen, J.H. & Lee, J.J. (1977). Effect of meiofauna on incorporation of aged eelgrass, *Zostera marina*, detritus by the polychaete *Nephthys incisa*. *J. Fish. Res. Board Can.*, **34**, 563–567.

Waringer, J.A. (1987). Spatial distribution of Trichoptera larvae in the sediments of an Austrian mountain brook. *Freshwater Biology*, **18**, 469–482.

Whitman, R.L. & Clark, W.J. (1984). Ecological studies of the sand-dwelling community of an East Texas stream. *Freshwater Inver. Biol.*, **3**, 59–79.

Williams, D.D. & Hynes, H.B.N. (1974). The occurrence of benthos deep in the substratum of a stream. *Freshwater Biology*, **4**, 233–256.

Zar, J.H. (1984). *Biostatistical analysis*. Prentice Hall.

5 Interstitial fauna along an epigean-hypogean gradient in a Rocky Mountain river

J. V. WARD* & N. J. VOELZ*

* Department of Biology, Colorado State University, Fort Collins, Colorado 80523, USA

** Department of Biology, St. Cloud State University, St. Cloud, Minnesota 56301, USA

ABSTRACT Interstitial animals were collected during spring, summer, and autumn from eight sampling sites along a Rocky Mountain river to examine small scale patterns of diversity, abundance, and faunal composition across the groundwater/surface water ecotone. At each site samples were taken from benthic habitats (superficial bed sediments), hyporheic habitats (underflow 30 cm below the bed surface), and phreatic habitats (30 cm below the water table of adjacent alluvial bars). Total taxa decreased markedly along the epigean-hypogean gradient. Crustaceans and insects (the majority of animals in all three habitats) progressively increased (Crustacea) or progressively declined (Insecta) in relative abundance along the epigean-hypogean gradient. Most of the 142 common taxa exhibited one of four types of distribution patterns along the gradient (hypogean, epigean, transitional, eurytopic). This study demonstrated a marked faunal gradient on a scale of meters across the groundwater/surface water ecotone. Faunal similarity coefficients (e.g., Jaccard) and detrended correspondence analysis of faunal distributions may provide comparative measures of ecotone permeability between physically contiguous interstitial habitats.

INTRODUCTION

Riverine sediments form habitat patches of various sizes and complexity, reflecting the interactions of fluvial dynamics and biogeochemical features (Schumm, 1977; Richards, 1982; Rust, 1982). The distribution patterns of animals inhabiting porous alluvia reflect gradients in environmental conditions that occur at a variety of spatial scales (reviewed by Gibert et al., 1994; Ward & Palmer, 1994). Interstitial animals residing in the water-filled spaces between substrate particles are distributed along a gradient of habitat conditions from superficial bed sediments (benthic biotope), to deeper sediments beneath the channel (hyporheic biotope=underflow), to alluvial deposits situated some distance laterally from the active channel (phreatic biotope).

The purpose of this paper is to describe distribution patterns of interstitial fauna along such an epigean-hypogean gradient on a scale of meters across the surface water/groundwater ecotone (Gibert et al., 1990) of an alluvial river. This investigation is part of a comprehensive study of the groundwater fauna of a Rocky Mountain river system.

The geomorphic setting and faunal distribution patterns along the extensive elevation gradient have been reported elsewhere (Ward & Voelz, 1990; 1994).

SITE DESCRIPTION AND METHODS

The South Platte River heads at 3700 m a.s.l. in the Rocky Mountains of Colorado, descends 2000 m in about 200 km as a mountain stream, then becomes a Great Plains river (Fig. 1). Alluvial aquifers of the mountain stream segment consist of shallow deposits of coarse-grained sediment lying directly on crystalline bedrock. The unconfined alluvial aquifer of the plains river consists of sediment deposited in broad valleys cut into underlying sedimentary bedrock. The width of the alluvial plains aquifer ranges from 1.6 to 16 km and the thickness of the saturated zone varies between <10 and >75 m.

Interstitial fauna was collected from three habitat types, designated phreatic, hyporheic, and epigean. The term 'phreatic' is used to be consistent with habitat designations in Ward et al. (1994), but does not imply that the interstitial

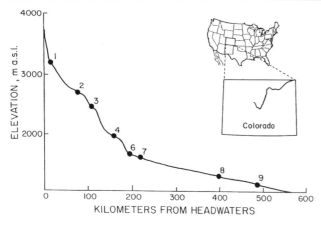

Fig. 1 Longitudinal profile of the South Platte River in Colorado (impoundments omitted) showing sampling locations. Site 5 excluded.

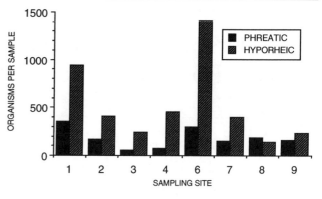

Fig. 2 Mean abundance levels, all dates combined, of the total fauna in phreatic and hyporheic habitats.

habitat of the alluvial bars is exempt from the influence of surface water. Phreatic and hyporheic samples were collected with a Bou-Rouch pump (Bou, 1974). To reduce contamination from soil fauna when sampling the phreatic habitat, the overlying alluvial deposits (*c.* 30–50 cm) were removed by digging a hole to the water table prior to inserting the standpipe. Six 5–liter samples were collected during spring, summer, and autumn at each site, three 30 cm below the stream bed (hyporheic biotope=underflow) and three 30 cm below the water table in adjacent (2–20 m from water's edge) alluvial bars (phreatic biotope). A third habitat, the benthic biotope was sampled by taking multiple cores (8 cm diameter) of finer surficial sediments (mainly gravel and sand) from the top 10 cm of the stream bed until a composite sample of 1.5–3.0 liters of sediment was obtained.

Sampling was conducted at nine sites (riffles), eight of which are included herein (site 5 is excluded because the epigeic biotope was not sampled). Rubble substratum predominated at Sites 1–7, whereas the bed of the plains river (Sites 8 and 9) consisted largely of sand and gravel. Mineral substratum up to 64 mm diameter was collected from the bed (hyporheic) and alluvial bars (phreatic) during autumn base flow conditions. The hyporheic substratum was similar at Sites 1–7, with pebble and gravel constituting the majority of sediment <64 mm diameter. In the plains river (Sites 8 and 9) pebble is less abundant and sand is more abundant than in the mountain stream. At mountain stream sites sand was generally more abundant in phreatic than in hyporheic habitats. Silt and clay each constituted <2% of the substratum in all samples. See Ward & Voelz (1994) for a more detailed description of geomorphology, geological history, sample collection and processing, and physico-chemical conditions.

Faunal data were subjected to two analytical techniques. The Jaccard coefficient, a simple measure of faunal similarity between two habitat types, is the ratio of twice the number

of taxa common to both habitats, to the sum of the number of taxa in each of the habitats. Detrended correspondence analysis (DCA), an ordination technique (Gauch, 1982), was employed to further examine faunal gradients. Abundance data were log transformed to account for the extremely high densities of a few taxa and the downweighting option was applied to reduce the influence of rare species.

RESULTS

Over 200 animal taxa were identified from the interstitial habitats. For the purposes of this paper, taxa that were rare (<10 organisms per 5-liter Bou-Rouch sample or per liter of surface gravel) and occurred at only one site are excluded, reducing the number of taxa to 142. Most of these were identified to species. However, Nematoda (common) and calanoid copepods (infrequent) were not identified further. These two taxa were excluded from detrended correspondence analysis. See Ward & Voelz (1994) for a complete list of taxa.

At Sites 1–7 densities of the total fauna were 2.5 to 6.2 times greater in hyporheic than in phreatic habitats (Fig. 2). Only at the two plains locations (Sites 8 and 9) were phreatic and hyporheic densities similar. Densities per sample were highest in the benthic habitat; however, because sampling units differed, abundance values cannot be directly compared with those of the other two habitat types.

Eight groups collectively accounted for >98% of total organisms collected from each habitat (Fig. 3). Crustaceans and insects, which constituted the majority of the organisms in all three habitats, exhibited opposing patterns across the epigean-hypogean gradient. Whereas crustaceans exhibited progressively higher relative abundances from benthic to phreatic biotopes, insects declined dramatically (Fig. 3). Although the strength of these patterns varied at individual sites (not shown), the general patterns across the gradient prevailed with few exceptions (e.g., the near absence of crustaceans in the phreatic at Site 2).

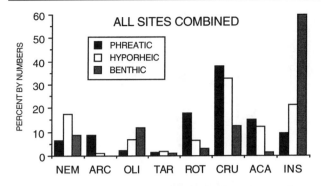

Fig. 3 Relative abundances of the eight major faunal groups in each habitat. NEM=nematodes, ARC=archiannelids, OLI=oligochaetes, TAR=tardigrades, ROT=rotifers, CRU=crustaceans, ACA=acarines, INS=insects.

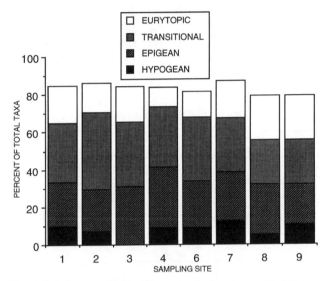

Fig. 4 Relative contribution of taxa to the four types of habitat distribution patterns defined in the text.

Habitat distribution types

The majority of taxa exhibited one of the following types of habitat distribution patterns: (1) hypogean forms are largely restricted to phreatic and/or hyporheic biotopes, rarely if ever occurring in surficial bed sediments; (2) epigean forms are largely restricted to surficial bed sediments, rarely if ever occurring in hyporheic biotopes and never colonizing the phreatic; (3) transitional forms commonly occur in both surficial bed sediments and the hyporheic, but rarely if ever occupy the phreatic biotope; (4) eurytopic forms commonly occur in all three biotopes.

The relative contribution of the four types of habitat distributions are remarkably similar between sites (Fig. 4). Hypogean forms were absent from Site 3 where oxygen levels were <1.0 mg l^{-1} in both hyporheic and phreatic biotopes. Examples of taxa that clearly demonstrate the four types of distribution are shown in Fig. 5.

Biodiversity and abundance patterns

The total number of taxa progressively increased, often markedly so, along the hypogean-epigean gradient, except at Site 9 where a slightly higher level was recorded in phreatic than hyporheic habitats (Fig. 6). Considering the low levels of oxygen, a fairly high faunal diversity was recorded from hyporheic and phreatic biotopes at Site 3, although densities were low. In Fig. 7, the number of taxa collected from each habitat type all sites combined, is shown within the rectangles. The numbers within the circles indicate how many taxa are common to different combinations of habitats.

At each site, Jaccard similarity is lowest between phreatic and benthic habitats, as expected, although values vary considerably between sites (Fig. 8). At most sites, similarity between hyporheic and benthic habitats is greater than between hyporheic and phreatic habitats, but this relationship is reversed at the two plains sites.

Although the large elevation range encompassed by the study sites complicates the two-dimensional array of the detrended correspondence analysis (Fig. 9), there is a strong tendency within a site for benthic habitats to plot to the left (lower values) and for phreatic habitats to plot to the right (higher values) along the primary axis. Based on proximity in two-dimensional space, the strength of the epigean-hypogean gradient varies in an apparently consistent manner along the elevation gradient, as discussed subsequently. DCA plots based on presence/absence data and plots based on abundance data without downweighting produced results similar to the ordination diagram of Fig. 9.

DISCUSSION

Other studies have examined the distribution of interstitial animals in alluvial aquifers over a range of spatial scales (e.g., Bretschko, 1981; Dole, 1985; Pennak & Ward, 1986; Rouch, 1988; Stanford & Ward, 1988; Williams, 1989; Creuzé des Châtelliers, 1991; Danielopol, 1991; Boulton et al., 1992; Marmonier et al., 1992; Ward & Palmer, 1994). Gradient analysis techniques have been employed to analyze faunal distribution patterns along elevation profiles (Ward & Voelz, 1994) and at the floodplain scale (Creuzé des Châtelliers & Reygrobellet, 1990; Ward et al., 1994).

A variety of interacting factors structure the community composition, biodiversity, and abundance patterns of interstitial faunal assemblages, although few definitive data are available to assess the specific role of many of them in field situations. For example, the rather high diversity (but low density) and lack of hypogean forms in the poorly-oxygenated biotopes of Site 3 is difficult to explain and is in opposition to the findings of Rouch (1988). In a review of spatial distribu-

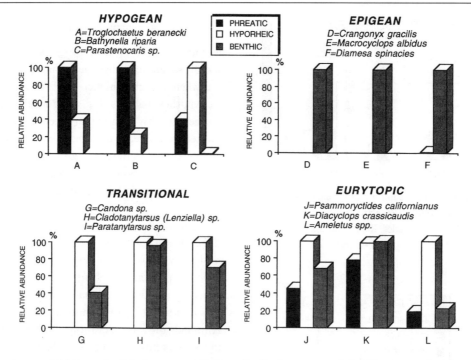

Fig. 5 Examples of taxa exhibiting each of the four types of habitat distribution patterns defined in the text, based on relative abundance by density in the three biotopes, all sites and dates combined.

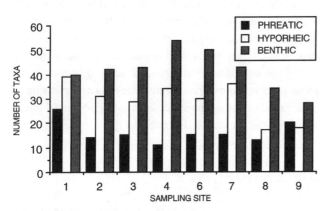

Fig. 6 Number of animal taxa (excluding rare taxa – see text) identified from each habitat at each site.

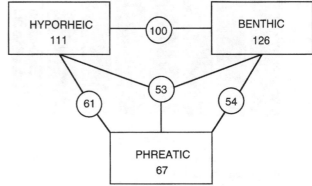

Fig. 7 Number of animal taxa collected from each habitat (values in rectangles) and the number of taxa shared between habitats (values in circles) all sites combined. Excluding rare taxa – see text.

tion patterns of interstitial fauna in alluvial river-aquifer systems, Ward & Palmer (1994) identified eight categories of factors that may influence such patterns: (1) characteristics of the alluvium, (2) exchange properties, (3) disturbance, (4) hypogean affinity, (5) food resources, (6) biotic interactions, (7) reproductive patterns, and (8) age distribution. Site specific geomorphic or hydrogeologic processes are major determinants of all the abiotic factors listed above.

The present study demonstrated a marked faunal gradient on a scale of meters across the groundwater/surface water ecotone at various elevations along the course of an alluvial river system. The DCA plots (Fig. 9), although difficult to interpret, exhibit an interesting pattern in the strength of the epigean-hypogean gradient along the longitudinal gradient. In the high elevation headwaters (Site 1), habitat plots are more-or-less equidistant, with a slightly greater proximity between benthic and hyporheic habitats. In the montane zone (Sites 2–4), hyporheic habitats exhibit a much greater proximity to benthic than to phreatic habitats. At Sites 6 and 7 in the foothills, the habitat plots are equidistant. In the plains river (Sites 8 and 9), the three habitats are more tightly clustered with the greatest proximity between the hyporheic and phreatic. The significance of this superimposed longitudinal pattern is difficult to discern, however, because we know of no other study of interstitial faunal assemblages that has been conducted over such an extensive elevation gradient.

Results suggest that faunal similarity coefficients (e.g., Jaccard similarity) and DCA plots of faunal distribution patterns may provide a means of roughly quantifying ecotone

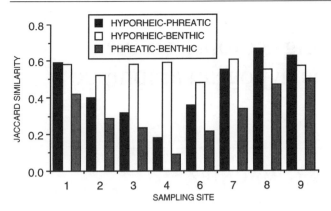

Fig. 8 Jaccard coefficients of faunal similarity between habitat pairs at each site.

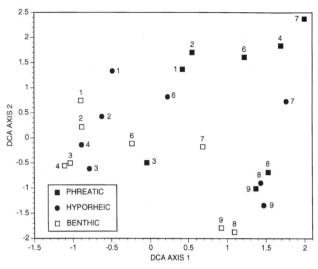

Fig. 9 DCA ordination diagram for the faunxa of each habitat at each site (CANOCO, Ter Braak, 1988). Eigenvalues are 0.409 (axis 1) and 0.354 (axis 2).

permeability between physically contiguous interstitial habitats and could contribute to the development of a relative measure of ecotone permeability at different locations and at different spatial scales.

ACKNOWLEDGEMENTS

The following specialists graciously provided taxonomic assistance: Dr J. W. Reid and Dr R. L. Whitman (Copepoda), Dr M. J. Wetzel (Oligochaeta), Dr P. Marmonier (Ostracoda), Dr L. C. Ferrington (Chironomidae), Dr G. W. Krantz, Dr R. A. Norton, Dr B. P. Smith, and Dr D. E. Walter (Acarina), Dr B. C. Kondratieff (Insecta), and Dr R. D. Waltz (Collembola). The Colorado State University Soil Testing Laboratory conducted particle size analyses and determined the organic content of substrate samples. We thank Professor J. Gibert and two anonymouse reviewers for comments that improved the paper and Mrs Nadine Kuehl for typing the manuscript. Support provided in part by the Colorado Water Resources Research Institute.

REFERENCES

Bou, C. (1974). Les méthodes de récolte dans les eaux souterraines interstitielles. *Ann. Spéléol.*, **29**, 611–619.

Boulton, A. J., Valett, H. M. & Fisher, S. G. (1992). Spatial distribution and taxonomic composition of the hypoheos of several Sonoran Desert streams. *Arch. Hydrobiol.*, **125**, 1–37.

Creuzé des Châtelliers, M. (1991). Geomorphological processus and discontinuities in the macrodistribution of the interstitial fauna. A working hypothesis. *Verh. Internat. Verein. Limnol.*, **24**, 1609–1612.

Creuzé des Châtelliers, M. & Reygrobellet, J. L. (1990). Interactions between geomorphological processes, benthic and hyporheic communities: First results on a by-passed canal of the French upper Rhône River. *Regulated Rivers*, **5**, 139–158.

Danielopol, D. L. (1991). Spatial distribution and dispersal of interstitial Crustacea in alluvial sediments of a backwater of the Danube at Vienna. *Stygologia*, **6**, 97–110.

Dole, M.-J. (1985). Le domaine aquatique souterrain de la plaine alluvial du Rhône à l'est de Lyon. 2. Structure verticale des peuplements des niveaux supérieurs de la nappe. *Stygologia*, **1**, 270–291.

Gauch, H. G. (1982). *Multivariate Analysis in Community Ecology*. New York, Cambridge Univ. Press.

Gibert, J., Dole-Olivier, M.-J., Marmonier, P. & Vervier, P. (1990). Surface water-groundwater ecotones. In *The Ecology and Management of Aquatic-Terrestrial Ecotones*, ed. Naiman, R. J. & Décamps, H., pp. 199–225. Parthenon, Casterton Hall, England.

Gibert, J., Stanford, J. A., Dole-Olivier, M.-J. & Ward, J. V. (1994). Basic attributes of groundwater ecosystems and prospects for research. In *Groundwater Ecology*, ed. Gibert, J., Danielopol, D. & Stanford, J.A., pp. 7–40. San Diego, Academic Press.

Marmonier, P., Dole-Olivier, M. J. & Creuzé des Châtelliers, M. (1992). Spatial distribution of interstitial assemblages in the floodplain of the Rhône River. *Regulated Rivers*, **7**, 75–82.

Pennak, R. W. & Ward, J. V. (1986). Interstitial faunal communities of the hyporheic and adjacent groundwater biotopes of a Colorado mountain stream. *Arch. Hydrobiol. Suppl.*, **74**, 356–396.

Richards, D. S. (1982). *Rivers: Form and Process in Alluvial Channels*. Methuen, London.

Rouch, R. (1988). Sur la répartition spatiale des Crustacés dans le sous-écoulement d'un ruisseau des Pyrénées. *Annls. Limnol.*, **24**, 213–234.

Rust, B. R. (1982). Sedimentation in fluvial and lacustrine environments. *Hydrobiologia*, **91**, 59–70.

Schumm, S. A. (1977). *The Fluvial System*. New York, Wiley.

Stanford, J. A. & Ward, J. V. (1988). The hyporheic habitat of river ecosystems. *Nature*, **335**, 64–66.

Ter Braak, C. J. F. (1988). CANOCO-a FORTRAN program for canonical community ordination by [partial] [detrended] [canonical] correspondence analysis, principal components analysis and redundancy analysis (version 2.2). Agricultural Mathematics Group, Wageningen, The Netherlands.

Ward, J. V. & Palmer, M. A. (1994). Distribution patterns of interstitial freshwater meiofauna over a range of spatial scales, with emphasis on alluvial river-aquifer systems. *Hydrobiologia*, **287**, 1, 147–156.

Ward, J. V., Stanford, J. A. & Voelz, N. J. (1994). Spatial distribution patterns of Crustacea in the floodplain aquifer of an alluvial river. *Hydrobiologia*, **287**, 1, 11–17.

Ward, J. V. & Voelz, N. J. (1990). Gradient analysis of interstitial meiofauna along a longitudinal stream profile. *Stygologia*, **5**, 93–99.

Ward, J. V. & Voelz, N. J. (1994). Groundwater fauna of the South Platte River system, Colorado. In *Groundwater Ecology*, ed. Gibert, J., Danielopol, D. & Stanford, J. A., pp. 391–423. San Diego, Academic Press.

Williams, D. D. (1989). Towards a biological and chemical definition of the hyporheic zone in two Canadian rivers. *Freshwat. Biol.*, **22**, 189–208.

6 Filter effect of karstic spring ecotones on the population structure of the hypogean amphipod *Niphargus virei*

F. MALARD*, M.-J. TURQUIN* & G. MAGNIEZ **

* *Université Lyon 1, URA CNRS 1974, Ecologie des Eaux Douces et des Grands Fleuves, Hydrobiologie et Ecologie Souterraines, 43 Bd du 11 novembre 1918, 69622 Villeurbanne cedex, France*

** *Laboratoire de Biologie Animale et Générale, Université de Bourgogne, 6 Bd Gabriel, 21000 Dijon, France*

ABSTRACT Faunal sampling was carried out in 1990–91 on three springs of a karstic system located at 15 km north-east of the city of Dijon (France) to investigate the filter effect of these outlets on the drifting population of the hypogean amphipod *Niphargus virei*. Results showed that, in places where limestones outcropped, the karstic spring ecotone did not modify the structure of the drifting population. In that case, faunal sampling of the springs during the floods was an effective method to study the dynamics of the population living in the aquifer. However, this method could not be applied in places where deep karstic groundwater circulated through water-saturated superficial deposits prior to emerge at the land surface. Indeed, in that case, the size frequency histograms of the drifting population of *N. virei* were characterised by abnormally low percentages of small size individuals and/or to a lesser extent by low percentages of large size individuals. These changes in the population structure were probably caused by two types of filters exerted by the karstic spring ecotone: 1) a mechanical filter which corresponded to the change of matrix between the deep karstified limestone aquifer and the shallow aquifer; 2) a biological filter probably due to the predation exerted by benthic invertebrates which have colonised the shallow aquifer.

INTRODUCTION

Surface water/groundwater ecotones (i.e. interaction zones occurring between surface water and groundwater systems) regulate the flow of matter, energy, information and organisms between two contrasted ecological systems (UNESCO, 1980; Pennak & Ward, 1986; Vanek, 1987; Ford & Naiman, 1989; Gibert, 1991a; Sharley, 1994). This regulation is dependent upon the four main functional characteristics of the ecotone, namely the elasticity, the permeability, the biodiversity and the connectivity (Gibert *et al.*, 1990). Out of these four characteristics, the ecotone permeability which corresponds to the degree to which the boundary deflects the movements of various 'vectors' (i.e. physical forces or animals that actively move materials or energy in the system) (Wiens *et al.*, 1985; Forman & Moore, 1992), is probably the most important one to consider. Indeed, the boundary permeability influences ecological flows and thus the patch dynamics, but it also modifies the propagation of disturbances from one patch to another (Hansen *et al.*, 1988).

Surface water/groundwater ecotone permeability is governed by three different types of filter: the mechanical filter, the photic filter, and the biochemical filter (Vervier *et al.*, 1992). Interstitial ecotone impedes the propagation of disturbances such as floods and may act as either permanent or temporary sinks since abiotic and biotic fluxes can be removed or temporarily lost from adjacent resource patches (see review in Gibert *et al.*, 1990). On the contrary, karstic spring and sinkstream ecotones transmit directly the disturbances and have a very low filtering efficacy (Rouch, 1970; Gibert, 1991b; Vervier & Gibert, 1991). It is usually considered that karstic spring ecotone has no effect on biological fluxes conveying from the underground to the surface environment. For example, faunistical samples collected at the spring during a flood are used to determine the structure of invertebrate communities (Rouch, 1977; Barthélémy, 1984; Gibert, 1986) or populations (Turquin, 1984a; Turquin & Barthélémy, 1985) living in the aquifer.

This paper aims to provide a counter-example of the so-called 'null-filtering-efficacy' of karstic spring ecotones with regard to biotic fluxes. To this end, we describe the results of

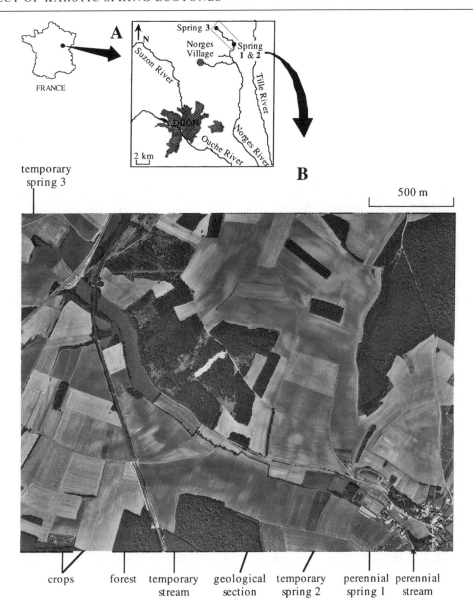

Fig. 1 Location (A) and aerial view (B) of the study area showing location of the three studied springs.

an ecological survey carried out on three springs of a karstic system. First, the springs are differentiated on the basis of physico-chemical characteristics of groundwater and second, on the basis of the structure of their populations of the hypogean amphipod *Niphargus virei* Chevreux, 1896. Differences in the population structures are then related to the mechanical and biological filter effects exerted by the ecotone.

MATERIALS AND METHODS

The study area

The three studied artesian springs are located in a flat area, 15 km north-east of the city of Dijon (France) (Fig. 1). They drain the groundwaters of a 31 km² catchment area which is mainly covered by crops (wheat, barley, and rape). Spring 1 is the only perennial exsurgence of the karstic system. Its discharge, which was measured by means of a staff gage, ranges from 20 to 200 l.$^{-1}$. Spring 2 is one of the many little temporary springs, which emerge at about 200 m north-west of the spring 1, in a swampy meadow located on the left bank of a temporary stream. This spring, whose discharge does not exceed 10 l.s^{-1}, flows during most of the year and is dry only during the summer months (from June to August), when discharge of the perennial spring is low (< 40 l.s^{-1}). Temporary fracture spring 3, whose discharge can reach 1000 l.s^{-1} during high-rainfall periods, lies at 2.5 km NW of spring 1. It comprises several outlets located at the bottom of a 4–m deep water-filled doline, which flow only about three months a year when discharge of the perennial spring is high (> 180 l.s^{-1}).

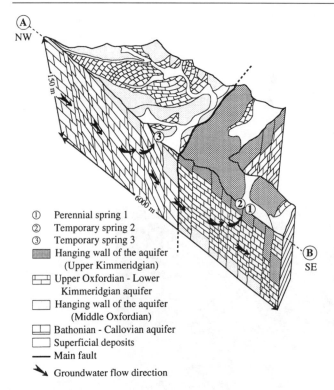

① Perennial spring 1
② Temporary spring 2
③ Temporary spring 3
▨ Hanging wall of the aquifer
   (Upper Kimmeridgian)
⊞ Upper Oxfordian - Lower
   Kimmeridgian aquifer
☐ Hanging wall of the aquifer
   (Middle Oxfordian)
⊟ Bathonian - Callovian aquifer
☐ Superficial deposits
— Main fault
↘ Groundwater flow direction

Fig. 2 Simplified geological section of the studied area showing location of the main karstified limestone aquifers.

Hydrogeological settings

Fig. 2 shows a geological section of the studied area. Two karstic aquifers may be distinguished (BRGM, 1978). A semi-confined aquifer develops in the Bathonian and Callovian limestones, which are partly overlain by Middle Oxfordian argillaceous limestones. A confined aquifer, develops in the Upper Oxfordian and Lower Kimmeridgian limestones overlain by Upper Kimmeridgian argillaceous limestones. Due to a major fault oriented NE-SW, the Bathonian-Callovian and Upper Oxfordian-Lower Kimmeridgian limestone aquifers are hydraulically connected. North of the major fault, the limestones dip gently to the SE (2–5°), while they dip to the SW, south of the fault. The local direction of groundwater flow is to the S-E. Artesian springs are located in the vicinity of faults which allow the groundwater to rise at the surface. Groundwater rising in these faults may emerge directly at the land surface when limestones outcrop locally (e.g. temporary spring 3) but in most cases it circulates through a variety of superficial deposits. This is more particularly the case for the perennial spring 1 and the temporary spring 2 which are located in an area where Upper Kimmeridgian argillaceous limestones are covered by Albian sand (Cretaceous), calcareous conglomerate of lacustrine origin (Tertiary), colluvial and alluvial deposits (Quaternary).

Physico-chemical sampling of the springs

From October 1990 to June 1991, a physico-chemical survey was carried out on springs 1, 2 and 3. About twice a week, temperature and conductivity were measured at the springs with a mercury thermometer and a Merck CM 85 conductivity meter. Groundwater samples were collected and analysed for bicarbonate, calcium and nitrate. Bicarbonate was determined by potentiometric titration with sulphuric acid to end point pH (A.P.H.A., 1985). The calcium content of the samples were measured by titration with EDTA, the end point being detected with the Patton-Reeder indicator (Patton & Reeder, 1956; Aminot, 1974). The cadmium reduction method using a Merck-Clévenot spectrophotometer with Hach reagents was used for the analysis of nitrate concentration.

Faunal sampling of the springs

Faunal sampling of springs 1, 2, and 3 was carried out to study the population structure of the subterranean amphipod *Niphargus virei*. Groundwater of springs 1 and 2 was filtered through a 300 μm-mesh net from February to May, 1991. Specific equipment had to be designed to filter the groundwater of spring 3. A PVC pipe 0.1 m in diameter and 5 m long was connected with concrete to one of the fractures located at the bottom of the water-filled doline. When a flood occurred, groundwater rose at a maximum discharge of 3.5 $l.s^{-1}$ within this pipe to the surface of the water-filled doline. Using such a sampling technique, groundwater of spring 3 was filtered through a 300 μm-mesh net during four floods, which occurred in October 1990, November, 1990, January 1991, and March, 1991. Because the discharge of filtered groundwater at the three sampled springs was quite low (< 25 $l.s^{-1}$ for spring 1, < 10 $l.s^{-1}$ for spring 2, and < 3.5 $l.s^{-1}$ for spring 3), the nets were removed only every 3 to 4 days. Faunal samples were collected in plastic containers and transported immediately to the laboratory. Individuals of *N. virei* were then collected alive and preserved in 70% alcohol with glycerine while the remaining part of the sample was preserved with formalin solution.

Data analysis

One-way analysis of variance (ANOVA) were used to test the differences between the means of temperatures, conductivities, bicarbonate concentrations, calcium concentrations, and nitrate concentrations of springs 1, 2, and 3. The relationships between physico-chemical parameters measured in groundwater of spring 2 were then analysed via Pearson-product moment correlation.

The life-cycle of *N. virei* is known thanks to Ginet's (1960) and Turquin's (1981, 1984a, 1988) studies. These authors demonstrated that this hypogean amphipod, which has a very long life-span (at least 14 years), grows very slowly. For example, its maximum growth rate is only 0.06 mg.day^{-1} (Turquin, 1988), while that of *Gammarus pulex* may reach 0.5 mg.day^{-1} (Sutcliffe *et al.*, 1981). Individuals of *N. virei* are 0.3 cm long at birth, about 0.5 cm long at 1.5 years old, and 1.5 cm long at 5 years old. The longest individuals do not exceed 4 cm long.

In this study, the body length of *N. virei* was used to study the population structure. The animals were placed on a ruler with a 0.5 mm scale and the length between the base of the first antenna and the base of the telson was measured by means a binocular lens. As many as 13103 individuals were measured during this survey. The size frequency histograms for different faunal samples were then represented using a class interval of 2 mm.

RESULTS

Physico-chemical characteristics of groundwaters

Fig. 3 shows the conductivities, temperatures, bicarbonate concentrations, calcium concentrations, and nitrate concentrations through time of groundwater samples collected at springs 1, 2 and 3 from October 1990 to June 1991. Numbers of samples, means, and standard deviations for these five physico-chemical parameters are given in Table 1. We also indicated the results of the one-way analysis of variance used to differentiate the three studied springs on the basis of their mean physico-chemical characteristics. On the whole, results showed that the water physico-chemistry of springs 1, 2, and 3 were statistically different (Fig. 3, Table 1). Significant differences ($p<0.05$) between the mean concentrations of the three studied springs were observed for most of the parameters. Only the comparison between mean conductivities of springs 1 and 2 and the comparison between mean conductivities of springs 2 and 3 did not yield significant differences ($p>0.05$). Particular attention was given to physico-chemical characteristics of spring 2 because they differed markedly from those of springs 1 and 3. Indeed, groundwater of spring 2 was characterised by low temperatures and low nitrate concentrations (Fig. 3 and Table 1).

Fig. 4 shows the conductivities, temperatures, bicarbonate concentrations, calcium concentrations, and nitrate concentrations through time of groundwater samples collected at spring 2 from November 1990 to May 1991. Air temperature values were also represented. They correspond to the averages of the mean daily temperatures recorded during the four days prior to the collection of the groundwater samples. Because the groundwater discharge of spring 2 was not recorded, only the perennial spring discharge has been indicated in Fig. 4. The Pearson-product moment

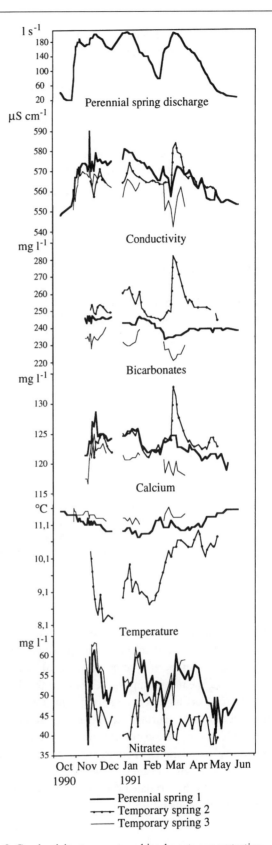

Fig. 3 Conductivity, temperature, bicarbonate concentration, calcium concentration, and nitrate concentration measured in groundwaters of springs 1, 2, and 3 from October 1990 to June 1991.

Table 1. *Differences between the average temperatures, conductivities, bicarbonate concentrations, calcium concentrations, and nitrate concentrations of the three studied springs. Results of analysis of variance: numbers of samples (Nb), means, standard deviations (S.D.), F-factors and levels of significance (p<0.05 significant; p<0.01 highly significant; p<0.001 very highly significant) are given*

	Temp.	Cond.	HCO$_3^-$	Ca^{2+}	NO$_3^-$
	°C	μScm^{-1}	mg l^{-1}	mg l^{-1}	mg l^{-1}
Nb of samples					
spring 1	82	79	63	61	62
spring 2	49	49	49	49	49
spring 3	44	44	32	32	32
Mean					
spring 1	11.1	567.5	241.1	123.1	52.3
spring 2	9.7	565.5	253.4	123.9	44.0
spring 3	11.4	562.8	231.6	120.6	55.9
S.D.					
spring 1	0.2	8.8	3.8	1.9	5.1
spring 2	0.8	5.6	8.4	2.2	4.1
spring 3	0.1	8.0	5.0	2.0	4.5
p					
spring 1/2	<0.001	0.1507	<0.001	<0.05	<0.001
spring 2/3	<0.001	0.0661	<0.001	<0.001	<0.001
spring 1/3	<0.001	<0.01	<0.001	<0.001	<0.01
F-ratio					
spring 1/2	252.20	2.0904	106.68	4.3297	84.874
spring 2/3	194.07	3.4600	174.14	47.109	149.44
spring 1/3	35.523	8.5235	105.55	35.145	11.000

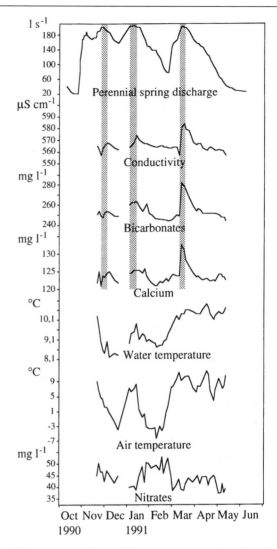

Fig. 4 Temporal variations of air temperature and conductivity, temperature, bicarbonate concentration, calcium concentration, nitrate concentration measured in groundwater of temporary spring 2 from November 1990 to May 1991.

correlation coefficients between the physico-chemical parameters are given in Table 2. There were very highly significant positive correlations ($p<0.001$) between conductivity, bicarbonate concentration and calcium concentration. For these three parameters, an increase could be observed each time a flood occurred, the maximum values being recorded during the flood crests of the perennial spring. This increase was particularly obvious during the February flood which occurred after a long period of low-water. Very highly significant positive correlations were also observed between air and groundwater temperatures. Finally, nitrate concentrations were negatively correlated with groundwater temperature and with air temperature.

Macro-invertebrate communities and population structure of the hypogean amphipod *N. virei*

Filtering of the water of perennial spring 1 from February to May, 1991, only provided a few individuals of *N. virei*, whose

Table 2. *Correlation coefficients between physico-chemical parameters measured in groundwater of temporary spring 2 (n=49). Values given are the Pearson-product moment correlation coefficients and their levels of significance (* significant p<0.05; ** highly significant p<0.01; *** very highly significant p<0.001)*

	Conductivity	HCO$_3^-$	Ca^{2+}	Water temp.	Air temp.
HCO$_3^-$	0.869***				
Ca^{2+}	0.804***	0.875***			
Water temp.	0.115	0.262	0.279		
Air temp.	0.108	0.340*	0.412**	0.827***	
NO$_3^-$	−0.158	−0.353*	−0.441**	−0.522***	−0.675***

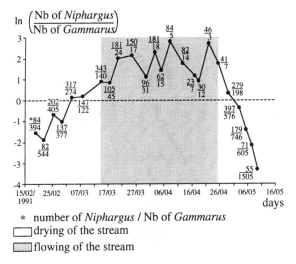

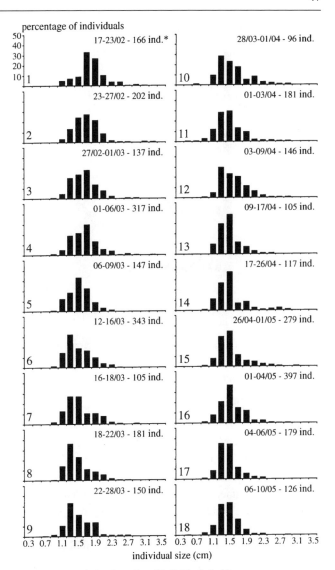

* sampling period and number of individuals (ind.)

Fig. 5 Temporal changes in the logarithmic value of the ratio 'number of individuals of *N. virei*/number of individuals of *G. fossarum*' measured for 24 samples collected at the temporary spring 2 from February 17th to May 10th, 1991. Pale and dark grey areas indicate respectively the drying and flowing periods of the temporary stream.

body length never exceeded 0.8 cm. Any of the samples contained epigean macro-invertebrate. In contrast, as many as 3374 individuals of *N. virei* were collected at temporary spring 2 from February 17th to May 10th, 1991. This hypogean amphipod was associated with epigean macro-invertebrates which mainly consisted of the amphipod *Gammarus fossarum*, some Coleoptera, Heteroptera, and larvae of Trichoptera. In this paper, we only considered the number of individuals of the epigean amphipod *G. fossarum*. For each sample, we calculated the logarithm of the ratio 'number of individuals of *N. virei*/number of individuals of *G. fossarum*'. Negative values for this index mean a greater number of *G. fossarum* than *N. virei*, while positive values mean a greater number of *N. virei* than *G. fossarum*. Values of the index for 24 samples collected from February 17th to May 10th, 1991, are given in Fig. 5. The drying (pale grey areas) and flowing (dark grey area) periods of the temporary stream were also indicated (Fig. 5). It clearly appeared that the index tended to be positive when the stream was flowing and negative when the stream was dry.

Measurement of the body length of the 3374 individuals of *N. virei* collected at spring 2 allowed us to plot 18 size frequency histograms (Fig. 6). The numbers of *N. virei* used to calculate these histograms usually correspond to those collected in each of the sample indicated in Fig. 5. However, when less than 100 *N. virei* were recovered in a sample, we considered the sum of the numbers of individuals collected in several successive samples (i.e. histograms 1, 12, 13, 14 and 18). The distribution of size classes was gaussian and quite similar from one sample to another. From 80.1 to 95.2 % of

Fig. 6 Length frequency distribution of *N. virei* recovered in 18 samples collected at the temporary spring 2 from February 17th to May 10th, 1991.

individuals collected in a sample occurred in only 5 size classes (i.e. classes 1.0–1.2 mm; 1.2–1.4 mm; 1.4–1.6 mm; 1.6–1.8 mm; 1.8–2.0 mm). These 5 classes contained 3034 individuals out of 3374 collected during all the sampling period. A surprising result was the almost total absence of individuals whose body length was less than 0.8 cm (only 5 individuals out of 3374).

In 1990–91, faunal sampling of the temporary spring 3 provided a total of 9729 *N. virei*. Two other hypogean macro-invertebrate species were also collected (the stygobitic isopod *Caecosphaeroma burgundum* Dolfus, 1898, and the stygophilic leech *Trocheta bykowskii* Gedroyc, 1913), but no epigean macro-invertebrate species was observed. Filtering of the groundwater, which was carried out during four floods, provided 8 samples during the first flood (from October 29th

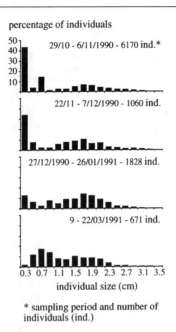

* sampling period and number of
individuals (ind.)

Fig. 7 Length frequency distribution of *N. virei* recovered during four floods (October, November 1990, January, and March 1991) of the temporary spring 3.

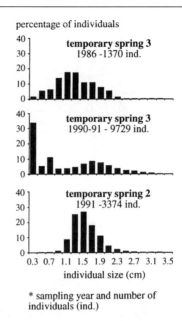

* sampling year and number of
individuals (ind.)

Fig. 8 Length frequency distribution of all *N. virei* collected at temporary spring 3 in 1986 and 1990–91 and at temporary spring 2 in 1991.

to November 6th, 1990), 5 samples during the second (from November 22nd to December 7th, 1990), 8 samples during the third (from December 27th, 1990 to January 26th, 1991), and 4 samples during the fourth (from March 9th to March 22nd, 1991). The numbers of *N. virei* collected in the samples of each flood were summed to calculate four size frequency histograms (Fig. 7). The distribution of size classes was plurimodal and differed markedly from the gaussian distribution observed for the temporary spring 2. The small size classes (i.e. classes 0.2–0.4 mm; 0.4–0.6 mm; 0.6–0.8 mm) represented a significant component of the total number of individuals (from 61.3 % in October 1990 to 21.9 % in January 1991). The first size class (i.e. 0.2–0.4 mm) even contained the highest number of individuals in October and November 1990.

Finally, to show clearly the difference between the population structures *of N. virei* of temporary springs 2 and 3, the total numbers of individuals collected in 1990–91 were taken into account to plot the size frequency histograms (Fig. 8). The 1990–91 histograms were also compared with those obtained by Turquin (unpublished data, 1986) for the temporary spring 3. This author measured the body length of 1370 individuals of *N. virei* collected during three floods, which occurred in February, March, and May 1986. Two relevant observations could be made. First, it clearly appeared that the 1990–91 distribution of spring 3 had a high percentage of small size individuals (49.9 % of individuals with body length less than 0.8 mm). Second, the 1986 and 1990–91 distributions of spring 3 were both very different from that of spring 2. Distribution of spring 2 was compressed in the middle size classes (i.e. 1.0–1.2 mm; 1.2–1.4 mm; 1.4–1.6 mm; 1.6–1.8

mm; 1.8–2.0 mm; 2.0–2.2 mm) and was thus characterised by abnormally low percentages of small size individuals and to a lesser extent by a low percentage of large size individuals.

DISCUSSION

Physico-chemical differentiation of the three studied springs

Although springs 1, 2, and 3 drain the groundwater of the same hydrogeological system, our results showed that they had different physico-chemical characteristics. Differences observed between main physico-chemical parameters of springs 1 and 3 (e.g. higher temperature and lower conductivity, bicarbonate concentration and calcium concentration for spring 3) could simply be due to the fact that their groundwaters originated from two aquifers: the Upper Oxfordian-Lower Kimmeridgian limestone aquifer for the perennial spring 1 and the Bathonian-Callovian limestone aquifer for the temporary spring 3. However, the results obtained for the temporary spring 2 also showed that the water physico-chemistry could be modified when deep artesian groundwater flowed through superficial deposits. Indeed, the high groundwater thermal variability and the high correlation between groundwater and air temperatures observed at the temporary spring 2 clearly demonstrated that the deep groundwater circulated in a shallow aquifer prior to flow at the land surface. Because of the great variety of tertiary and quaternary aged deposits occurring in the area, it is very difficult to define precisely the structure of this superficial aquifer. The thickness of coarse alluvium layer observed on the stream bottom is too low (less than 30-cm thick) to hold significant quantity of groundwater. Moreover, the increase in conductivity, bicarbonate concentration, and calcium concentration during the floods indicated that groundwater was discharged at the surface by piston-flow mechanism. Such a particular flow type might occur in solution channels developed in calcareous conglomerates of lacustrine origin (Tertiary), which were tapped to a depth of 8 m by a well drilled at only 20 m SW of the perennial spring 1. This geological formation consists of calcareous pebbles linked by a salmon-pink clayed matrix (50 to 60 % of $CaCO_3$). Finally, low nitrate concentrations observed in groundwater of temporary spring 2 might be due to nitrate adsorption by plant-cover of the swampy meadow. Although, supplementary data are needed to confirm this assumption, it is reinforced by the fact that high nitrate concentrations were observed in groundwater of spring 2 during the coldest months of January and February 1991 (i.e. significant negative correlation between air temperature and nitrate concentration, see Table 2), when the plant metabolism was lower.

Filter effects of the karstic spring ecotones on the population structure of *N. virei*

Great differences were observed between the population structures of *N. virei* of karstic springs 1, 2, and 3. These differences did not probably reflect spatial changes in the structure of the populations of *N. virei* living in the underlying karstic aquifers. They were rather due to the presence of different filter effects which might locally occur in the superficial deposits observed in the vicinity of the springs. At temporary spring 3, where limestones outcrop locally, individuals of *N. virei* were swept along with the current during the floods. There was no selective filtration, and although we could not be sure that the drift was similar for all size classes, the size frequency histograms probably reflected the population structure of *N. virei* living in the Bathonian-Callovian limestone aquifer. Such an approach has already been developed by Turquin (1984b) and Turquin & Barthélémy (1985) to compare the population structures of *N. virei* of several French karstic systems and to study their evolutions throughout the years. Filtering carried out in 1986 and 1990–91 at temporary spring 3 showed that all the size classes were represented. Chauvin (1987), who also studied the population structure of *N. virei* of the temporary spring 3, found similar results. The high number of 0.3-cm long individuals (i.e. body length at hatching) during the floods of October and November 1990 were in good agreement with the biology of *N. virei*. This hypogean amphipod has a definite breeding season in summer, that is followed by an annual hatching period in early autumn (Ginet, 1960;

Turquin & Barthélémy, 1985). In 1986, such a high percentage of small-size individuals was not observed probably because filtering of the spring only started in February.

The population of *N. virei* of temporary spring 2, which most only consisted of 3 to 6 year old individuals (i.e. body length ranging from 1 to 2 cm), should not maintain over time. However, individuals of *N. virei* were regularly collected at the spring in 1992, 1993 and 1994 (Malard, unpublished data). The almost total absence of young individuals is thus believed to be due to a 'biological filter' occurring in the water-bearing calcareous conglomerates drained by the spring. Filtering of the spring water showed that this superficial aquifer was colonised by various epigean macro-invertebrates (e.g. Amphipoda, Coleoptera, Heteroptera, Trichoptera larvae). The evolution from February 17th to May 10th, 1991, of the logarithmic value of the *N. virei*/*G. fossarum* ratio also suggested that the downward migration of epigean organisms was even more pronounced when the temporary stream was dry. In that case, the surface water fauna probably found refuge within the groundwater. Benthic invertebrates, which have colonised the superficial aquifer, might prey on young individuals of *N. virei* and thus modify the structure of the population of *N. virei* originating from the deep Upper Oxfordian-Lower Kimmeridgian limestone aquifer. The influence of predation by epigean organisms on the presence within the sediments of karst outflows of groundwater invertebrates drifting from the karst, has also been emphasised by Essafi *et al.* (1992). These authors studied the colonisation of the interstitial layers of two karstic streams by the hypgean amphipod *Niphargus rhenorhodanensis*. They suggested that this colonisation was partly dependent upon the presence in the interstitial layer of the streams of many *Gammarus* which could compete and prey on *N. rhenorhodanensis*.

Finally, the fact that only a few small size individuals of *N. virei* were collected at perennial spring 1 strongly suggested that, in some cases, a mechanical filter could also impede the movement of karstic macro-invertebrates towards surface waters. With regard to the downward migration of *N. virei*, this mechanical filter probably corresponded to the change of matrix between deep karstified limestone aquifers and superficial aquifers with small substrate pore size. For example, karstic groundwater of perennial spring 1 probably circulated through a lens of sand prior to emerging at the surface. Indeed, the nets used to filter the spring water always contained sand particles which were identified as Albian sand of Cretaceous age.

CONCLUSION

Many authors have studied the way by which the ecotone may modify the nature and intensity of biological fluxes between groundwater and surface water sytems (Danielopol, 1976; Dole, 1983, Marmonier & Dole, 1986; Schmidt *et al.*, 1991; Plénet *et al.*, 1992; Chafiq *et al.*, 1992). However, most studies were based on the examination of the spatio-temporal distribution of aquatic invertebrate communities. Our results showed that changes in the population structure of hypogean species might also occur in the surface water/groundwater interactions zones. These results are all the more interesting as they concern karstic spring ecotones which have often been considered as highly permeable interfaces with regard to groundwater biological fluxes. The permeability of karstic spring ecotones is probably very high when groundwater is discharged directly at the land surface as it often the case in mountain karst areas. However, in some cases, karstic groundwaters may circulate in various geological deposists prior to emerge at the land surface (e.g. colluvial, alluvial, morainic deposits). These superficial deposits, which are less permeable than the underlying karstified limestones but may also contain some epigean and/or hypogean invertebrates, are interactions zones between the karstic groundwater and surface water environments. In these transition zones, mechanical and biological filters may regulate the flow of karstic groundwater organisms towards the surface.

REFERENCES

A.P.H.A. (1985). *Standard methods for the examination of water and wastewater.* 16th edn. Washington: A.P.H.A., A.W.W.A. & W.P.C.F.

Aminot, A. (1974). Géochimie des eaux d'aquifères karstiques. II. – Les analyses chimiques en hydrogéologie karstique. *Annls. Spéléol.*, 29, 461–483.

Barthélémy, D. (1984). *Impact des pollutions sur la faune stygobie karstique: approche typologique sur seize émergences des départements de l'Ain et du Jura.* Thesis, Univ. Lyon I, France.

BRGM (1978). *Notice et carte géologique de la France: feuille Mirebeau XXXI-22.* BRGM (ed.), Orléans, France.

Chafiq, M., Gibert, J., Marmonier, P., Dole-Olivier, M.-J. & Juget, J. (1992). Spring ecotone and gradient study of interstitial fauna along two floodplain tributaries of the River Rhône, France. *Regulated Rivers*, 7, 103–115.

Chauvin, C. (1987). Une application de la méthode de filtrage aux exsurgences ascendantes. *Bull. Sci. Bourg.*, 40, 61–65.

Danielopol, D.L. (1976). The distribution of the fauna in the interstitial habitats of riverine sediments of the Danube and the Piesting (Austria). *Int. J. Speleol.*, 8, 23–51.

Dole, M.-J. (1983). *Le domaine aquatique souterrain de la plaine alluviale du Rhône à l'est de Lyon: écologie des niveaux supérieurs de la nappe.* Thesis, Univ. Lyon I, France.

Essafi, E., Mathieu, J. & Beffy, J.-L. (1992). Spatial and temporal variations of *Niphargus* populations in interstitial aquatic habitat at the karst/floodplain interface. *Regulated Rivers*, 7, 83–92.

Ford, T.E. & Naiman, R.J. (1989). Groundwater-surface water relationships in boeral forest watersheds: dissolved organic carbon and inorganic nutrient dynamics. *Canadian Journal of Fisheries and Aquatic Sciences*, 46, 41–49.

Forman, R.T.T. & Moore, P.N. (1992). Theoretical foundations for understanding boundaries in landscape mosaics. In *Landscape boundaries, consequences for biotic diversity and ecological flows*, ed. A.J. Hansen & F. Di Castri, pp. 236–258, Ecological Studies 92, Springer-Verlag, New York.

Gibert, J., Dole-Olivier, M.J., Marmonier, P. & Vervier, Ph. (1990). Surfacewater/groundwater ecotones. In *Ecology and management- of aquatic-terrestrial ecotones*, ed. R.J. Naiman & H. Décamps, pp. 199–225, Partenon Publ., London.

Gibert, J. (1986). Ecologie d'un système karstique jurassien. Hydrogéologie, dérive animale, transits de matières, dynamique de la population de *Niphargus* (Crustacé Amphipode). *Mém. Biospéol.*, 13, 1–379.

Gibert, J. (1991a). Groundwater systems and their boundaries: conceptual framework and prospects in groundwater ecology. *Verh. Internat. Verein. Limnol.*, 24, 1605–1608.

Gibert, J. (1991b). Les écotones souterrains/superficiels: des zones d'échanges entre environnements souterrain et de surface. *Hydrogéologie*, 3, 233–240.

Ginet, R. (1960). Ecologie, éthologie et biologie de *Niphargus* (Amphipodes Gammaridés hypogés). *Annls Spéléol.*, 15, 1–254.

Hansen, A.J., Di Castri, F. & Naiman, R.J. (1988). Ecotones: what and why ? *Biology International, Special Issue*, 17, 9–46.

Marmonier, P. & Dole, M.-J. (1986). Les Amphipodes des sédiments d'un bras court-circuité du Rhône: logique de répartition et réaction aux crues. *Sciences de l'Eau*, 5, 461–486.

Patton, J. & Reeder, W. (1956). New indicator for titration of calcium with (ethylenedinitrilo) tetraacetate. *Anal. Chem.*, 28, 1026.

Pennak, R. & Ward, J. (1986). Interstitial fauna communities of the hyporheic and adjacent groundwater biotopes of a Colorado mountain stream. *Archiv für Hydrobiologie, Supplement*, 74, 356–396.

Plénet, S., Gibert, J. & Vervier, P. (1992). A floodplain spring: an ecotone between surface water and groundwater. *Regulated Rivers*, 7, 93–102.

Rouch, R. (1970). Recherches sur les eaux souterraines. 12. Le système karstique du Baget. 1. Le phénomène d''Hémorragie' au niveau de l'exutoire principal. *Annals Spéléol.*, 25, 665–690.

Rouch, R. (1977). Le système karstique du Baget. VI – La communauté des Harpacticides. Signification des échantillons récoltés lors des crues au niveau de deux exutoires du système. *Annls Limnol.*, 13, 227–249.

Schmidt, C.M., Marmonier, P., Plénet, S., Creuzé des Châtelliers, M. & Gibert, J. (1991). Bank filtration and interstitial communities. Example of the Rhône River in a polluted sector (downstream of Lyon, Grand-Gravier, France). *Hydrogéologie*, 3, 217–223.

Sharley, T. (1994). *The influence of exchange between surface and groundwater on ecotone processes.* In Proc. of Ecotones Regional Workshop, Barmera, South Australia, 12–15 October 1992. Ecotones at the river basin scale – global land/water interactions, ed. A. E. Jensen, pp. 86–91, South Australian Department of Environment and Natural Resources Publ., Adelaide.

Sutcliffe, D.W., Carrick, T.R. & Willoughby, L.G. (1981). Effects of diet, body size, age and temperature on growth rates in the amphipod *Gammarus pulex*. *Freshwat. Biol.*, 11, 183–214.

Turquin, M.-J. (1981). Profil démographique et environnement chez une population de *Niphargus virei* (Amphipode troglobie). *Bull. Soc. Zool. Fr.*, 106, 457–466.

Turquin, M.-J. (1984a). Age et croissance de *Niphargus virei* (Amphipode pérennant) dans le système karstique de Drom: méthodes d'estimation. *Mém. Biospéol.*, 11, 37–49.

Turquin, M.-J. (1984b). Un cas de transition démographique dans le milieu souterrain. *Verh. Internat. Verein. Limnol.*, 22, 1751–1754.

Turquin, M.-J. (1988). *Les âges de Niphargus virei (Crustacé stygobie).* In Actes du colloque 'Aspects récents de la biologie des Crustacés', Caen, France, August 1988, ed. IFREMER, pp. 195–198.

Turquin, M.-J. & Barthélémy, D. (1985). The dynamics of a population of the troglobitic amphipod *Niphargus virei* Chevreux. *Stygologia*, 1, 109–117.

UNESCO (1980). *Surface water and groundwater interaction.* Studies and Report in Hydrology 29, Paris, France.

Vanek, V. (1987). The interactions between lake and groundwater and their ecological significance. *Stygologia*, 3, 1–23.

Vervier, Ph. & Gibert, J. (1991). Dynamics of surface water/groundwater ecotones in a karstic aquifer. *Freshwat. Biol.*, 26, 241–250.

Vervier, Ph., Gibert, J., Marmonier, P. & Dole-Olivier, M.-J. (1992). A perspective on the permeability of the surface freshwater-groundwater ecotone. *J. N. Am. Benthol. Soc.*, 11, 93–102.

Wiens, J.A., Crawford, C.S. & Gosz, J.R. (1985). Boundary dynamics: a conceptual framework for studying landscape ecosystems. *Oikos*, 45, 421–427.

7 Community respiration in the hyporheic zone of a riffle-pool sequence

M. PUSCH

Limnological Institute, University of Constance, D – 78434 Konstanz, Germany

Present address: Institut für Gewässerökologie und Binnenfischerei, Müeggelseedamm 260, D – 12587 Berlin, Germany

ABSTRACT   A portable incubation device was used to study community respiration activity in hyporheic sediments of a mountain stream. Substantial heterotrophic activity was measured in these sediments, which contribute significantly to total organic matter processing in stream ecosystems. Hyporheic community respiration (HCR), which was mainly attributable to microbial metabolism, decreased with sediment depth in a riffle, with a minimum value at 20 cm. At all depths, HCR activity, loosely-associated particulate organic matter (LAPOM), and the LAPOM protein content were all much lower in the pool sediments than in the riffle sediments. This implies that the hyporheic zone investigated here was probably not supplied with significant amounts of nutrients from downwelling water in the pool, as implied by current concepts of the hydraulics in riffle-pool sequences.

INTRODUCTION

In running waters, organic matter (OM) is subjected to biological transformation as well as physical transportation due to the flowing water. In streams with a moderate or high gradient, the residence time of water is short and the biological activity in the open water is low (e. g. Meyer *et al.*, 1990). However, the sediment interstices act as effective retention sites for particulate organic matter (POM) (e. g. Schwoerbel, 1961; Metzler & Smock, 1990; Leichtfried, 1991). Thus most biological activity in streams is associated with the sediments (Naiman *et al.*, 1987).

In many cases, the stream channel cuts into an alluvial floodplain, and surface waters are in contact with phreatic groundwater. In the transition zone between superficial sediments and deeper layers of alluvium, which is known as the hyporheic zone, characteristics of both habitats occur (Schwoerbel, 1961). This ecotone (cf. Gibert *et al.*, 1990) is linked hydrologically to phreatic as well as surface waters by various degrees of exfiltration/infiltration and downwelling/upwelling, respectively (e. g. Vaux, 1968; White *et al.*, 1987; Valett *et al.*, 1990; 1994, Stanford & Ward, 1993). Occurrence of the latter processes often appears to be associated with longitudinal changes in the stream-bed and water table gradients, for example within riffle-pool sequences (Vaux, 1968;

White *et al.*, 1987). Riffles are defined here as stream sections where the water table gradient is steeper than the mean gradient of the stream, pools are defined by the opposing situation (cf. Leopold *et al.*, 1964; Figs. 1, 2).

In riffle-pool sequences, distinct spatial patterns of downwelling and upwelling processes have been reported (Fig. 1; Vaux, 1968; White *et al.*, 1987; Hendricks, 1993). The water table is somewhat higher than the mean gradient line at the lower end of pools, while it is relatively low at the lower end of riffles. Thus in accordance with the respective hydraulic heads, downwelling should occur at the lower end of pools while upwelling should occur in riffles. Dissolved and fine particulate organic matter should be transported into the hyporheic zone of pools along with the downwelling water. On the other hand, riffle sections are supplied with DOM by groundwater discharged into areas with a relatively low hydraulic head.

Little is known about the functional relationships between abiotic and biological processes in the hyporheic zone (Valett *et al.*, 1993). In particular, the extent of organic matter degradation in the hyporheic zone of stream sediments is poorly understood (Hendricks, 1993). Respiration measurements have been used to study heterotrophic activity in the benthic layer of stream sediments (Bott *et al.*, 1985). The only studies on community respiration in the hyporheic zone that I

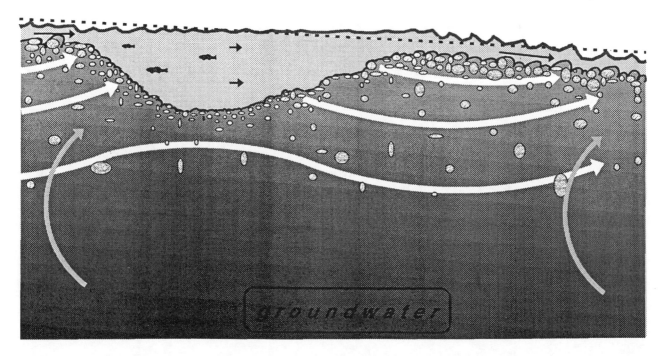

Fig. 1 Longitudinal section through an idealized riffle-pool sequence. Arrows show water flow according to current concepts of the hydraulics in this environment (e.g. Vaux, 1968; White *et al.*, 1987). The dotted line indicates the mean gradient of the water table.

am aware of have been conducted in a desert stream in Arizona (Sycamore Creek, Grimm & Fisher, 1984; Valett *et al.*, 1990). These reported a relatively high degree of heterotrophic metabolism in the hyporheic zone of this warm-water stream, and this was similar to the benthic community respiration.

The objective of this study was to measure hyporheic community respiration (HCR) at several depths in the sediments of a temperate stream. This was carried out at a riffle site and a corresponding pool site of a riffle-pool sequence in order to compare the pattern of HCR with predicted hydrological inputs of organic matter as well as a number of other biological parameters which were measured simultaneously.

STUDY SITE AND METHODS

The study was conducted in a third-order reach of the Steina, a soft-water mountain stream in the Black Forest, Germany. At this site at *c.* 700 m a.s.l., the stream is *c.* 4.5 m wide with a gradient of 1.6%. The catchment of 19 km² is dominated by

coniferous forest and meadows, and is scarcely populated. Sediments consist mainly of gneiss and sandstone and are badly sorted, so that even boulders are frequent everywhere in the stream bed. The median grain size of the <63 mm fraction is 9.0 mm, the lower quartile (25%) is 2.4 mm, and the upper quartile (75%) is 21.5 mm (M.H.E. Pusch, 1987). For further details on the study site see Meyer *et al.*, (1990), Pusch (1993) and Pusch & Schwoerbel (1994).

In a typical riffle-pool sequence (Fig. 2), water depth was measured every 2 m along the stream, and relative levels of the water table were compared every 10 m using a transparent plastic hose.

Interstitial flow rates were estimated using pulses of a sodium chloride solution as a conservative tracer (cf. Bencala *et al.*, 1984). The most frequent hyporheic transport velocity was derived from the peak conductivity value observed in the downstream hyporheic probe. The mean hyporheic transport velocity was derived from the median of the transition curve of the tracer pulse. Measurements were carried out at a sediment depth of 30 cm in the permanent stream bed in riffle sections at mean flow conditions.

Sediments from the Steina were dried and sieved into two grain sizes (1–2 mm, 2–4 mm). Plastic pipes (15 cm in length, 0.754 dm³ volume) were filled with a uniform weight of one or the other of the two types of sediment, and closed with a

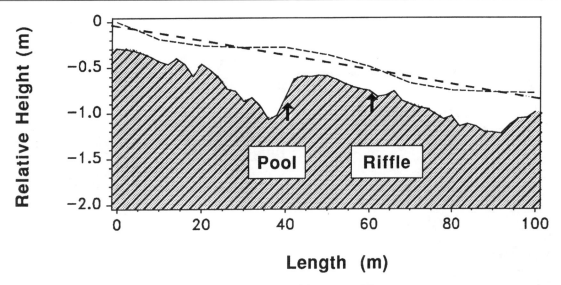

Fig. 2 Longitudinal section through the riffle-pool sequence studied in the Steina. Relative heights of the stream bed and the water table are shown along with the mean gradient (regression line) of the water table. Arrows indicate locations of the vertical incubation containers in the riffle and pool sites.

gauze of 1 mm mesh size. These sediments were used as quasi-natural substrates and incubated within the hyporheic zone for 16–20 months for colonization by stream biota. Faunal colonization of sediment pipes in the hyporheic zone reaches equilibrium after about four months of incubation (M. H. E. Pusch, 1987). Vertical incubation containers made of plastic were installed in the stream bed so that hyporheic sediments did not have to be disturbed when retrieving the pipes for measurements. Each vertical incubation container had lateral openings at sediment depths of 10, 20, 30 and 40 cm. These corresponded to the depths of the incubated sediment pipes, so that they were in direct contact with the surrounding sediments at both ends. Six vertical incubation containers were installed in the riffle and pool of a typical riffle-pool sequence, respectively. Thus a total of 48 sediment pipes were used, with 24 of each grain size. Half of the pipes were retrieved between 21. 3. 92 and 5. 5. 92 at mean stream discharge (spring samples), while the remainder were retrieved between 15. 8. 92 and 11. 9. 92 under base flow conditions (summer samples). Unfortunately, on the bottom of several containers in riffle some fine POM accumulated in spring, which could have influenced conditions within the lowermost samples by a small slit.

On the stream bank, sediment pipes were then inserted into an incubation device which simulated *in situ* conditions in the hyporheic zone (Pusch & Schwoerbel, 1994). This recirculating system was designed to measure of hyporheic community respiration and consisted of two conical screw adapters which accomodated a sediment pipe, an oxygen probe, a heat exchange unit for heating or cooling, and a pre-

cision pump. These components were connected together by rubber and plastic tubing (Fig. 3). In this gas-tight system, the pump generated a flow rate equivalent to natural water velocities in the hyporheic zone. Flow velocity in the hyporheic zone of riffle sections (cf. results) was used for measurements of both riffle and pool sediments. Incubation temperatures always corresponded to current stream temperatures in spring and summer, respectively.

On termination of an HCR measurement, sediments were scoured, in order to separate macro- and meiofauna >100 μm from particulate organic matter (POM) <100 μm. The latter comprised >90 % of the total POM which could be rinsed out. Hyporheic fauna were counted and their length measured; subsequently, their biomass and respiration rates were calculated (Pusch & Schwoerbel, 1994). Ash-free dry mass of POM was determined by combustion. POM was classified into two fractions: (A) POM which was only loosely associated with the sediment (LAPOM) was obtained by scouring the sediments and included microinvertebrates <100 μm, and (B) the POM fraction which remained on rinsed sediments, i. e. the POM strongly associated with sediment (SAPOM). Further details are given in Pusch & Schwoerbel (1994).

LAPOM content and macrofaunal density in benthic sediments were estimated tentatively on 21. 9. 92. An iron frame and a simple suction device were used for LAPOM sampling. Nine areas (7.5 cm² each) were sampled in riffle and pool sections (cf. Fig. 2), respectively. For benthic macrofauna, one sample using a square-foot Surber sampler with a 300 μm net was taken at the riffle and pool sites.

RESULTS

The most frequent flow rate of water parallel to the direction of stream flow in the hyporheic sediments was 1.43 m h⁻¹

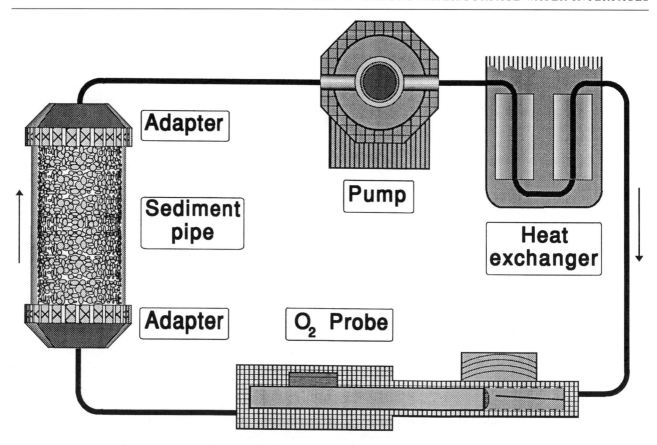

Fig. 3 Incubation system for measuring hyporheic community respiration (HCR) in sediment pipes at *in situ* temperatures.

(s. d.=0.34, n=5). The mean velocity was 0.74 m h^{-1} (+/−0.40). The corresponding values for the pool were 0.37 and 0.19 m h^{-1}, respectively (single observations).

There was nearly six times more LAPOM in the hyporheic zone of the riffle than in the hyporheic zone of the pool (Fig. 4a). While the LAPOM had an even depth distribution in the pool site, there was a distinct LAPOM – minimum at a depth of 20 cm in the riffle sediment. This spatial pattern was apparent at low discharge in summer as well as at mean discharge in spring. The content of LAPOM at the lowermost sediment level (40 cm) in spring was probably slightly over-estimated due to an artifact in the vertical incubation containers (see methods). In contrast to the hyporheic zone, LAPOM content was not higher in the benthic layer of the riffle during base flow conditions: a mean of 72 g m^{-2} of LAPOM were measured in the riffle, compared to 132 g m^{-2} in the pool.

The proportion of protein within the LAPOM fraction was relatively constant, both spatially and temporally. It was about 7% in pool samples and about 16% in riffle samples. Thus the protein content (Fig. 4b) had a similar vertical distribution to the LAPOM itself. However, average protein contents in the riffle and pool samples differed by a factor of 16, which was higher than the difference for LAPOM. The

distribution of SAPOM in the hyporheic zone (Fig. 4c) did not differ markedly between the riffle and pool sites, but did show some decrease between spring and summer.

Hyporheic community respiration ranged between 0.03 mg O$_2$ dm^{-3} h^{-1} in the pool sediments and 1.3 mg O$_2$ dm^{-3} h^{-1} in the riffle sediments. The spring values were, on average, 7.2 times higher at the riffle site (Fig. 5a); this difference was reduced to a factor of 3.4 in the summer. The activity at 20 cm was consistently less than half that at 10 cm, where HCR reached its maximal values. Thus HCR followed the same general spatial pattern as the protein content of LAPOM as well as LAPOM itself in the sediments. Again there was a distinct gradient at the riffle site, while there were no clear differences between sediment depths at the pool site. When respiration rates were standardized for temperature, HCR was an average of 5 times higher in the riffle sediments than the pool sediments. HCR in spring was slightly higher than in summer, but this was not statistically significant (Wilcoxon test p >0.5, Fig. 5b).

Respiration activity of meio- and macrofauna >100 µm was calculated as a mean of 7.9 % of the total community respiration in both the riffle and the pool samples. Hence 92.1 % of community respiration in the hyporheic zone was attributed to micro-organisms, including protozoa, which could not be quantified directly.

Hyporheic fauna were more than thirteen times more abundant in riffle sediments than in pool sediments, corre-

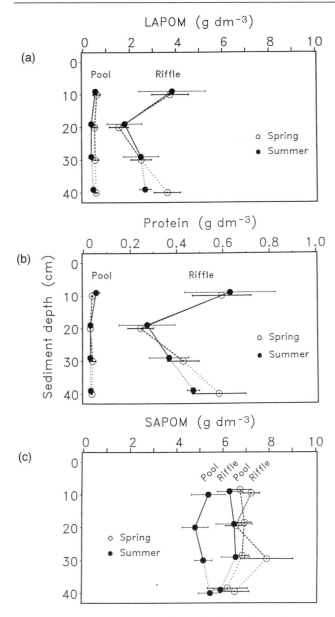

Fig. 4 Vertical distribution of different fractions of organic matter in the hyporheic zones of the riffle and pool sites, which were sampled once in spring and summer, respectively (cf. text). Values at the 40 cm depth were probably slightly overestimated due to an artifact (dotted lines). a) Loosely associated particulate organic matter (LAPOM), b) protein within the LAPOM fraction, and c) strongly associated particulate organic matter (SAPOM).

sponding to a factor of 7 in terms of biomass. Benthic macrofauna had a density of c. 2000 indiv. $(0.1\ m^2)^{-1}$ in the riffle sediments and 260 indiv. $(0.1\ m^2)^{-1}$ in the pool sediments.

DISCUSSION

Heterotrophic activity is a central process in the functioning of every ecosystem. Its level is crucial for the extent to which

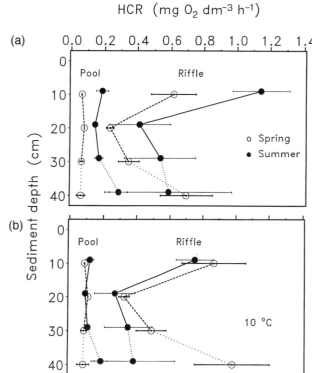

Fig. 5 Vertical distribution of hyporheic community respiration (HCR) (+/– 1 SE, n= 5) a) at ambient temperatures (5 °C in spring, 16 °C in summer), b) standardized to a uniform temperature of 10 °C using a Q10–value of 2.0. Symbols as in Fig. 4.

organic substances can be utilized by a given biocenosis. In the hyporheic sediments of the mountain stream studied here, a surprisingly high level of heterotrophic activity was detected. Related to an areal basis, this level of HCR is similar to levels of benthic community respiration reported in other streams (Bott et al., 1985).

Despite standardization of the substrata by sieving, there was still a relatively high degree of variability in the HCR measured in in the hyporheic zone of the riffle site (cf. Fig. 5). Nevertheless, temporally and spatially consistent patterns of HCR activity were still apparent. HCR decreased with depth in the riffle site, with a conspicuous minimum at 20 cm. By contrast, HCR in the pool hyporheic zone had an almost uniform vertical profile, with a low level of activity. This pattern was reflected in the temporal and spatial distribution of the LAPOM protein content as well as the spatial distribution of the LAPOM fraction itself. This similarity, which also appeared in a study of seasonal variability (Pusch & Schwoerbel, 1994), suggests that these parameters were connected functionally. Higher protein content in organic matter is probably an indicator for high-quality particulate organic matter, which forms a suitable substrate for sediment bacteria and other micro-organisms. Due to the method used, the biomass of these micro-organisms would also have contributed to the estimated protein content of POM. In a more general way, it appears that, in hyporheic zone sediments, areas with a high content of protein are probably correlated with 'hot spots' of community respiration.

Parameters measured in the hyporheic zone of the riffle contrasted markedly with those measured in the pool. For hyporheic fauna, similar spatial heterogeneity has been observed by Rouch (1991), which he explained by differing hydraulic situations. It is plausible that differences in LAPOM content and HCR activity in the Steina would have corresponded to parallel differences in the supply of organic matter, either from the benthic zone or shallow groundwater. As the amount of LAPOM stored in the benthic zone did not correspond to this pattern, it would appear that there was no significant downward transport of LAPOM with down-welling surface water at that time in the pool studied here. This finding is contrary to current concepts of the hydraulics in riffle-pool sequences (e.g. White et al., 1987; Hendricks, 1993).

On the other hand, there does seem to be an effective mechanism by which large amounts of POM can become entrained in the hyporheic zone of riffles, either from the surface or deeper strata. Local differences in hydraulic head within a rough stream bed can generate intensive vertical exchange (Williams, 1993), and there might also be a substantial input of material from exfiltrating groundwater.

Speculatively the consistent minimum in LAPOM values at the 20 cm depth in the riffle site might have been due to episodic sedimentation processes occuring during spates (cf. Metzler & Smock, 1990). However, the characteristic profiles might also have been formed by the interplay of continious processes such as the supply of different forms of organic matter from the surface layer or groundwater, and microbial heterotrophic activity. It is therefore necessary to include transport rates and pathways of organic matter, as well as seasonal variability of hyporheic processes (cf. Valett et al., 1994) and historical aspects of hyporheic sediments formation, in conceptual models aimed at improving our understanding of metabolism in the hyporheic zone.

ACKOWLEDGEMENTS

I wish to thank Professor J. Schwoerbel for inspiring discussions, M.H.E. Pusch for valuable help, the students participating in the limnological summer course of 1992 for some field data, and Dr D. Fiebig for linguistic correction and valuable comments.

REFERENCES

Bencala, K.E., Kennedy, V.C., Zellweger, G.W., Jackman, A.P. & Avanzino, R.J. (1984). Interactions of solutes and streambed sediment. 1. An experimental analysis of cation and anion transport in a mountain stream. Wat. Resour. Res.,, 20, 1797–1803.
Bott, T.L., Brock, J.T., Dunn, C.S., Naiman, R.J., Ovink, R.W. & Petersen, R.C. (1985). Benthic community respiration in four temperate stream systems: An inter-biome comparision and evaluation of the river continuum concept. Hydrobiologia, 123, 3–45.
Gibert, J., Dole-Olivier, M.J., Marmonier, P. & Vervier, P. (1990). Surface water-groundwater ecotones. In Ecology and management of aquatic-terrestrial ecotones, ed. R.J. Naiman & H. Decamps, pp. 199–225, Man and the Biosphere Series, Vol. 4, Parthenon Publ., Park Ridge, New Jersey.
Grimm, N.B. & Fisher, S.G. (1984). Exchange between interstitial and surface water: implications for stream metabolism and nutrient cycling. Hydrobiologia, 111, 219–228.
Hendricks, S.P. (1993). Microbial ecology of the hyporheic zone: a perspective integrating hydrology and biology. J. N. Am. Benthol. Soc., 12, 70–78.
Leichtfried, M. (1991). POM in bed sediments of a gravel stream (Ritrodat-Lunz study area, Austria). Verh. Internat. Verein. Limnol., 24, 1921–1925.
Leopold, B.L., Wolman, M.G. & Miller, J.P. (1964). Fluvial processes in geomorphology. San Francisco, Freeman & Co.
Metzler, G.M. & Smock, L.A. (1990). Storage and dynamics of subsurface detritus in a sand-bottomed stream. Can. J. Fish. Aquat. Sci., 47, 588–594.
Meyer, E., Schwoerbel, J. & Tillmans, G.C. (1990). Physikalische, chemische und hydrographische Untersuchungen eines Mittelgebirgsbaches: Ein Beitrag zur Typisierung kleiner Flieflgewässer. Aquat. Sci., 52, 236–255.
Naiman, R.J., Melillo, J.M., Lock, M.A., Ford, T.E. & Reice, S.R. (1987). Longitudinal patterns of ecosystem processes and community structure in a subarctic river continuum. Ecology, 68, 1139–1156.
Pusch, M. H. E. (1987). Die Besiedlung des hyporheischen Interstitials in einem Schwarzwaldbach. – Diploma thesis Univ. Freiburg.
Pusch, M. (1993). Heterotropher Stoffumsatz und faunistische Besiedlung des hyporheischen Interstitials eines Mittelgebirgsbaches (Steina, Schwarzwald). Doct. thesis Univ. Freiburg, 123 pp.
Pusch, M. & Schwoerbel, J. (1994). Community respiration in hyporheic sediments of a mountain stream (Steina, Black Forest). Arch. Hydrobiol., 130, 35–52.
Rouch, R. (1991). Structure du peuplement des Harpacticides dans le milieu hyporhéique d'un ruisseau des Pyrénées. Annls Limnol., 27, 227–241.
Schwoerbel, J. (1961). Über die Lebensbedingungen und die Besiedlung des hyporheischen Lebensraums. Arch. Hydrobiol. Suppl., 25, 182–214.
Stanford, J.A. & Ward, J.V. (1993). An ecosystem perspective of alluvial rivers: connectivity and the hyporheic corridor. J. N. Am. Benthol. Soc., 12, 48–60.
Valett, H.M., Fisher, S.G., Grimm, N.B. & Camill, P. (1994). Vertical hydrologic exchange and ecological stability of a desert stream ecosystem. Ecology, 75, 548–560.
Valett, H.M., Fisher, S.G. & Stanley, E.H. (1990). Physical and chemical characteristics of the hyporheic zone of a Sonoran Desert stream. J. N. Am. Benthol. Soc., 9, 201–215.
Valett, H.M., Hakenkamp, C.C. & Boulton, A.J. (1993). Perspectives on the hyporheic zone: integrating hydrology and biology. Introduction. J. N. Am. Benthol. Soc., 12, 40–43.
Vaux, W. G. (1968). Intragravel flow and interchange of water in a streambed. U.S. Fish. Wildl. Serv. Fish. Bull., 66, 479–489.
White, D.S., Elzinga, C.H. & Hendricks, S.P. (1987). Temperature patterns within the hyporheic zone of a northern Michigan River. J. N. Am. Benthol. Soc., 6, 85–91.
Williams, D.D. (1993). Nutrient and flow dynamics at the hyporheic/groundwater interface and their effects on the interstitial fauna. Hydrobiologia, 251, 185–198.

8 Diversity, connectivity and variability of littoral, surface water ecotones in three side arms of the Szigetköz Region (Danube, Hungary)

A. BOTHAR & B. RATH

Hungarian Danube Research Station, Hungarian Academy of Sciences, H-2131 God, Hungary

ABSTRACT The 137 km long Szigetköz side-arm system extends on the right side of the river Danube in Hungary and shows the unique characteristics of river, shore, riparian forest, and groundwater ecotones. The investigated three side arms are in different alluvial stages, their surface sediment is variable: sand, sandy mud, mud, mud covered with detritus. The composition, life forms, and cover of aquatic macrophyte stands are good describers of the different side arms. On the surface of the sediment, the habitats are numerous and varied, and linked to the vegetation and open water region. Crustacean communities of these habitats are very diverse, mosaic-form, and are in dynamic connection with each other mainly through the developing stages, and even through certain stenoecious species.

INTRODUCTION

The Szigetköz Danube stretch in Hungary (1848–1806 river km) flows on its own alluvial cone. Fall is small, 3–10 cm km^{-1}. A 137 km long side-arm system extends on the right side of the river, with 22.2 million m^3 volume. The whole region is composed of five side-arm systems. Regulation activities began in the last century. The main goal of the engineering has been the assurance of the shipping on the Danube and the flood control. The present situation is the result of the regulation works carried out in 1966–1983 when numbers and heights of closing dams were increased. During low and mean water periods, 90 % of the water discharge flows in the Danube itself. Therefore, quantity and intensity of water supply of the side arms have decreased. The situation of the water table in the Szigetköz region did not changed drastically in this period. The future fate of the Szigetköz region will depend on the form of construction of the Gabcikovo power plant system. Present situation was valid only till October 1992 when the diversion of the river Danube was carried out (so called C variant of the power plant). There are only few papers dealing with limnological investigations referring to macrophyte vegetation and Crustacean communities in the Szigetköz region. Ráth (1987) described the macrophyte vegetation of a small branch-system at Dunaremete

(Szigetköz, river km 1826). Gulyàs (1987, 1994) carried out zooplankton investigations in the Szigetköz from 94 sampling sites. According to the Rotatoria, Cladocera and Copepoda fauna he could differentiate 7 types of water bodies on the protected area and on the alluvial plain. Bothár & Ráth (1992) informed about abundance dynamics of Crustaceans in the Szigetköz. Detailed zooplankton investigations were carried out by Vranovsky (1975, 1985, 1991) in different side-arms of the opposite, Slovakian bank of the River Danube.

STUDY AREA, TIME OF SAMPLING

The Szigetköz region consists of five arm systems. The investigated Cikola side-arm system (850 ha) is formed by several branches with 28 km length. As mentioned earlier, the system is regulated for mean water level by closing the upper part of the arms with overflow dams. When the water level of the Danube is lower than that of the overflow dam, the water supply for the side arms occurs only from the direction of their mouth and from the groundwater. In the last decade, the recession of the Danube-bed and that of the low water levels also had unfavourable effect on the groundwater level. Therefore an aging process can be observed in the whole system.

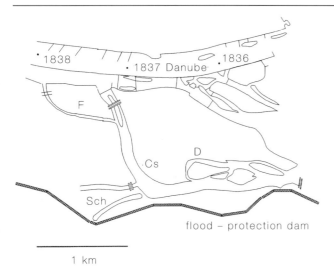

Fig. 1 Map of the sampling area, F: side-arm Forrásos, Sch: side-arm Schisler, D: side-arm Disznós, Cs: main side-arm Csákányi.

The length of the different arms are not big: 500–600 m, width: 14–40 m. The slope of the shore is steep, with *Salicetum albae-fragilis, Populetum canadensis* alongside the branches. Bottom material: several meters of alluvial gravel, covered by sand, sandy mud, or mud. Water depth is biggest in early summer (200–250 cm) when even the vegetation of the littoral region *(Rorippo-Agrostetum, Scirpo-Phragmitetum)* is inundated. During autumn low water periods, water surfaces are getting contracted and depths are decreasing to 0–60 cm. During high water periods, flow velocity and water levels are increasing rapidly in the side arms.

Investigations were carried out in three side arms: Disznós, Schiszler, Forrásos (Fig.1). Main differences of the three water bodies are based on their different extensions, in connection with their alluvial age. The direct water supply of the arms comes from the main side-arm of the system, Csákányi, where the height of the overflow is 460 cm water level on the Dunaremete gauge, 1826 river km (Fig. 2).

Times of sampling were as follows. In 1991: 5 June – during long-lasting low-water period; 18 July – after a short flood; 14 August – high-water period, after an extreme great flood; 15 October – extreme low-water period. In 1992: 9 June, 4 August, 9 October during the whole sampling period water level was low, direct water supply did not occur (Fig. 2).

METHODS

Botanical analysis was performed according to the Braun-Blanquet technique. Quantitative samples for Crustaceans were taken from different habitats: integrated samples, 5×10 litres of water from the open water region along the long axis

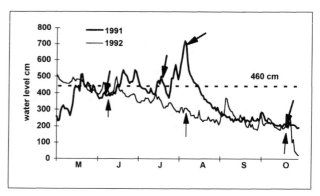

Fig. 2 Water-level graph for the Danube at the Dunaremete gauge (1826 river km). Arrows indicate the sampling times.

of the side arms, bottom samples with Ekman-Birge sampler (one sample consists from three random subsamples from 4–5 m area), 5–10 litre samples from water bodies among littoral vegetation and aquatic macrophytes (5–10 random subsamples from 2–3 m² area). Samples were conserved in 4% formaline.

RESULTS

Side-arm Disznós

This water body is in a moderate alluvial stage, lake-like, the flushing begins at the 500 cm water level from the Csákányi-arm which causes only slight bed-flow. The bed profile is u-shaped, width is 25 m, average depth is 150 cm, shore line is gently sloping. The alluvial sand and gravel is covered by sandy mud, and mud. Its thickness and composition is variable (2–15 cm) depending on the actual flowing velocity.

The water warms up earliest among the three investigated arms. Water temperature can achieve 23–24 °C already in May which is favourable for the phytoplankton. Dominant species of the macrovegetation are *Ranunculus circinatus* and *Myriophyllum spicatum*. Appearance of macrophyton stands is periodical, their cover is variable, 10–80%, caused by the phytoplankton/macrophyton competition. The reed belt is 3–4 m wide. The marshy foot of the branch is covered by *Rorippa amphibia* stands. Adventive *Impatiens glandulifera is* characteristic in the gallery forest *Populetum canadensis.*

During a long-lasting low-water period, at the beginning of the vegetation period, June 1991, large Crustacean populations were present in the water (10^2–10^6 m^{-3}). Greatest individual numbers could be found among the littoral inundated *Agrostis* stands in the 20–30 cm deep water. Copepods in copepodite and nauplius stages were abundant only in the plankton and in the water among the logs fallen into the water as a consequence of timbering. On the surface of the mud, copepodite IV resting stages of *Cyclops vicinus*

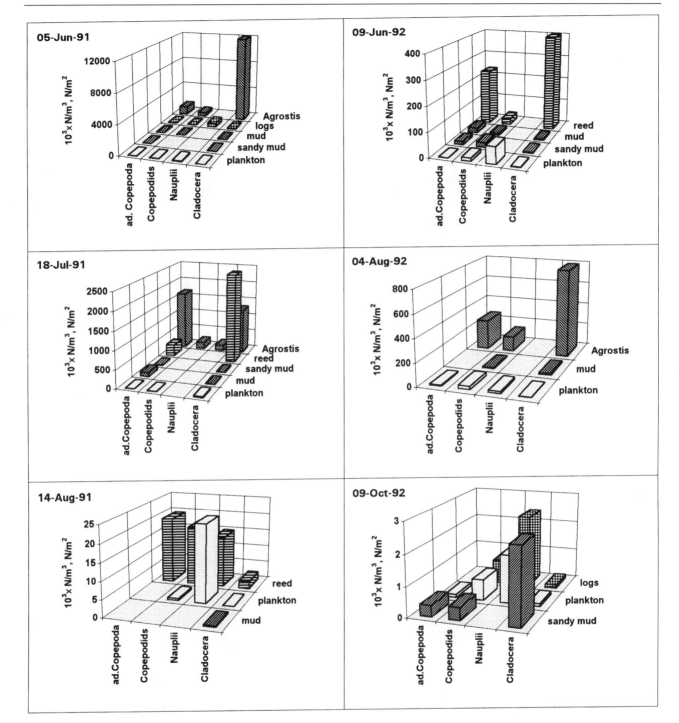

Fig. 3 Distribution of adult Crustaceans and larval stages of Copepods in the different habitats of the side-arm Disznós in 1991 and 1992 (N/m^2-values of the z-axis refer to the bottom samples).

were recorded (Fig. 3). Twentynine crustacean species were determined from the different habitats, which means an extremely high diversity. Species numbers in the plankton were small; it is remarkable that the euplanktonic *Bosmina longirostris* was abundant in the open water only in very few numbers, and among the *Agrostis* stand it was overcrowded (11×10^6 m^{-3}). Of the 29 species, only two (*Bosmina lon-*

girostris and *Acanthocyclops robustus*) were abundant in all the five habitats (Fig. 4).

In the declining period of the small flood, July 1991, a slight flow could be experienced in the branch. Crustacean communities were highly responsive to this water movement. In the open water area, a large Crustacean population was present with 13 species, all of them typical for the phytal and the bottom. Presumably they were washed out from these habitats to the plankton. The species composition of the different habitats was very variable, among the 28 registered species there was not a single one which was abundant in

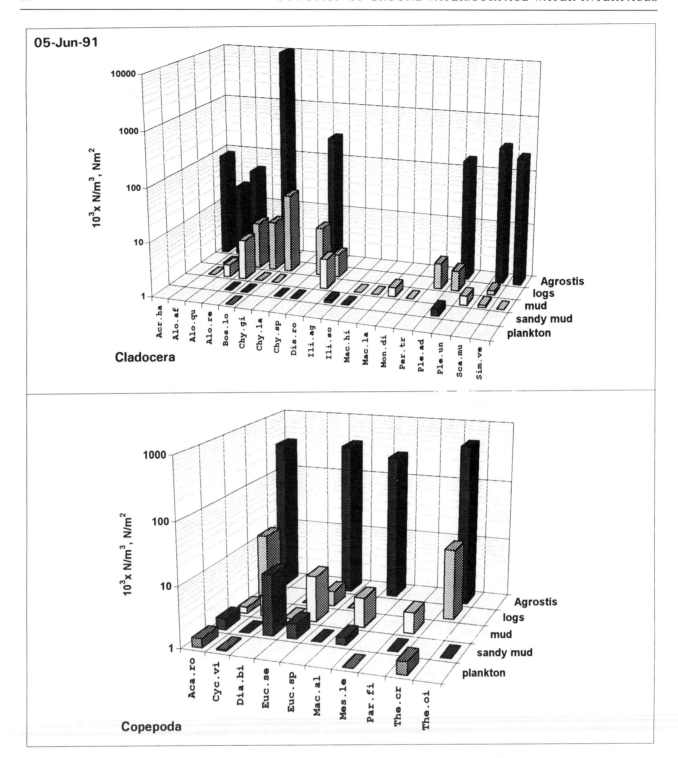

Fig. 4 Distribution of Cladocera and Copepoda species in the side-arm Disznos, 5 June 1991. Explanation of the abbreviations of species names in Table 1 (N/m^{2-} values of the z-axis refer to the bottom samples).

every habitat investigated. Euplanktonic *Bosmina longirostris* was abundant even on the surface of the bottom with 18×10 $N\,m^{-3}$, while in the plankton only with 3.2×10^3 $N\,m^{-3}$ individual numbers. In the water body among the reeds and among the *Agrostis* stand, a very diverse Crustacean community was present, but of the 16 species, only 8 were common in the two habitats (Fig. 5). *Nauplius* larvae disappeared from the plankton, they were found only in the more protected littoral area in the *Agrostis* stand.

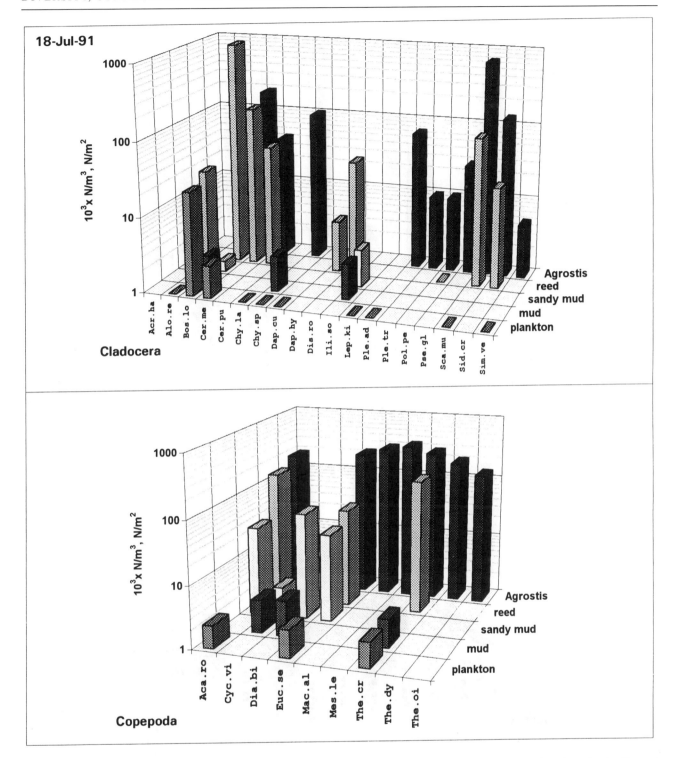

Fig. 5 Distribution of Cladocera and Copepoda species in the side-arm Disznós, 18 July 1991. Explanation of the abbreviations of species names in Table 1 (N/m^2-values of the z-axis refer to the bottom samples).

Copepodite larvae were also more abundant in the littoral region (Fig. 3).

During the extremely great flood in August 1991, Crustacean communities were practically washed out from the side-arm. The plankton consisted only of a single species, in few examples, mainly larval stages of Copepods were abundant in the open water. Adult Crustaceans found protection among the reed belt. Only 8 species could be registered, individual numbers were two orders of

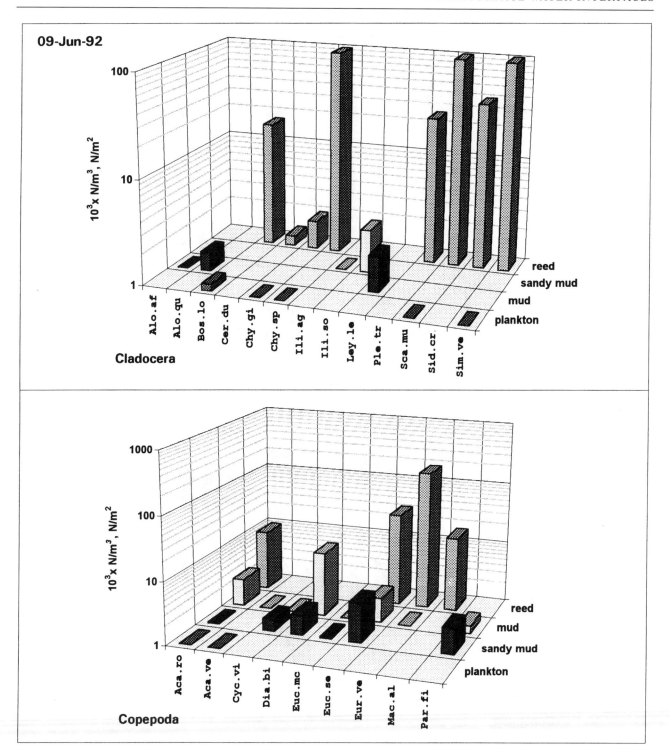

Fig. 6 Distribution of Cladocera and Copepoda species in the side-arm Disznós, 9 June 1992. Explanation of the abbreviations of species names in Table 1 (N/m^2-values of the z-axis refer to the bottom samples).

magnitude smaller than those in the earlier sampling time (Fig. 3).

In 1992, the distribution of adult Crustaceans and larval stages were similar to that of experienced in 1991 (Fig. 3). During the different sampling periods, registered species were less than in the former year, 22 and 21 in June and August respectively (Figs. 6 and 7). Individual numbers were also smaller, mainly in June, they reached only 10^4 m^{-3} order of magnitude. During the low-water period in October, there

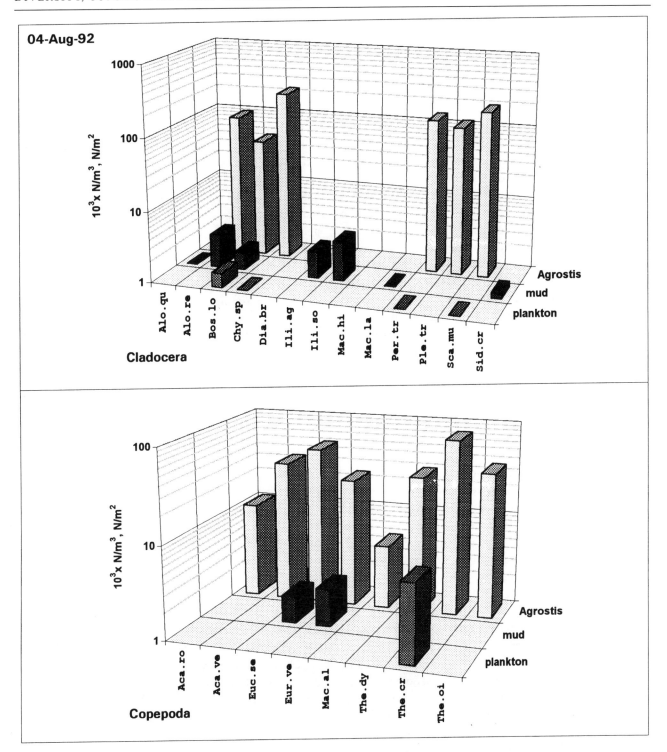

Fig. 7 Distribution of Cladocera and Copepoda species in the side-arm Disznós, 4 Aug. 1992. Explanation of the abbreviations of species names in Table 1 (N/m^2-values of the z-axis refer to the bottom samples).

remained a larger quantity of water in the river bed as compared with 1991. In the 30–40 cm deep water, adult Crustaceans were abundant on the surface of the bottom, with larval stages in the water body itself (Fig. 3).

Side-arm Schisler

The side-arm is in the latest alluvial stage among the investigated water bodies. Overflow begins only by the 510 cm water level on the Danube. The greatest water depth is 180 cm, the volume of water is the most stagnant among the three arms. The alluvial gravel is covered by thick (30–60 cm) mud, and muddy detritus.

The old alluvial age of this arm is proved by the expansion of *Ceratophyllum demersum*. In the middle, deep parts it creates homogenous stands, while in the littoral, it is mixed with *Myriophyllum spicatum, Ranunculus circinatus*. In the shallow, sandy parts, different *Potamogeton* species are characteristic *(P. perfoliatus, P. lucens, P. pectinatus)*. In summer, by mean-water level, cover is 70–80%. At the end of the vegetation period, in the shallow water, only turions of *Ceratophyllum demersum* and small stands of *Myriophyllum spicatum* can be found. The foot of the branch is marshy, its muddy bottom is covered with *Rorippa amphibia*, which is mixed with *Agrostis alba,* on the less humid parts. The dominant species of the reed belt is *Phragmites australis,* which is submitted with *Typha angustifolia* in the deeper sites.

During the low-water period, in June 1991, Crustacean populations with high individual numbers were registered in the open water region. Reed belt and littoral vegetation were not inundated at that time. During the next sampling period, in July, the increased water level reached the littoral and the marshy vegetation *(Agrostis, Rorippa* stands). Great Crustacean individual numbers could be registered from both the open water and the littoral region. Species segregation was strong: only 3 species among the registered 26 were abundant in all the four investigated habitats. Euplanktonic *Bosmina longirostris* was found in biggest individual numbers in the open water region. Larval stages were mainly abundant in the plankton. In August, during the great flood, individual numbers of Crustaceans decreased drastically. Larval stages and adult specimens found protection against the flushing effect in the reed belt and the *Rorippa* stand (Fig. 8). In contrast to the other two investigated arms, the flood caused here mainly the increasing of the water level, the water volume itself remained in the bed. The main cause of the decreasing of individual numbers could be the dilution effect of the flood, and not the washing out effect as in the case of the other two arms. In June 1992, in contrast to the former year, littoral vegetation was inundated and Crustacean populations with big individual numbers were abundant there. At the same time, in the open water, only a few specimens could be found. In August, dispersion of Crustaceans both in the littoral region and in the plankton was similar to that of in 1991. As a consequence of the uninterrupted low-water period, in October water level drastically decreased, water depth became 50–60 cm, water surface contracted, littoral vegetation remained dry. Open water was overcrowded by the larval stages of Copepods, adult specimens were found mainly on the surface of the bottom (Fig. 8).

Side-arm Forrásos

The side-arm lies nearest the Danube and is in the earliest alluvial stage. It is separated from the Danube by a wide embankment, built only a few years ago. It is in active connection with the Csákányi-arm, overflowing begins by the 460 cm water level on the Danube. It has several unique habitat characteristics: springlets are breaking up on the bottom, originating from the groundwater. The close connection with the groundwater table is proved by the fact that pH, dissolved O_2-N-values, and temperature are lower as compared with other arms. Bottom material consists of sand.

Cover of macrophyton communities is small, 5–25 %, but they are of high floristical value: *Ranunculus fluitans* and *Elode nutallii* are botanical rarities (Ráth, 1992). Also abundant here are *Ranunculus circinatus, Myriophyllum verticillatum,* and in moderately sedimented spots *Ceratophyllum demersum*. The narrow littoral region can be characterised by plants tolerating water level fluctuations *(Polygonum mite, Polygonum hydropiper, Myosotis palustris, Agrostis alba)*. *Phragmites australis is* mainly substituted by *Phalaris arundinacea*. Mass productions of *Urtica dioica* are frequent in the cleared *Salix alba* gallery forests.

In this arm, a unique habitat was also sampled for Crustaceans among those in the other two side-arms. At the upper part of the branch, a little lakelet was separated around a breaking up groundwater spring, covered with a dense *Cladophora* clump. In June 1991, during the low-water period, Crustaceans were less abundant in this arm as compared with the Disznós. Individual numbers reached only 10^3 N m^{-3} values. From the plankton, 8 species, and from the reed belt 9 species, could be registered. In the open water region, phytophyl species were dominant. Euplanktonic *Bosmina longirostris* avoided the open water. Larval stages also avoided this habitat, and the bottom region; they preferred the reed belt and the so called *Cladophora* lakelet. In July 1991, 17 Crustacean species were registered in the five investigated habitats. In the single habitats only 1–7 species occurred. Species segregation was very strong, only one species, *Bosmina longirostris*, was abundant in four habitats at the same time. On the surface of the sandy bottom, mainly psammophyl species were abundant. The extreme great flood in August washed out the arm, and Crustaceans were found mainly in the reed belt as in a protected area against flushing. The little lakelet was absolutely inundated at that time, and conditions were the same there as in the open water region. In October, during the long-lasting low-water period, water depth remained only 30–40 cm in the moderately dried out bed. Crustaceans accumulated in the plankton, both the adult specimens and the larval stages (Fig. 9).

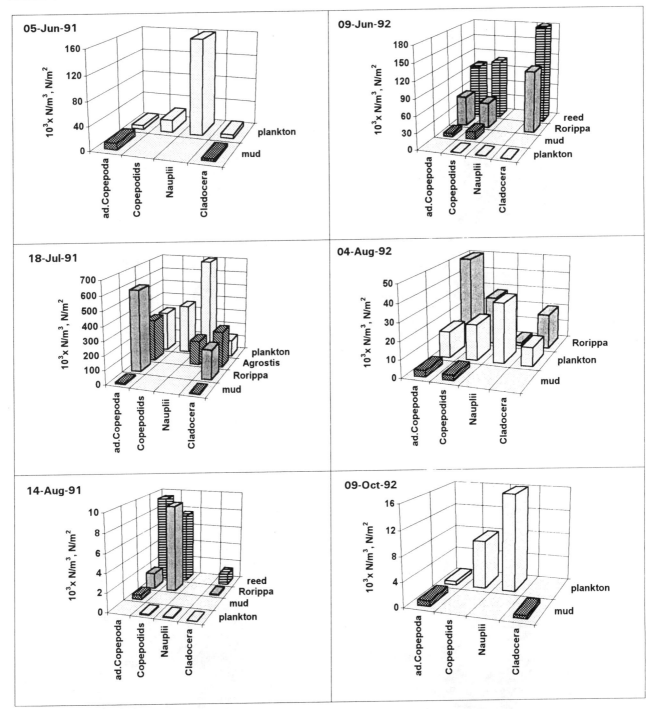

Fig. 8 Distribution of adult Crustaceans and larval stages of Copepods in the different habitats of the side-arm Schisler in 1991 and 1992 (N/m^2-values of the z-axis refer to the bottom samples).

Faunistical remarks

During 1991 and 1992, 39 Cladocera and 21 Copepoda species were registered (Table 1). According to our investigations, the Szigetköz region is the richest in species along the Hungarian Danube section. Most of the species are typical for the littoral region, there are only few which are typical

euplanktonic ones (*Bosmina longirostris, Daphnia cucullata, Cyclops vicinus*). Several species are considered as rarities for the Hungarian fauna (*Chydorus gibbus, Macrocyclops distinctus*). It is worth mentioning that *Eurytemora velox* was found first in 1992 in the three investigated side arms when it invaded nearly every habitats in them in great individual numbers. This species is new for the fauna of Hungary; it was registered first in 1990, in the backwaters of the Szigetköz region in the protected area (Forró & Gulyás, 1992).

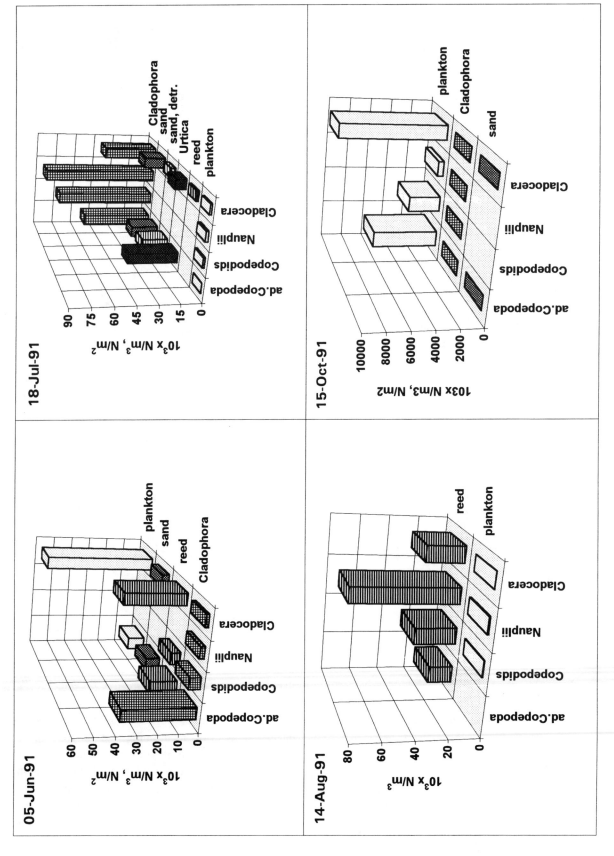

Fig. 9 Distribution of adult Crustaceans and larval stages of Copepods in the different habitats of the side-arm Forrásos in 1991 (N/m^2 - values of the z-axis refer to the bottom samples).

Table 1. *Species composition of Crustaceans in the three side arms*

CLADOCERA		1991	1992
Acroperus harpae (Baird)	Acr.ha	X	X
Alona affinis (Leydig)	Alo.af	X	X
Alona quadrangularis (O.F.M.)	Alo.qu	X	X
Alona guttata G.O.S.	Alo.gu		X
Alona rectangula G.O.S.	Alo.re	X	X
Alonella excisa (Fischer)	Alo.ex		X
Alonella nana (Baird)	Alo.na	X	X
Bosmina longirostris (O.F.M.)	Bos.lo	X	X
Ceriodaphnia dubia Richard	Cer. du		X
Ceriodaphnia megops G.O.S.	Cer.me	X	
Ceriodaphnia pulchella G.O.S.	Cer.pu	X	X
Chydorus gibbus Lilljeborg	Chi.gi	X	X
Chydorus latus G.O.S.	Chi.la	X	
Chydorus sphaericus (O.F.M.)	Chi.sp	X	X
Daphnia cucullata G.O.S.	Dap.cu	X	X
Daphnia hyalina Leydig	Dap.hy	X	
Diaphanosoma brachyurum (Lievin)	Dia.br	X	X
Disparalona rostrata (Koch)	Dis.ro	X	X
Eurycercus lamellatus (O.F.M.)	Eur.la	X	X
Graptoleberis testudinaria (Fischer)	Gra.te		X
Iliocryptus agilis Kurz	Ili.ag	X	X
Iliocryptus sordidus (Lievin)	Ili.so	X	X
Leptodora kindtii (Focke)	Lep.ki	X	
Leydigia leydigi (Schoedler)	Ley.le	X	X
Macrothrix hirsuticornis Norm.et Brady	Mac.hi	X	X
Macrothrix laticornis (Jurine)	Mac.la	X	X
Moina micrura Kurz	Moi.mi		X
Monospilus dispar G.O.S.	Mon.di	X	
Peracantha truncata (O.F.M.)	Per.tr	X	X
Pleuroxus aduncus (Jurine)	Ple.ad	X	X
Pleuroxus laevis G.O.S.	Ple.la		X
Pleuroxus trigonellus (O.F.M.)	Ple.tr	X	X
Pleuroxus uncinatus Baird	Ple.un	X	
Polyphemus pediculus (Linne)	Pol.pe	X	
Pseudochydorus globosus (Baird)	Pse.gl	X	
Scapholeberis mucronata (O.F.M.)	Sca.mu	X	X
Sida crystallina (O.F.M.)	Sid.cr	X	X
Simocephalus serrulatus (Koch)	Sim.se	X	
Simocephalus vetulus (O.F.M.)	Sim.ve	X	X

COPEPODA		1991	1992
Acanthocyclops robustus (G.O.S.)	Aca.ro	X	X
Acanthocyclops vernalis (Fischer)	Aca.ve		X
Cyclops vicinus Uljanin	Cyc.vi	X	X
Diacyclops bicuspidatus Claus	Dia.bi	X	X
Ectocyclops phaleratus (Koch)	Ect.ph		X
Eucyclops macruroides (Lilljeborg)	Euc.mc	X	X
Eucyclops macrurus (G.O.S.)	Euc.ma	X	X
Eucyclops serrulatus (Fischer)	Euc.se	X	X
Eucyclops speratus (Lilljeborg)	Euc.sp	X	

Table 1. (*cont.*)

COPEPODA		1991	1992
Eudiaptomus gracilis (G.O.S.)	Eud.gr	X	
Eurytermora velox (Lilljeborg)	Eur.ve		X
Macrocyclops albidus (Jurine)	Mac.al	X	X
Macrocyclops distinctus (Richard)	Mac.di		X
Macrocyclops fuscus (Jurine)	Mac.fu	X	
Megacyclops viridis (Jurine)	Meg.vi	X	X
Mesocyclops leuckarti (Claus)	Mes.le	X	X
Microcyclops varicans (G.O.S.)	Mic.va	X	X
Paracyclops fimbriatus (Fischer)	Par.fi	X	X
Thermocyclops crassus (Fischer)	The.cr	X	X
Thermocyclops dybowskii (Lande)	The.dy	X	X
Thermocyclops oithonoides (G.O.S.)	The.oi	X	X

REFERENCES

Bothár, A. & Ráth, B. (1992). *Abundance dynamics of Crustaceans in different littoral biotopes of the "Szigetköz" side arm system, River Danube, Hungary*. XXV International Congress of SIL Barcelona, Abstract, 509.

Forro, L. & Gulyás, P. (1992). *Eurytemora velox* (Lilljeborg, 1853) (Copepoda, Calanoida) in the Szigetköz region of the Danube. *Misc. Zool. Hung.*, **7**, 53–58.

Gulyás, P. (1987). Tägliche Zooplankton-Untersuchungen im Donau Nebenarm bei Ásványráró im Sommer 1985 (Diurnal zooplankton investigations in the side-arm Ásványráró in summer 1985). *Wissenschaftliche Kurzreferaten, 26. Arbeitstagung der IAD*, 123–126.

Gulyás, P. (1994). Studies on Rotatoria and Crustacea in the various water bodies of Szigetköz. *Limnologie Aktuell.*, **2**, 63–78.

Ráth, B. (1987). The macrophyte vegetation of a small branch system of the Danube at Dunaremete (Szigetköz, river km 1826). *Acta Botanica Hungaric*, **33**, 3–4, 187–197.

Ráth, B. (1992). Uj adventiv vizi növény Magyarországon: *Elodea nuttallii* (Planchon) st.John (A new aquatic plant in Hungary: *Elodea nuttallii* (Planchon) st.John). *Botanikai Közlemények*, **79**, 1, 35–40.

Vranovsky, M. (1975). Untersuchungen des Zooplanktons im Donaunebenarm "Zofin" (Str.km 1836). *Arbeitstagung der Internat Arbeitsgem. Donauforschung 14–20 Sept. 1975, Regensburg*, 261–278.

Vranovsky, M. (1985). Zooplankton of two side arms of the Danube at Baka (1820.5–1825.5 river km) (English summary). *Prace Lab. Rybar. Hydrobiol.*, **5**, 47–100.

Vranovsky, M. (1991). Zooplankton of a Danube side arm under regulated ichthyocoenosis conditions. *Verh. Internat Verein. Limnol.*, **24**, 4, 2505–2508.

9 Seasonal dynamics and storage of particulate organic matter within bed sediment of three streams with contrasted riparian vegetation and morphology

L. MARIDET*, M. PHILIPPE*, J.-G. WASSON* & J. MATHIEU**

* CEMAGREF, Division BEA, Hydroécologie Quantitative, BP 220, 69009 Lyon, France

** Université Lyon 1, URA CNRS 1974, Ecologie des Eaux Douces et des Grands Fleuves, Hydrobiologie et Ecologie Souterraines, 43 Bd du 11/11/1918, 69622 Villeurbanne cedex, France

ABSTRACT Seasonal dynamics and storage of particulate organic matter (POM) were examined at three sites in adjacent watersheds in the French granitic Massif Central mountains. The three study areas differed mainly by their streamside vegetation and morphology:

– an undisturbed site in a deciduous forest located in a V-shaped moderately incised valley,
– a site in a pasture area with narrow forested buffer strips along the banks,
– a site in a pasture with only isolated trees.

The last two streams flow through gently sloping plateau valleys.

For each season between July 1991 and April 1992, the freezing-core technique was used to extract three cores from different morphodynamic units at each site.

Riparian vegetation influenced the seasonal dynamics of POM inputs. Streambank vegetation and macrophytes (also influenced by the canopy) were sources of organic matter and they controlled its transport by modifying the retentiveness of the channel. The amount of POM buried in the bed sediment depended on interactive factors such as substratum composition and porosity. Porosity was closely linked with the percentage of grain size <1 mm that acts as a limiting factor. The duration of POM stored within bed sediment depended on the timing and magnitude of storms and on retention structures. During high discharge, channel and bank morphology (percentage of riffles or backwaters, slope of the bank) and streambank vegetation influenced transport and retention of POM.

INTRODUCTION

Many studies have shown interactions between channel, riparian and foodplain zones and nutrient cycling in stream ecosystems (Grimm & Fisher, 1984; Minshall et al., 1985; Décamps et al., 1988; Naiman et al., 1988; Essafi, 1990; Mathieu & Essafi, 1991). For small streams, riparian influences may predominante (Conners & Naiman, 1984; Cummins et al., 1984; Wasson, 1989). Danielopol (1982) recognized the importance of the hyporheic zone to stream ecology but most research to date has focused on the vertical distribution of invertebrates into bed sediment (Bretschko, 1981; Strommer & Smock, 1989) and on the potential importance of this zone as a refuge (Williams & Hynes, 1974; Dole-Olivier & Marmonier, 1992; Palmer et al., 1992; Griffith & Perry, 1993). Bed sediments can also be important to stream metabolism (Grimm & Fisher, 1984), but organic matter inputs and decomposition dynamics during storage in

bed sediment, and its subsequent export to the surface, are relatively unknown (Mathieu et al., 1991). Porosity and sediment stability, which are related to particle size (Maridet et al., 1992), influence entrainment of organic matter and retention time (Essafi et al., 1992; 1994) within bed sediments of streams.

The purposes of the present study were 1) to describe the spatial and temporal variability of particulate organic matter within the bed sediment, 2) to understand the influence of the structure of sediment on the vertical distribution of organic matter and 3) to examine the influence of streamside vegetation upon organic matter dynamics.

STUDY SITES AND METHODS

Study sites

The study was carried out between 1991 and 1992 in three streams in adjacent catchments of the French granitic

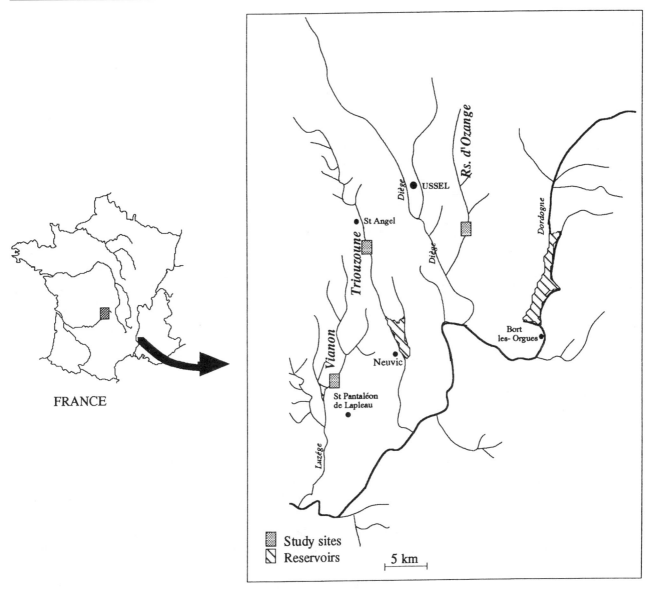

Fig. 1 Location of the three streams reaches selected for study.

Massif Central mountains (Fig. 1). The three streams and sites differed mainly by riparian and watershed vegetation. The Vianon was relatively undisturbed, dominated by beech (*Fagus silvatica*) forest. The study reach was located in a V-shaped and moderately incised valley (Cupp, 1989), and dense vegetation shaded most of the streambed. The reach was dominated by a rapid-riffle (Bisson *et al.*, 1982; Malavoi, 1989) of cobbles, and a sandy sidepool. The other two watersheds were partially deforested, with pastures and conifers or deciduous forest patches. The streams flow through gently sloping plateau valleys. The Ozange study site was located in a pasture with narrow forested buffer strips along the banks; the reach was made up of a riffle-pool sequence with an extensive sandy lentic channel (Malavoi, 1989). The Triouzoune was lined with isolated trees and the moderately sinuous channel was made up of riffle-pool sequences of pebbles and sandy gravel (see also

Table 1). The three streams were similar in hydrologic conditions, corresponding to a pluvio-nival regime with high flow from December to April and a minimum in August (Fig. 2 shows the daily discharge of the Triouzoune during the period of study).

Methods

The bed sediment was sampled with the freezing-core technique using liquid nitrogen preceded by the electro-positioning technique (Bretschko & Klemens, 1986). At each site, three cores were taken from different morphodynamic units (lotic, lentic, and depositional defined by Malavoi, 1989) down to a sediment depth of 60 cm. Each core was divided into 4 sediment layers (subsamples 0–15, 15–30, 30–45, 45–60 cm). The samples were transported frozen to the laboratory.

Subsamples were mechanically shaken through a column of sieves (12, 8, 5, 2, 1, 0.500, 0.250, 0.050 mm). The < 50 μm

Table 1. *Physical characteristics of the studied reach streams*

	Streams		
Variables	Ozange	Triouzoune	Vianon
Order (Strahler, 1957)	3	3	4
Catchment area (km²)	49	77	102
Altitude (m)	590	630	370
Mean discharge (m³·s⁻¹)	1.15	1.82	2.42
Mean width (m)	5.4	8.7	7.5
Slope (‰)	2.7	2.9	18.5

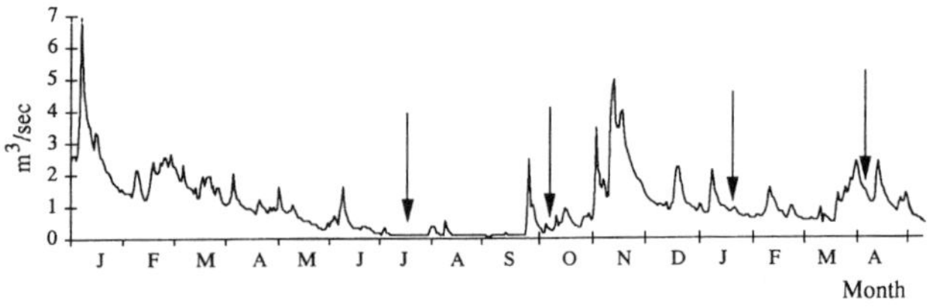

Fig. 2 Hydrograph ($m^3.s^{-1}$) for the study site on the Triouzoune River. Arrows indicate dates of sampling (July, October 1991; January, April 1992).

Table 2. *Annual means of quartiles (in mm, D 25, D 50, D 75%) in the 4 sediment layers for the 3 streams*

	Vianon				Ozange				Triouzoune			
Streams	0–15	15–30	30–45	45–60	0–15	15–30	30–45	45–60	0–15	15–30	30–45	45–60
D 25%	22.6	18.6	17.6	26.2	9.5	5.0	15.5	3.1	14.1	11.5	4.2	8.3
D 50%	58.7	57.4	72.6	65.8	30.2	34.3	45.5	19.7	34.2	49.0	26.6	43.7
D 75%	77.9	81.8	94.0	88.4	50.3	72.4	72.0	49.0	61.2	85.4	58.4	66.4

fraction was further condensed with a centrifuge. Each fraction size was dried, weighed and quantified as a percentage of total dry weight of sediment. The diameter of grains was calculated on semilogarithmic cumulatives curves for 25, 50, and 75 % (D25, D50, D75) (Vatan, 1967). The porosity was calculated as percentage (by volume) of interstitial water flowing out from sediment thawing at room temperature (Stocker & Williams, 1972)

Particulate organic matter (POM) was regrouped into three categories: coarse (CPOM > 1 mm), fine (FPOM 1mm-50 µm) and ultra-fine particulate organic matter (UPOM < 50 µm). Macroinvertebrates larger than 500 µm, sorted by hand, were not included in the POM. The organic matter from each sample was oven-dried at 60 °C to constant weight before ashing at 550 °C for one hour to determine ash-free dry weight (AFDW). Particulate organic carbon (POC) and particulate organic nitrogen (PON) were analysed on UPOM to calculate a C/N ratio to estimate the nutritional quality of POM (Hynes & Kaushik, 1969). This ratio was only determined on the < 50 µm fraction which generally contains most of the organic matter. POC was determined with a CHNS-O analyzer (EA 1108, Carlo Erba instruments) and PON by distillation and colorimetric measuring with a Büchi analyzer.

RESULTS

Substrate particle size distribution of the three rivers varied with time and between the different morphodynamic units.

But, overall the grain-size assemblage was very different in the three river sites (Table 2). In the Vianon, the size of sediment grains (quartile values) was greater and did not vary with depth. Its porosity was also homogeneous with depth (Fig. 3 B). In the Ozange and Triouzoune, porosity was much higher in the first 15 cm than below. In these two streams a greater heterogeneity of sediment grain size was found with depth and in the four layers Q25 % was lower (quartiles values, Table 2). A layer composed of fine sediment particles seems to limit vertical exchanges at a depth of 15–30 cm in the Ozange and between 30–45 cm in the Triouzoune.

The relationship between grain-size assemblage of the substrate and porosity was more complex. The graph of porosity *versus* percentage of sediment particle size <1mm shows a triangular shape (Fig. 4).

POM varied between the three rivers and during the study. The distribution of CPOM in the stream was aggregated or patchy; so variances were high and the differences were not significant. CPOM were much more abundant in the top 15 cm of sediment. Their concentration in this surface layer (Fig. 5) was higher ($p > 0.01$) in July in the Triouzoune. The three sites behaved similarly with an increase of about one third between October and January. In April there was an increase in the Triouzoune and Ozange, and a decrease in the Vianon.

The vertical distribution of the three fractions constituting FPOM was relatively homogeneous with depth in the three sites (Fig. 3 A). Deviations from this pattern were an increase of the larger fraction in the deeper layer of the Vianon, and a regular decrease of the smallest one in the Triouzoune.

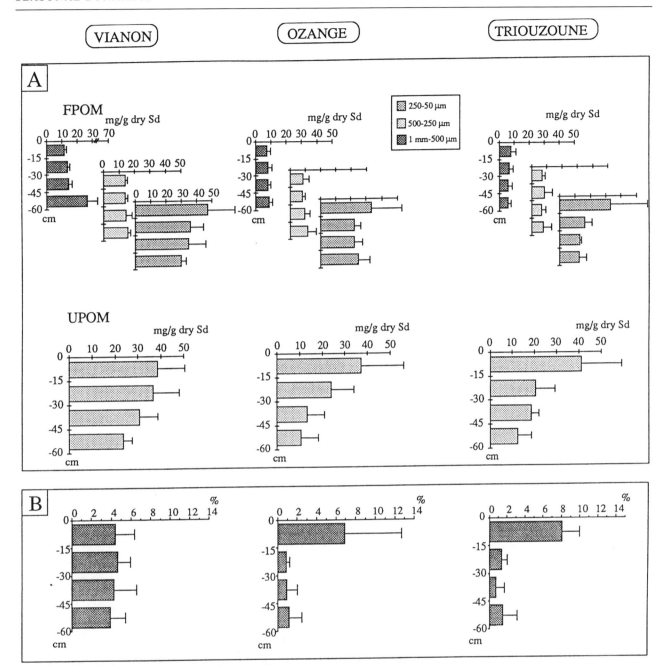

Fig. 3 (A): Vertical distribution (cm) of annual means of FPOM (AFDW, mg g^{-1} dry sediment) and UPOM (POC, mg.g^{-1} dry sediment) at each sites (with S. D.). (B): Annual means of depth distribution (with S. D.) of porosity (%) at each sites.

However, the global amount of FPOM was noticeably greater in the Vianon ($p<0.01$).

The UPOM variables, AFDW and POC, were strongly correlated (t-test; $r=0.97$; $p<0.01$), with a AFDW/POC ratio of 1.93. Thus only POC data were used in Fig. 3. In the Ozange and Triouzoune the vertical distribution of POC followed the same pattern, decreasing with depth (Fig. 3 A). In the Vianon, the distribution was more uniform and its amount decreased only below 30 cm. As for FPOM, the

Vianon contained globally more UPOM than the other two sites ($p<0.01$).

The C/N ratio did not vary significantly between streams and with depth, but varied significantly between seasons ($p<0.01$). The POC was highly correlated with PON ($r=0.94$; $p<0.01$). Annual mean C/N for Vianon, Ozange and Triouzoune were 10.8±0.96; 9.5±1,94 and 9.8±1.12, respectively.

DISCUSSION

CPOM dynamics

Spatial distribution and temporal changes in the stock of organic matter are important factors affecting trophic

Porosity

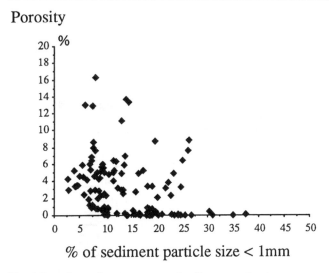

% of sediment particle size < 1mm

Fig. 4 Porosity against percentage of sediment smaller than 1 mm, all data pooled.

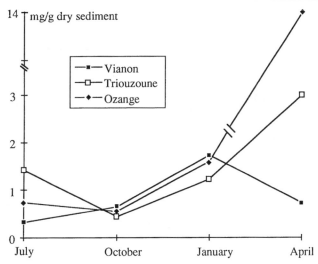

Fig. 5 Mean detritus storage >1 mm (i. e., CPOM as AFDW) in the top 15 cm of sediment in the three streams between July 1991 and April 1992.

dynamics in stream ecosystems. In a temperate stream, these fluctuations result from the balance between autumnal inputs of leaves and senescent macrophytes, and exports controlled by high discharges and physical retentiveness.

The growing season for macrophytes at our sites was May through September and maximum standing stock occurred in July–August. During summer, the high density of macrophytes in the Triouzoune probably accounts for the higher CPOM stock observed in this unshaded stream. Macrophytes can be a relatively important source of autochthonous organic matter and also entrain or trap allochthonous detritus drifting downstream. During low flow, macrophytes significantly increase hydraulic roughness, thus leading to POM settling (Angradi, 1991). Macrophytes were more abundant in the unshaded stream (Triouzoune) and in the partially shaded stream (Ozange) than in the forest stream (Vianon) where only some sparse bryophytes grew, because of the shade inhibiting their development (Dawson & Kern-Hansen, 1978; 1979).

In all sites, CPOM increased between October and January. The greatest amounts of allochthonous organic matter enter the streams following autumnal leaf fall (Bilby & Bisson, 1992). Hense, inputs of CPOM were assumed to be greater in the forest stream than in the open one. But the stock of CPOM was not significantly higher in the Vianon in January. Thus, CPOM was probably exported rapidly during the November flood. Snaddon *et al.* (1992) showed that riffles decreased in retention efficiencies with increasing discharge; conversely, backwaters were more efficient at higher discharge. In the Vianon, riffles occupied the highest proportion of the reach and retention structures were sparse, thus leading to a lower retentiveness at high flow. In this stream large woody debris dams, which act as primary retainers, were very sparse.

Riparian vegetation is a source of CPOM but also creates retention structures by the mean of roots and backwaters (Ehrman & Lamberti, 1992). During receding flows, CPOM stored on the banks is also a source for the channel. April samples were obtained during high flow in a receding stage (Fig. 2). In the Ozange stream, a greater area of the streambed was submerged and the roots and submerged branches created many small backwaters where leaves were probably trapped. Thus, organic matter was stored in the subsurface and the amount of CPOM increased markedly. In the Vianon, such vegetal structures are much less developed; the canopy is very high and roots are not apparent, thus banks are much less retentive during high flow. These first results and hypothesis will be completed by more important samples of the surface CPOM.

POM vertical distribution

The amount of particulate organic matter stored in the bed sediment is the result of two processes: burial and release (Metzler & Smock, 1990). The quantities of detritus trapped within the streambed may be greatly influenced by the substrate composition (Rabeni & Minshall, 1977; Culp & Davies, 1983). In a sandy stream, Smock (1990) showed that 82% of the CPOM stored in the channel was buried, 15% was lying on the sediment surface (0–2 cm) and 3% was trapped in debris dams. Exchanges of CPOM between surface and deep sediments occur only during spates (Metzler & Smock, 1990) and the release of buried detritus depends on the depth of sediment scouring during spates (Jones & Smock, 1991). But exchanges of FPOM and UPOM may occur throughout the year through the water exchanges between surface and hyporheic flows. In our streams, the vertical distribution of

POC and the porosity exhibited similar patterns, except in the Ozange where the relatively high concentration of FPOM in the deeper layers was due to the presence of roots of the adjacent trees. The higher porosity of the Vianon also seems to allow a deeper penetration of the largest FPOM (1–0.5 mm) in this site.

Thus, effective porosity seems to be a primary factor controlling the concentration of the finest POM fractions (<250 μm) in the bed sediment.

Leichtfried (1988; 1991) suggested that POM distribution in the bed sediment is positively correlated with the distribution of the grain size class <1 mm. But the percentage of the fine fraction in the sediment also strongly influences the porosity, although the relationship between fine fractions and porosity is complex. The results of the present study (Fig. 4) confirmed the observation of Maridet *et al.* (1992) showing that fine particles act as a limiting factor: the porosity may be high or low when the percentage of fine particles is low, but porosity is always low when a large quantity of fine particles is present. Porosity is closely linked with sizes and distribution of bed sediments interstices; a large quantity of fine particles fills these spaces and decreases the porosity. Partial deforestation of Ozange and Triouzoune watersheds increased erosion of soils. Thus, in these slow slope streams, the amount of fine sediment particles increased.

The C/N results of the present study did not agree with those of Bretschko & Leichtfried (1988) who found stratification with depth and a maximum between 20 and 40 cm. The depth distribution of ratio in our study was uniform, because both POC and PON changed in the same direction.

CONCLUSION

The amount of organic matter deposited on the sediment surface or buried within the bed sediment depends on many interactive components like riparian vegetation, density of aquatic plants, channel and bank morphology, substrate composition and porosity, while the duration of CPOM stored within bed sediment is dependent on the timing and magnitude of floods and the quantity and nature of the retention structures. The forest streams received major POM inputs, from litter fall, and stored more fine POM due to a more porous substrate, but in the Vianon less coarse detritus in the surface layer was retained because of a higher slope and less efficient retention structures. Surface CPOM distribution is patchy with complex dynamics and cannot be adequately characterized by the few samples taken during this study. Another sampling method such as litter traps will be used.

Seasonal timing of inputs or outputs of detrital materials is very important for the availability of food resources for the aquatic fauna. Culp & Davies (1985) showed the importance of the interstitial organic matter in the substrate as a factor influencing the microdistribution of benthos. Conversely, many other studies (e. g. Cummins, 1974; Petersen & Cummins, 1974; Chergui & Pattée, 1988; Stewart, 1992) have clearly demonstrated the important role of macroinvertebrates and microorganisms in the processing of detrial materials. These aspects will also be investigated in forthcoming studies.

Our results further show the necessity of a comprehensive description not only of riparian vegetation, but also of stream morphology including substrate granulometry to understand and further predict POM dynamics.

ACKNOWLEDGEMENTS

Financial support was provided by 'Ministère Français de l'Environnement' (D.R.A.E.I.). Many thanks to P. Roger and many others for their essential contribution in the field, to G. Rofes, B. Motte and G. Hamelin for laboratory assistance. We thank also Professor Dr J. Stanford (The University of Montana, USA) for his critical help and assistance in editing the English text.

REFERENCES

Angradi, T.R. (1991). Transport of coarse particulate organic matter in an Idaho river, USA. *Hydrobiologia*, **211**, 171–183.

Bilby, R.E. & Bisson, P.A. (1992) Allochtonous versus autochtonous organic matter contributions to the trophic support of fish populations in clear-cut and old-growth forested streams. *Can. J. Fish. Aquat. Sci.*, **49**, 540–551.

Bisson, P.A., Nielsen, J.L., Palmason, R.A. & Grove, L.E. (1982). A system of naming habitat types in small streams, with examples of habitat utilization by salmonids during low stream flow. In *Acquisition and Utilization of Aquatic Habitat Inventory Information*, ed. Armantrout N.B., pp. 62–73, Portland Oregon, American Fisheries Society.

Bretschko, G. (1981). Vertical distribution of zoobenthos in an alpine brook of the Ritrodat-Lunz study area. *Verh. Internat. Verein. Limnol.*, **21**, 873–876.

Bretschko, G. & Klemens, W. (1986). Quantitative methods and aspects in the study of the interstitial fauna of running waters. *Stygologia*, **2**, 4, 279–316.

Bretschko, G. & Leichtfried, M. (1988). Distribution of organic matter and fauna in a second order alpine gravel stream (Ritrodat-Lunz study area, Austria). *Verh. Internat. Verein. Limnol.*, **23**, 1333–1339.

Chergui, H. & Pattee, E. (1988). The dynamics of Hyphomycetes on decaying leaves in the network of the river Rhône (France). *Arch. Hydrobiol.*, **114**, 3–20.

Conners, M.E. & Naiman, R. (1984). Particulate allochthonous inputs: Relationships with stream size in an undisturbed watershed. *Can. J. Fish. Aquat. Sci.*, **41**, 1473–1484.

Culp, J.M. & Davies, R.W. (1983). *An assessment of the effects of streambank clear-cutting on macroinvertebrate communities in a managed watershed*. Canadian Technical Report of Fisheries and Aquatic Sciences, 1208, 116 pp.

Culp, J.M. & Davies, R.W. (1985). Responses of benthic macroinvertebrate species to manipulation of interstitial detritus in Carnation Creek, British Columbia. *Can. J. Fish. Aquat. Sci.*, **42**,139–146.

Cummins, K.W. (1974). Structure and function of stream ecosystems. *BioScience*, **24**, 631–641.

Cummins, K.W., Minshall, G.W., Sedell, J.R., Cushing, C.E. & Petersen, R.C. (1984). Stream ecosystem theory. *Verh. Internat. Verein. Limnol.*, **22**, 1818–1827.

Cupp, C.E. (1989). *Stream corridor classification for forested lands of Washington.*, Olympia, Washington, Washington Forest Protection Association, 46 p.

Danielopol, D.L. (1982). Phreatobiology reconsidered. *Pol. Arch. Hydrobiol.*, **29**, 375–386.

Dawson, F.H. & Kern-Hansen, U. (1978). Aquatic weed management in natural streams, the effect of shade by the marginal vegetation. *Verh. Internat. Verein. Limnol.*, **20**, 1451–1456.

Dawson, F.H. & Kern-Hansen, V. (1979). The effect of natural and artificial shade on the macrophytes of lowland streams and the use of shade as a management technique. *Int. Revue Ges. Hydrobiol.*, **64**, 4, 437–455.

Décamps, H., Fortuné, M., Gzelle, F. & Pautou, G. (1988). Historical influence of man on riparian dynamics of a fluvial landscape. *Landscape Ecology*, **1**, 163–173.

Dole-Olivier, M. J. & Marmonier, P. (1992). Effects of spates on the vertical distribution of the interstitial community. *Hydrobiologia*, **230**, 49–61.

Ehrman, T.P. & Lamberti, G.A. (1992). Hydraulic and particulate matter retention in a 3rd-order Indiana stream. *J. N. Am. Benthol. Soc.*, **11**, 4, 341–349.

Essafi, K. (1990). *Structure et transfert des peuplements aquatiques souterrains à l'interface karst-plaine alluviale.* Thèse de Doctorat, Lyon, Fr., 102 pp.

Essafi, K., Chergui, H., Pattée, E. & Mathieu, J. (1994). The breakdown of dead leaves buried in the sediment of a permanent stream in Marocco. *Arch. Hydrobiol.*, **130**, 105–112.

Essafi, K., Mathieu, J. & Beffy, J. L. (1992). Spatial and temporal variations of *Niphargus* populations in interstitial aquatic habitat at the karst/floodplain interface. *Regulated Rivers: Reasearch & Management*, 7, 83–92.

Griffith, M.B. & Perry, S.A. (1993). The distribution of macroinvertebrates in the hyporheic zone of two small Appalachian headwater streams. *Arch. Hydrobiol.*, **126**, 373–384.

Grimm, N.B. & Fisher, F.G. (1984). Exchanges between interstitial/surface water: Application for stream metabolism and nutrient cycling. *Hydrobiologia*, **111**, 219–228.

Hynes, H.B.N. & Kaushik, N.K. (1969). The relationship between dissolved nutrient salts and protein production in submerged autumnal leaves. *Verh. Int. Ver. Limnol.*, **17**, 95–103.

Jones, Jr J.B. & Smock, L.A. (1991). Transport and retention of particulate organic matter in two low-gradient headwater streams. *J. N. Am. Benthol. Soc.*, **10**, 2, 115–126.

Leichtfried, M. (1988). Bacterial substrates in gravel beds of a second order alpine stream. *Verh. Internat. Verein. Limnol.*, **23**, 1325–1332.

Leichtfried, M. (1991). POM in bed sediments of a gravel stream (Ritrodat-Lunz study area, Austria). *Verh. Internat. Verein. Limnol.*, **24**, 1921–1925.

Malavoi, J.R. (1989). Typologie des faciès d'écoulement ou unités morphodynamiques des cours d'eau à haute énergie. *Bulletin Français de pêche et pisciculture*, **315**, 189–210.

Maridet, L., Wasson, J.G. & Philippe, M. (1992). Vertical distribution of fauna in the bed sediment of three running water sites: Influence of physical and trophic factors. *Regulated Rivers: Research & Mangement*, 7, 45–55.

Mathieu, J. & Essafi, K. (1991). Changes in abundance of interstitial populations of *Niphargus* (stygobiont amphipod) at the karst/floodplain interface. *C. R. Acad. Sci. Paris*, **312**, 489–494.

Mathieu, J., Essafi, K. & Doledec, S. (1991). Dynamics of particulate organic matter in bed sediments of two karst streams. *Arch. Hydrobiol.*, **122**, 199–211.

Metzler, G.M. & Smock, L.A. (1990). Storage and dynamics of subsurface detritus in a sand bottomed stream. *Can. J. Fish. Aquat. Sci.*, **47**, 588–594.

Minshall, G.W., Cummins, K.W., Petersen, R.C., Cushing, C.E., Bruns, D.A., Sedell, J.R. & Vannote, R.L. (1985). Developments in stream ecosystems theory. *Canadian Journal of Fisheries and Aquatic Science*, **42**, 1045–1055.

Naiman, R.J., Holland, M.M., Décamps, H. & Risser, P.G. (1988). A new UNESCO program: research and management of land/inland water ecotone. *Biology International Special Issue*, **17**, 107–136.

Palmer, M.A., Bely, A.E. & Berg, K.E. (1992). Response of invertebrates to lotic disturbance: a test of the hyporheic refuge hypothesis. *Oecologia*, **89**, 182–194.

Petersen, R.C. & Cummins, K.W. (1974). Leaf processing in a woodland stream. *Freshwater Biology*, **4**, 343–368.

Rabeni, C.F. & Minshall, G.W. (1977). Factors affecting microdistribution of stream benthic insects. *Oikos*, **29**, 33–43.

Smock, L.A. (1990). Spatial and temporal variation in organic matter storage in low-gradient, headwater streams. *Arch. Hydrobiol*, **118**, 2, 169–184.

Snaddon, C.D., Stewart, B.A. & Davies, B.R. (1992). The effect of discharge on leaf retention in two headwater streams. *Arch. Hydrobiol.*, **125**, 1, 109–120.

Stewart, B.A. (1992). The effect of invertebrates on leaf decomposition rates in two small woodland streams in southern Africa. *Arch. Hydrobiol.*, **124**, 1, 19–33.

Strahler, A. N. (1957). Quantitative analysis of watershed geomorphology. *Trans. Amer. Geophys. Union*, **38**, 913–920.

Stocker, G. & Williams, D. D. (1972). A frezzing core method for describing the vertical distribution of sediments in a streambed. *Limnol. Oceanogr.*, **17**, 1, 136–138

Strommer, J.L. & Smock, L.A. (1989). Vertical distribution and abundance of invertebrates within the sandy substrate of a low-gradient head water stream. *Freshwater Biology*, **22**, 263–274.

Vatan, A. (1967). *Manuel de sédimentologie*. Paris, Technip, 397 p.

Wasson, J.G. (1989). Eléments pour une typologie fonctionnelle des eaux courantes, 1. Revue critique de quelques approches existantes. *Bull. Ecol.*, **20**, 2, 109–127.

Williams, D.D. & Hynes, H.B.N. (1974). The occurrence of benthos deep in the substratum of a stream. *Freshwater Biolology*, **4**, 233–256.

10 Bedsediments: protein and POM content (RITRODAT-Lunz study area, Austria)

M. LEICHTFRIED

Dept Biological Station Lunz, Institute for Limnology, Austrian Academy of Sciences, A – 3293 Lunz am See, Austria

ABSTRACT   The availability of organic matter to animal consumers is very dependent on its protein content. C/N relationships can therefore be used as food quality indicators, although C/N ratios are not only dependent on the actual protein content. The present study analyses the distribution in space and time of protein, TON and TOC in the bedsediments of a second order gravel stream (Oberer Seebach, RITRODAT – Lunz). All three parameters were measured in the same samples. The validity of C/N ratios as food quality indicators is confirmed for sample means but not for individual values.

INTRODUCTION

The energy basis of low order streams is mainly allochthonous organic matter. Above surface imports are bank run off and aerial drift. This organic matter must be processed by the microbial community to become available to animal consumers and the food quality depends on the intensity of microbial activity. The processing of the organic material takes place partly on the sediment surface and partly in the bedsediments. The bedsediments are defined as channel forming sediments quantitatively dominated by epigeic faunal elements (Bretschko, 1992). They are therefore the topmost layer of the hyporheic zone, the extent of which is usually not clearly defined (Bretschko & Moser, 1993; Schwörbel, 1961). The distributions of bacteria (Kasimir, 1990 and in press), meiofauna (Schmid-Araya, 1994) and macrofauna (Bretschko, 1981; Bretschko & Klemens, 1986) indicate a very high metabolic rate in the bedsediments. Organic matter is measured as total organic bound carbon (TOC) and nitrogen (TON). The spatial/temporal distribution of POM is known for a period of some years (Leichtfried, 1985; 1986; 1988; 1991a,b).

Organic matter appears in the system in three basic forms: as plant tissue, as sediment fauna and as biofilm. Plant tissue is largely dominated by cellulose molecules. Based on the total organic matter, up to 90% of the TON content and up to 80% of the TOC content of the bedsediments is attached to the grain size class <1 mm in diameter. Of this, about one quarter can be attributed to plant tissue (Leichtfried, 1991a). Even smaller is the contribution of the sediment fauna > 100 µm in body length: the ratio of total organic carbon to the carbon content of the macrobenthic community is of the order of 10^5:1 (Leichtfried, 1991a). Most of the POM content must therefore be attributed to biofilms, defined by Marshall (1984) as biomass of microorganisms and their organic excretions (exocellular polymers) attached to surfaces. Because more than three quarters of the total POM content of the bedsediments is attached to small grains, the study of these size classes should be sufficient to characterize the organic matter of the bedsediment.

The protein content is supposed to be least in plant tissues and greatest in the fauna and in biofilms. Since the quantity of fauna larger than 100 µm is small relative to that of the microbiocoenosis, most of the protein has to be attributed to the biofilm with its attached micro- and meiofauna. C/N ratios are supposed to reflect the protein content of organic matter (e.g. Atkinson, 1983; Hutchinson, 1957; Hyne, 1978; Taylor & Roff, 1984) and therefore describe the food availability to consumers. Low C/N ratios are correlated with high protein contents, indicating high food quality. This study sets out to prove this assumption by directly confronting measured protein concentrations with C/N ratios, derived from TOC and TON measurements.

Fig. 1 Geographic location of the study area.

Table 1. *Oberer Seebach, RITRODAT study area. Long term means±95 % conf. intervals (Bretschko, 1991)*

Discharge ($l\,s^{-1}$)		Surface water chemistry	
MQ:	720	pH	8.1 ± 0.1
MMAX Q:	2400	Cond. ($\mu S\,cm^{-1}$):	216 ± 6
MMIN Q:	320	Alkali. (meg l^{-1}):	2.18 ± 0.07
MAX:	17500	Ca^{++} (meg l^{-1}):	2.04 ± 0.06
MIN:	320	Mg^{++} (meg l^{-1}):	0.052 ± 0.04
		P_{tot} (mg l^{-1}):	0.009 ± 0.012
Mean temperature (°C)		N_{tot} (mg l^{-1}):	1.01 ± 0.238
Annual mean:	6.8	*Channel*	
Mean Max:	11.1	Width (m):	15.8 ± 1.3
Mean Mini:	1.9	Slope (%):	0.41 ± 0.04

STUDY SITE

The study area (Lunz am See, Austria) is situated roughly 150 km south-west of Vienna, on the low lying slopes of the northern limestone fringe of the Eastern Alps, 600 m above sea level (Fig. 1). The RITRODAT study site is a 100 m long stretch of Oberer Seebach, a second order limestone gravel stream. The karstic catchment of about 20 km is uninhabitated and the stream is therefore unpolluted and only very slightly influenced by man. Characteristic abiotic data are set out in Table 1. Sediment grains larger than 10 mm in diameter dominate the grain size distribution. The grain size class smaller than 1 mm in diameter contributes less than 10% of the bulk but contains most of the organic matter (80% TOC and 90% TON, biannual means, Leichtfried, 1988). The porosity is high (22–30 vol.%) with large and well interconnected interstices, allowing high throughflow rates. Oxygen saturation is therefore high down to a sediment depth of 1 m or even more (Leichtfried, 1986, 1988, 1991b). The macrozoobenthic stream community is typical of alpine streams and very dense, with abundances around 100 000 individuals per 1 m^2. The stream fauna (>100 μm body length) colonizes the sediments down to a depth of 0.5 m (Bretschko, 1992).

METHODS

Sampling was done on the 20th of each month during 1992 using permanently installed stand pipe traps. Stand pipes trap sediment grains drifting freely through the interstices. Most of these floating interstitial sediments were smaller than 1 mm. The actual sample consisted of sediment water and fine interstitial grains. For a detailed description see Bretschko & Klemens (1986) or Leichtfried (1986). A total of 24 stand pipes were sampled, 12 in a riffle and 12 in a pool area of the RITRODAT study site. In each of the 6 sediment depths sampled (5, 10, 25, 35, 45 and 55 cm) two pipes were installed. The sampling depths were actually the depths of the portholes of the pipes. Because the sediment grains float

into the pipes, samples were not taken from one plane, but from a layer of about 10 cm. Quantification is based on the sediment weight in one dm^3 of sample (g dm^{-3}), which is a fine interstitial sediment-sediment water complex, sucked out from the stand pipe with a pump. Immediately after sampling, the sediment-sediment water complex was deep frozen in the laboratory (224 °C) and later freeze dried to save the protein molecules. For the chemical analysis, the freeze dried sample was homogenized and pulverized, using a mortar mill (Fa.Retsch, Type MS).

Chemical analysis

ORGANIC CARBON (TOC) AND NITROGEN (TON)

The inorganic bound carbon was removed by acidification with 0.1N hydrochloric acid. The developing CO_2 was removed by blowing pressurized air through the sample. The latter was filtered through a nitrogen and carbon free, premuffeled glass fibre filter (Whatman GF/F). The dried filter, together with the organic matter sticking to it, was analyzed for carbon and nitrogen in a LECO CHN 600 Analyzer. The organic nitrogen was determined by measuring the total nitrogen and deducting inorganic bound nitrogen, determined separately (Bretschko & Leichtfried, 1987).

PROTEIN

Proteins were eluted from the freeze dried and pulverized sample with diluted sodium hydroxide according to the method published by Rausch (1981), modified by Pusch (1987). The eluted proteins were analyzed according to the method originally published by Lowry et al. (1951) as modified by Schmid-Araya (1989), using Folin-Ciocalteu reagent.

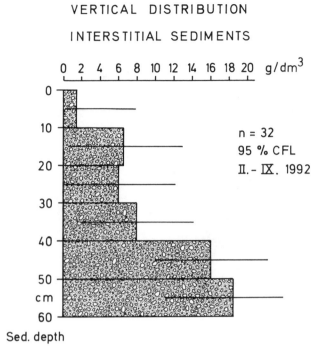

Fig. 2 Vertical distribution of mobile interstitial sediments in Oberer Seebach (RITRODAT – Lunz).

Table 2. *ANOVA: effect of the depth on concentration (mg g^{-1}) and content (g dm^{-3} orig.sample) of proteins TOC, TON in the bedsediments of Oberer Seebach, Austria*

	DF	F	P
Concentrations			
TON	5	0.429	0.828
TOC	5	1.129	0.346
PROTEIN	5	1.728	0.130
Contents			
TON	5	32.782	0.000
TOC	5	32.839	0.000
PROTEIN	5	35.601	0.000

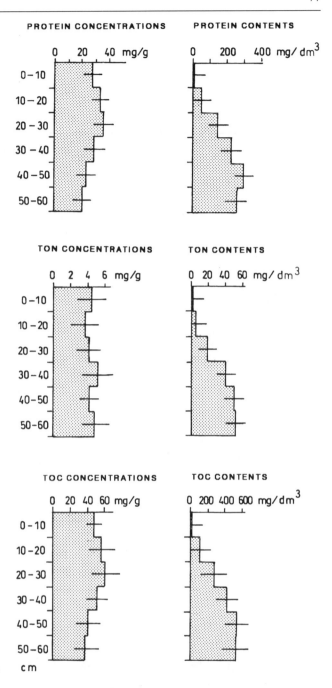

Fig. 3 Vertical distributions of protein, TON and TOC in the bedsediments of Oberer Seebach (RITRODAT – Lunz). Concentrations and contents.

RESULTS

Because of the permanently installed stand pipe traps, the hydraulic regime is influenced at the actual site by the fact that pool and riffle are becoming increasingly similar. The original aim, to disclose possible differences in protein quantities and distributions between the pool and riffle areas, became impeded. As expected, no significant differencies were discernable (ANOVA: $DF=1$; $F=0.290$; $P=0.593$) and in all further considerations, the two sites are therefore pooled.

The results of the analytical methods applied here are always concentrations, e.g. x mg carbon in y g sample ana-

lyzed. The amount of organic matter in stream sediments is closely correlated with the amount and size of sediment grains in a certain volume of the original sample. The actual organic matter content of a certain volume of original sample depends therefore on the concentration and on the amount of sediment grains. Consequently, it is necessary to differentiate between concentration and content.

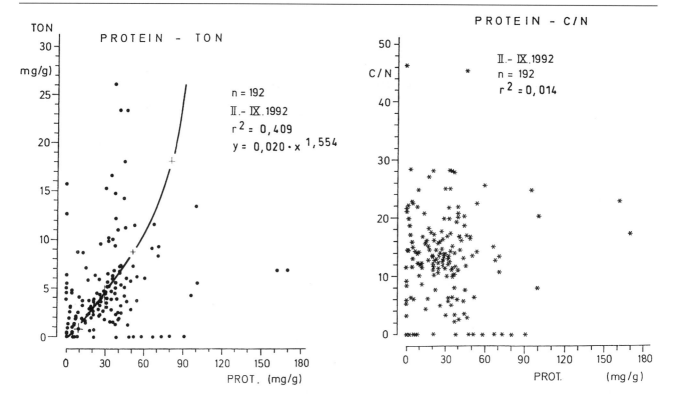

Fig. 4 Protein – TON regression and protein – C/N scatter of individual values in the bedsediments of Oberer Seebach (RITRODAT – Lunz).

CONCENTRATION: the measured parameter in milligramms per one gramme dry weight of the grain size class smaller than 1 mm.

CONTENT: The concentration of the parameter analyzed times the amount of sediment grains <1 mm in 1 dm^3 of the original sediment-sediment water sample.

Vertical distribution

The pattern of vertical distributions of the analyzed parameters (protein, TON, TOC) were not significantly different over the sampling period (II. – IX 1992) therefore all the time series were lumped together for description of the vertical distribution.

The amount of mobile sediment grains in the interstitial water increased significantly with sediment depth (ANOVA: $DF=5$; $F=3.847$; $P=0.002$; Fig.2). In the topmost layer (0–10 cm), the mean amount was only 1.5 g dm^{-3} interstitial water, in the bottom layer (50–60 cm) it was 18.5 g dm^{-3} interstitial water (Fig.2). The difference was more than one order of magnitude and therefore greatly influenced the vertical content distribution of all the parameters analysed.

Protein concentrations (mg g^{-1}) were evenly distributed

over the whole sediment column, with mean values from 21 to 36 mg g^{-1} (Fig.3). There were no significant differences between protein concentrations in different sediment layers, not even between the topmost and the deepest layers (Fig.3, Table 2). The protein content (mg dm^{-3} original sample) was dominated by the distribution of mobile interstitial sediment. Consequently, protein contents increased significantly with depth (Fig.3, Table 2). Mean values varied between 14 (topmost layer) and 304 (bottom layer) mg protein dm^{-3} original sample which is a very wide range, up to a 20 fold difference.

Similar vertical distribution patterns were found in organically bound nitrogen (TON) and carbon (TOC). Concentrations of TON and TOC (mg g^{-1}) were evenly distributed, showing no significant differences with sediment depth (Fig.3, Table 2). Mean values for TON concentrations varied between 3.8 and 5.2 mg g^{-1} and those for TOC between 38 and 61 mg g^{-1}. TON and TOC contents (mg l^{21} original sample) increased significantly with sediment depth (Fig.3, Table 2), reflecting the distribution of the fine interstitial sediment (Fig.2). The mean TON content of the topmost sediment layer was 2.7 mg dm^{-3} and in the bottom layer 51.1 mg dm^{-3} original sample. Mean TOC content was 24 mg dm^{-3} original sample in the upper sediment layer. The maximum content of 548 mg TOC dm^{-3} was found in the 40–50 cm sediment depth.

Thus, all three measured parameters of organic matter (protein, TON and TOC) were twenty times higher in the deepest sediment layers than near the surface.

Protein and C/N ratios

As described above, the C/N ratio is frequently used as a food quality indicator for consumers. Low C/N ratios are supposed to indicate high food quality caused by larger amounts of protein. Consequently, high C/N ratios are supposed to indicate low food quality and low amounts of protein. The amounts of cellulose in relation to C/N ratios are theoretically complementary to proteins (Atkinson, 1983; Hutchinson, 1957; Hyne *et al.*, 1978; Taylor & Roff, 1984). However, although C/N ratios are widely used and taken as food quality indicators, no attempt has been made to correlate C/N ratios with actual protein measurements in the sediments.

The relationships between the concentrations of proteins and nitrogen or carbon in the 192 samples of this study are just statistically significant ($r=0.409$ and $r=0.596$, respectively) but the scatter is extremely great (Fig.4). The correlation between proteins and C/N ratios is not even significant ($r=0.014$; Fig.4). These findings are contrary to expectation but the prediction is fulfilled as soon as the correlation analysis is based on sample means (integrating spatial distributions) rather than individual samples. Positive and negative linear regressions correlate protein concentrations with nitrogen ($r=0.851$) and C/N ratios ($r=0.779$), respectively. Both correlations are significant at a probability level greater than 99.9% (Fig.5). The strong relationships of the mean values are not only caused by a possible correlation between C/N and protein concentrations, but also by the protein, nitrogen and carbon regressions on time. Over the study period, protein and TON concentrations were significantly correlated with time by positive linear regressions ($r= 0.951$ and $r=0.408$, respectively; Fig.6).

TOC concentrations were variable in time, with minimum values in April and September and maximum in July as has already been shown by Leichtfried (1986, 1991a,b; Fig.6). C/N ratios were also significantly correlated with time by a negative linear regression ($r=0.858$; Fig.6). These correlations, of protein concentrations and C/N ratios on time, reappear in the regressions between protein concentrations and TON concentrations and C/N ratios, respectively (Fig.5).

DISCUSSION

Only a certain part of the sediment-sediment water complex has been used for analysis: the fine grained fraction floating freely through the interstitial voids. This sediment fraction covers most of the smaller grain size classes, and in previous studies, it has been shown that up to 90% of total organic matter is connected with these grain size classes (Leichtfried, 1985, 1986, 1991a,b). It is therefore reasonable to extrapolate from this fraction to the total.

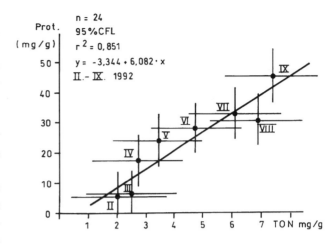

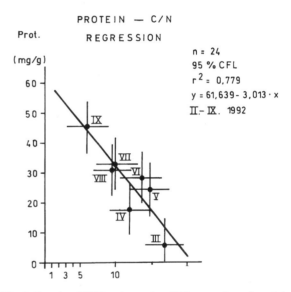

Fig. 5 Protein – TON and protein – C/N regression of spatial averages in the bedsediments of Oberer Seebach (RITRODAT – Lunz).

The central question of this study is the realibility of C/N ratios as indicators of food digestibility for faunal consumers. The null-hypothesis is therefore the existence of an inverse relationship between C/N ratios and protein concentrations. The null-hypothesis has to be rejected on the basis of the data in this study (Fig.4 right) because a significant correlation exists between protein and TON concentrations, although with a very large scatter (Fig.4 left). Both concentrations showed a steady and linear increase over the study period. This is in good agreement with previous studies at the same sampling site (Leichtfried, 1986).

The interrelationship between organic carbon concentrations and time is a more complex curvilinear one. Again, the distribution pattern is comparable with those described in

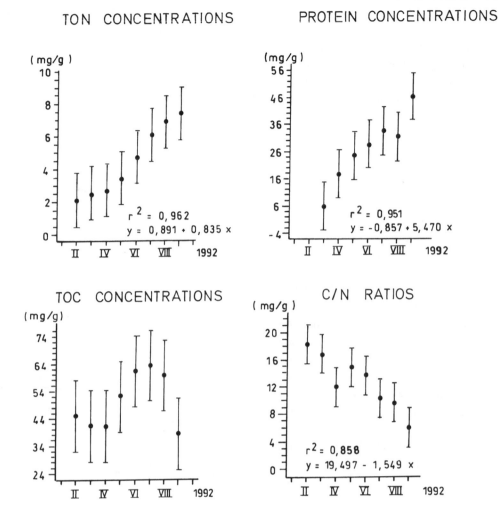

Fig. 6 Temporal distribution (spatial averages) of protein, TON, TOC concentrations and C/N ratios in the bedsediments of Oberer Seebach (RITRODAT – Lunz).

earlier studies (Fig.6, Leichtfried, 1986, 1991a,b) and, in spite of the curvilinear regression of organic carbon concentrations, the C/N ratios are significantly correlated with time in the form of a negative linear regression (Fig.6).

The strong relationships between organic nitrogen and carbon on one side and time on the other result in a highly significant correlation between protein concentrations and C/N ratios in the form of a negative linear regression (Fig.5). This regression fits perfectly the prediction of the theoretical concept but because the null-hypothesis had to be rejected on the basis of the very same data pool, the protein – C/N regression in Fig. 5 cannot be taken as an argument in favour of C/N ratios as food quality indicators because the regression is probably the result of the correlations between nitrogen and carbon and time.

Measurements of both proteins and C/N ratios in one sample are extremely scarce in the literature. Pick (1987) measured both parameters in seston samples from Lake Ontario and Jacks Lake. Although his sample size was very small ($n=7$), the results were the same as in this study: a significant correlation existed between protein and nitrogen concentrations, and no correlation was found between protein concentrations and C/N ratios.

Nearly all of the relevant studies deal only with protein measurements. Most of them analyze the protein contents of animals and plants. Only very few studies deal with protein contents of sediments. Brewer & Pfaender (1979) studied the proteins in the wetlands associated with lakes but, although they employed the same analytical methods, their findings are not comparable because they related the results to the wet weight of the samples. The same holds for Cunningham & Wetzel (1989), who studied the degradation of proteins in the sediments of *Typha latifolia* wetlands near Lake Lawrence (SW Michigan). The protein content of stream sediments was studied by Pusch (1987) in the low order river Steina in the Black Forest (Germany) but he confined his studies to benthic organic matter and related it to the organic ashfree dry weight per m, which does not allow any comparisons with the results of this study.

It has been shown in various studies that the bedsediments are an integral part of the stream ecosystem (Bretschko, 1992; Bretschko & Klemens, 1986; Harvey & Bencala, 1993). Following the ecotone definition of Holland (1988), the stream/groundwater ecotone is below the bedsediments. Whenever studied in detail, the lower border of the bedsediments proved to be stable and marked by a few hydrological parameters and by qualitative and quantitative change of the community. Some sort of influence by the latter on the vertical distribution of protein concentration is to be expected. As has already been demonstrated with organic carbon and nitrogen in earlier studies, there was also no marked change in the vertical distribution of protein concentrations around the lower border of the bedsediments (40–50 cm, Fig.3 left; Leichtfried, 1986, 1988, 1991a,b).

In contrast to the concentrations, contents showed a marked increase in all three parameters, proteins, TON and TOC, near the lower bedsediment border (Fig.3 right). This increase is caused by a significant increase in the frequency of freely floating fine grained sediments (Fig.2). The higher protein contents below the bedsediments are therefore simply the result of increased availability of colonizable surface areas. Like organic nitrogen and carbon distributions, protein distributions also mainly reflect the amount of biofilm present. Any possible influence of the macrofauna is probably too small to be detected (Bretschko & Leichtfried, 1988).

ACKNOWLEDGEMENTS

I have to thank Dr Peter E. Schmid for mathematical help and Dr Ing. Josef Toth for helping me with computer programs. For technical help I have to thank: Ing. Walpurga Fahrner, Christian Hödl, Ernestine Kronsteiner, Arnold Leichtfried, Melitta Reitbauer.

REFERENCES

Atkinson, M.J. (1983). C:N:P: ratios of benthic marine plants. *Limnol. Oceanogr.*, **28**, 568–574.

Bretschko, G. (1981). Vertical distribution of zoobenthos in an alpine brook of the Ritrodat-Lunz study area. *Verh. Internat. Verein. Limnol.*, **21**, 873–876.

Bretschko, G. (1991). The limnology of a low order alpine stream (Ritrodat-Lunz study area, Austria). *Verh. Internat. Verein. Limnol.*, **24**, 1908–1912.

Bretschko, G. (1992). Differentation between epigeic and hypogeic fauna in gravel streams. *Regulated Rivers*, **7**, 17–22.

Bretschko, G. & Klemens, W.E. (1986). Quantitative methods and aspects in the study of the interstitial fauna of running waters. *Stygologia*, **2**, 279–316.

Bretschko, G. & Leichtfried, M. (1987). The determination of organic matter in river sediments. *Arch. Hydrobiol. Suppl.*, **68**, 403–417.

Bretschko, G. & Leichtfried, M. (1988). Distribution of organic matter and fauna in a second order, alpine gravel stream (Ritrodat-Lunz study area, Austria). *Verh. Internat. Verein. Limnol.*, **23**, 1333–1339.

Bretschko, G. & Moser, H. (1993). Transport and retention of matter in riparian ecotones. *Hydrobiologia*, **251**, 95–101

Brewer, W.S. & Pfaender, F.K. (1979). The distribution of selected organic molecules in freshwater sediment. *Water Research*, **13**, 237–240.

Cunningham, H.W. & Wetzel, R.G. (1989). Kinetic analysis of protein degradation by a freshwater wetland sediment community. *Appl. Environ. Microbiol.*, **55**, 1963–1967.

Harvey, J.W. & Bencala, K.E. (1993). The effect of streambed topography on surface-subsurface water exchange in mountain catchments. *Water Resources Research*, **29**, 89–98.

Holland, M.M. (1988). SCOPE/MAB technical consultations on landscape boundaries: report of a SCOPE/MAB workshop on ecotones. *Biology International, Special Issue*, **17**, 47–106.

Hutchinson, G.E. (1957). *A treatise on limnology.* Vol.1, London, John Wiley & Sons, Inc., 1015 pp.

Hyne, N.J. (1978). The distribution and source of organic matter in reservoir sediments. *Environm. Geol.*, **2**, 279–285.

Kasimir, G.D. (1990). *Die mikrobiellen Biozönosen eines alpinen Baches.* V. Int. Hydromikrobiol.Symp., Smolenice, 5. – 8. Juni 1990, 245–250.

Kasimir, G.D. in press: Microbial biomass and activities in a second order mountain brook. *Verh. Internat. Verein. Limnol.*, **25**.

Leichtfried, M. (1985). Organic matter in gravel streams (Project Ritrodat-Lunz). *Verh. Internat. Verein. Limnol.*, **22**, 2058–2062.

Leichtfried, M. (1986). *Räumliche und zeitliche Verteilung von particulärer organischer Substanz (POM – Particulate Organic Matter) in einem Gebirgsbach als Energiebasis der Biozönose.* Dissertation, Univ. Wien, 360 pp.

Leichtfried, M. (1988). Bacterial substrates in gravel beds of a second order alpine stream (Project Ritrodat – Lunz, Austria). *Verh. Internat. Verein. Limnol.*, **23**, 1325–1332.

Leichtfried, M. (1991a). POM in bed sediments of a gravel stream (Ritrodat-Lunz study area, Austria). *Verh. Internat. Verein.Limnol.*, **24**, 1921–1925.

Leichtfried, M. (1991b). Organische Substanz in Bettsedimenten des Oberen Seebaches in Lunz/See, Niederösterreich. *Mitt. österr. geol. Ges.*, **83**, 229–241.

Lowry, O., Rosebrough, N.J., Farr, A.L. & Randall, H.J. (1951). Protein measurement with the Folin phenol reagent. *J. Biol. Chem.*, **193**, 265–275.

Marshall, K.E. (ed.) (1984). *Microbial adhesion and aggregation.* Life Sciences Research Report 31, Springer Verlag.

Pick, F.R. (1987). Carbohydrate and protein content of Lake Seston in relation to plankton nutrient deficiency. *Can. J. Fish. Aquat. Sci.*, **44**, 2095–2101.

Pusch, M. (1987). *Qualitative and quantitative Untersuchungen an abgelagertem partikulärem Material in einem Mittelgebirgsbach.* Diplomarbeit, Universität Freiburg, 101 pp.

Rausch, T. (1981). The estimation of micro-algal protein content and its meaning for the evaluation of algal biomass. 1. Comparison of methods for extracting protein. *Hydrobiologia*, **78**, 237–251.

Schmid-Araya, J.M. (1989). *Comparison of the characteristics of two rotifer species, Brechionus plicatilis (O.F. Müller) and Encentrum linnhei Scott 1974, as food organisms and their effect on growth and survival of newlyhatched larvae of the turbot Scophalmus maximus L.* Ph. D. Thesis, London Univ.

Schmid-Araya, J.M. (1994). Spatial and temporal distribution of micro-meiofauna groups in an alpine gravel stream. *Verh. Internat. Verein. Limnol.*, **25**, 1649–1655.

Schwoerbel, J. (1961). Über die Lebensbedingungen und die Besiedlung des hyporheischen Lebensraumes. *Arch. Hydrobiol. Suppl.*, **25**, 182–214.

Taylor, B.R. & Roff, J.C. (1984). Use of ATP and carbon: nitrogen ratio as indicators of food quality of stream detritus. *Freshwater Biology*, **14**, 195–201.

11 Dynamics and vertical distribution of particulate organic matter in river bed sediments (Morava River, Czech Republic)

M. RULIK

Department of Ecology, Faculty of Natural Sciences, Palacky University, Svobody 26, 771 46 Olomouc, Czech Republic

ABSTRACT Vertical and temporal distribution of particulate organic matter (POM) in river bed sediments was observed in two different stations in the Morava River. Samples of river sediments from 0–70 cm depths were taken by inserting a steel cylinder with a special grab. The main type of organic matter found in bed sediments was fine particulate organic matter (FPOM <1 mm). The organic matter had a variable distribution in the hyporheic zone. Generally, the concentration of organic matter was higher in the top 0–40 cm of sediment. The annual mean storage of buried organic matter (0–70 cm depth) was 5352.13 g AFDW.m^{-3}. The exchange of organic matter between surface sediments and deep river bed sediments is influenced mainly by flooding activity. The maximum storage of organic matter in the hyporheic zone appears in the spring when floods are decreasing.

INTRODUCTION

Allochthonous organic matter which is imported in running waters at different times and in different ways (Boling *et al.*, 1975; Cummins, 1974; Moser, 1991) is an important source of energy for sediment fauna (Fisher & Likens, 1973; Cummins, 1974). Fresh imported allochthonous organic matter is not available directly for benthic consumers and must first be processed by the microbial communities, especially fungi and bacteria (Rossi & Fano, 1979; Arsuffi & Suberkropp, 1985).

An exchange of water exists between the surface and hyporheic zone (Grimm & Fisher, 1984; Thibodeaux & Boyle, 1987; White *et al.*, 1987). In these exchange processes, particulate and dissolved organic material are transported into the bed sediments (Welton, 1980; Carling & McMahon, 1987; Mayack *et al.*, 1989; Metzler & Smock, 1990; Smock, 1990; Bretschko & Moser, 1993; Sear, 1993). Moreover, the hyporheic zone may be supplied by dissolved organic matter (DOM) from groundwater discharging through the stream bed (Fiebig & Lock, 1991; Fiebig, 1992; Vervier & Naiman, 1992; Vervier *et al.*, 1993). Physical adsorption and microbial utilization influence a part of both the dissolved nutrients and DOM in the sediments (Triska *et al.*, 1989). Biofilm, composed of microorganism biomass and their extra cellular polymers (Marshall, 1984; Characklis & Wilderer, 1989; Characklis & Marshall, 1990) which are attached to sediment grain surfaces and which are in turn a rich energy base of the zoobenthic hyporheic community, plays a main role in these processes (Bärlocher & Murdoch, 1989).

Flow through the sediments has important implications for nutrient cycling due to the greater surface area available for the retention of nutrients (Munn & Meyer, 1988) and bed sediments also play an important role in the self-purification processes of rivers (Fontvieille & Fevotte, 1981; Fontvieille & Cazelles, 1988; Sterba, 1990).

Particulate organic matter (POM >0.00045 mm; Boling *et al.*, 1975) appears in three main types in the bed sediments (Leichtfried, 1991): 1) pure organic particles like plant or animal tissue; 2) living organisms; and 3) biofilm. The biofilm is regarded as the most important of these three in both quantity and quality (Bretschko & Leichtfried, 1988). The POM buried within the sediments is an important source of interstitial FPOM and DOM (Crocker & Meyer, 1987). Subsurface organic matter storage can be separated into two types: shallow and deep (Metzler & Smock, 1990). The storage of organic matter in the hyporheic zone may be considerably higher than on the surface (Cummins *et al.*, 1983), partly because buried detritus has a slower

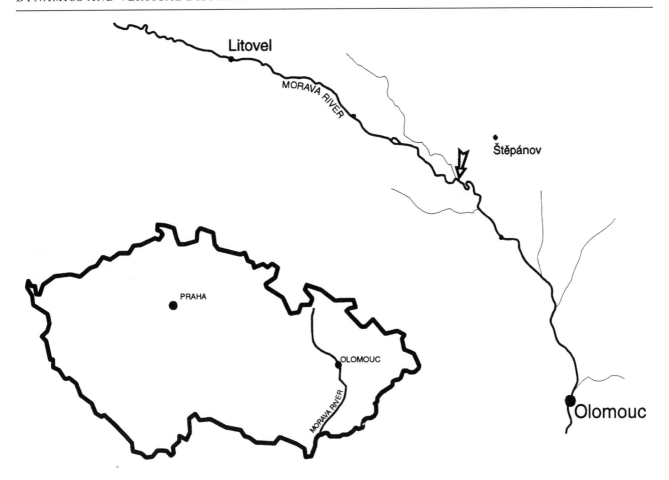

Fig. 1 Map of the Czech Republic showing the Morava River and Olomouc. The arrow indicates the sampling stations.

decomposition rate (Herbst, 1980; Rounick & Winterbourn, 1983) and a lower probability of downstream transport during spates (Metzler & Smock, 1990).

However, the vertical distribution and dynamic of particulate organic matter during storage in river bed sediments are relatively unknown. Thus, the aim of this study was to describe the composition, and the vertical and temporal distribution of particulate organic matter in river bed sediment in a lowland flood plain river.

STUDY LOCALITY

The study site is situated on the Morava River about 15 km north of Olomouc, in the vicinity of Stepánov (Fig. 1). Here, the Morava River is a multi-channel lowland river which meanders through an extensive complex of deciduous woodland. The river has an average flow of 20 m³.s⁻¹ ($Q_{min}=4$ m³.s⁻¹, $Q_{max}=350$ m³.s⁻¹). Two stations were chosen for the collection of samples. Station A lies above a gravel wave form riffle where surface water infiltration occurs. Station B is on the downstream face of a riffle, 20 m

below station A, at a place where the water upwells from the hyporheic zone (from sediments) and returns to the river (Fig. 2 – according to the basic hyporheic underflow model; Thibodeaux & Boyle, 1987; White, 1990, Hendricks & White; 1991, White *et al.*, 1992; White, 1993). The depth of the water during sampling was the same at both stations (30–50 cm).

MATERIALS AND METHODS

Twentysix field samples were collected (once or twice each month) between October 1989 – March 1991. Samples of river sediments for the qualitative and quantitative assessment of organic matter were taken with a 140 mm diameter steel cylinder inserted into the bed to a depth of 70 cm. A special grab (Sterba & Holzer, 1977), which consisting of two 110 mm diameter hemispheres, closed to enclose a sediment sample of 1583 cm³. Seven 10 cm layers (0–10, 11–20, 21–30, 31–40, 41–50, 51–60, 61–70 cm) were collected from each inserted cylinder. Each of these samples were transferred to a basket together with the sediment's interstitial water. The content of the basket was then filtered through a 1.0 mm pore size sieve. The ash-free dry weight (AFDW) of FPOM (fine particulate organic matter – particles smaller than 1 mm) and

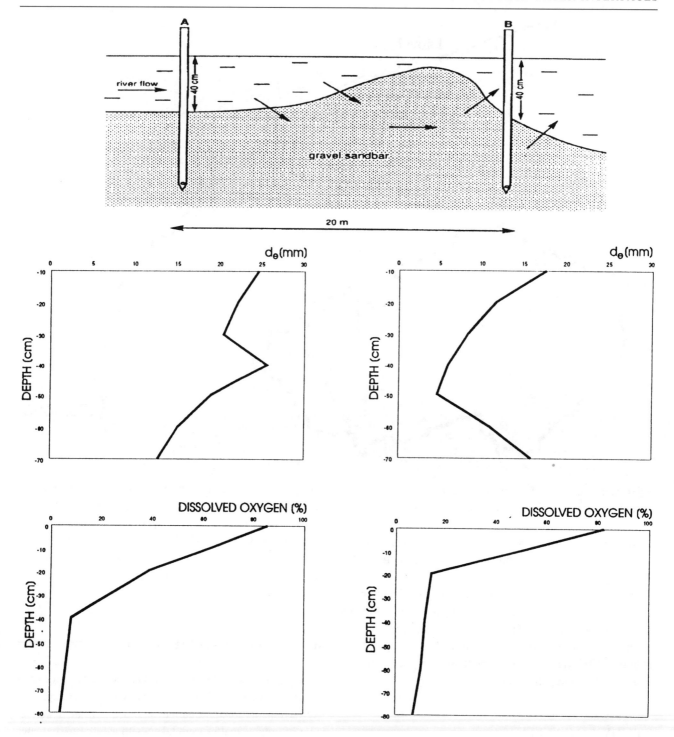

Fig. 2 The field location and the sites of the sampling stations A and B. Sediment particle size for each horizon at 10 cm intervals; d_e is the effective grain diameter, and average annual values of dissolved oxygen and organic matter at different horizons of the hyporheic zone.

CPOM (coarse particulate organic matter – particles larger than 1 mm) from horizons at 10 cm intervals was then determined by combustion at 550 °C for 5 h in a muffle furnace. The resulting quantity of organic matter was therefore expressed as g AFDW m^{-3} sediment. The annual variation of CPOM and FPOM in river bed sediments was estimated together for both stations because of the different number of samples taken at stations A and B.

The measurement of sediment particle size (granulometry) was determining using sieves of pore size 0.5, 1, 2, 4, 5.6, 8,

16, 22, 32, 45 and 63 mm. Size fractions were obtained for each horizon at 10 cm intervals, using the formula (Macura, 1966):

$$d_e = \frac{(\sum d_i p_i)}{100} \qquad (1)$$

where d_e is the diameter of effective grain, d_i=diameter of grains in mm, p_i=content of grains in separate grain size categories in %.

At the same time, interstitial water (for the sampling method see Sterba *et al.*, 1992) from up to 80 cm depth was analysed for dissolved oxygen. The oxygen content was determined by a portable oximeter (Oxi-96 WTW). The analysis of variance (ANOVA, Scheffer's multiple-range test) was used to test the significance of the vertical distribution.

RESULTS

Granulometry and oxygen content

The substrate at station A is composed of medium gravel (18–25 mm) up to 50 cm in depth. The river bed at station B is looser, as a consequence of fine particles being washed from the gravel/sand. The substrate here is composed of fine-medium gravel (4.6–17.5 mm) with sand (Fig.2). At both stations, generally, the medium-size class particles decreased with depth while smaller particles of the sediment became more abundant (Fig. 2).

The oxygen saturation of the surface water was balanced to an average of about 85 %. The maximum monthly average of 110 % was recorded in March 1991 when the river attained its high water mark. The minimum average value of 68 % was measured in October 1991. Oxygen saturation decreased rapidly with depth at both sites, but the decrease was most marked at station B (Fig.2).

Particulate organic matter

COMPOSITION AND DYNAMICS
FPOM was the predominant type of POM at both stations (Fig. 3). The average CPOM amount of 907.7 g AFDW.m^{-3} and 4 447. 43 g AFDW.m^{-3} for FPOM, respectively, was found for one sampling pipe (0–70 cm). Average values for one sampling core at station A were 1 722.97 g m^{-3} AFDW for CPOM and 19 405.98 g AFDW.m^{-3} for FPOM respectively. The average values for one sampling core at station B were higher for CPOM (5 054.54 g AFDW.m^{-3}) and smaller for FPOM (7 923.38 g AFDW.m^{-3}). Generally, the content of both types of organic matter was higher in the upper layer of the bottom sediment (0–40 cm) than in the lower layer (41–70 cm). The CPOM maximum amount in the upper

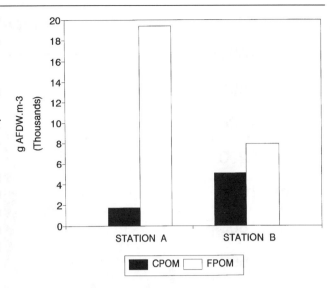

Fig. 3 Composition of organic matter found in 0–70 cm of river bed sediments at both stations.

layer was found in the summer (Fig. 4) while the storage of CPOM in lower parts of the bottom (41–70 cm) was the largest in September. The FPOM maximum was found shortly after spring flooding in April in both layers 0–40 and 41–70 cm (Fig. 5).

VERTICAL DISTRIBUTION
The amount of CPOM at station A significantly increased (F=4.466, P=0.05), up to 30 cm depth, while in the deeper horizons a decrease of CPOM was observed (Fig.6). At station B a similar situation was observed but the differences between horizons were not significant (F=1.104, P=0.05) (Fig. 7).

At station A the FPOM maximum was at 40 cm depth. A decrease in FPOM with depth was observed at station B (Fig. 7). In any case, no significant differences between horizons were found (F50.614, P50.05; F50.852, P50.05). The standard deviations for the annual mean storage of each horizon are probably very large due to the variability between individual samples and are shown in Table 1.

DISCUSSION

There exist two principal means of supplying bed sediments with particulate organic matter in rivers: 1) by infiltration of surface water into gravel bed sediments and, 2) detritus burial during spates and floods. Infiltration rates depend on the transport mechanism (i.e. suspended or bedload), local hydraulics, the dimensions of the interstices between the framework gravels, and scour and fill sequences during floods (Sear, 1993). Detritus exchange between surface and shallow sediments occurred throughout the year; exchange

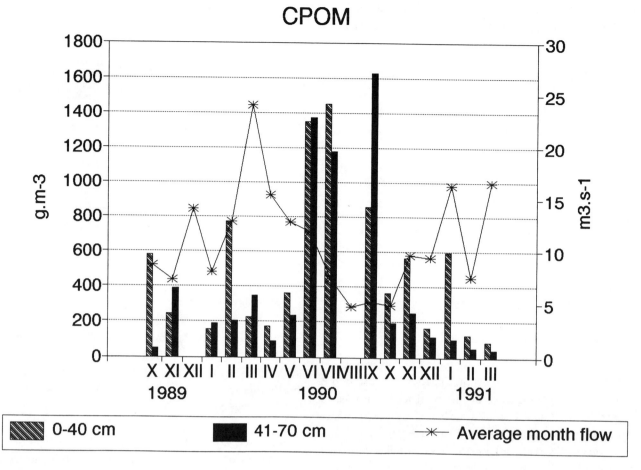

Fig. 4 Seasonal variation of water flow and amount of coarse particulate organic matter (CPOM) in river bed sediments.

between surface and deep sediments occurred only during spates (Metzler & Smock, 1990).

Fine particulate organic matter <1 mm was a dominant size fraction, composing 83 % of the total POM found in the Morava's bottom. FPOM as the dominant fraction of transported POM in the rivers was also found in some streams (e.g. Short & Ward, 1980; Jones & Smock, 1991). The maximum input of allochthonous CPOM is in the autumn as a consequence of leaf-fall (Lock & Williams, 1981; Jones & Smock, 1991; Moser, 1991). Metzler & Smock (1990) have reported that about 21% of autumnal leaf input to the stream is buried. Bretschko & Moser (1993) found that more than 80% of the annual input comes between April and August, the time with the most flashy hydrography, and less than 10% during defoliation. Although we did not estimate the transport of POM into the bed sediments, results from the Morava River indicated a similar situation to those mentioned above (Rulík, 1994). In autumn and winter (October – February) the *c*. 20 % of the annual in-bottom storage was found probably as a result of litter-fall (see Fig. 4.). The occurrence of this CPOM including pieces of bark, branches, stones of fruits (e.g. *Prunus* sp.) and the remains of leaves, was mainly

in the upper horizons. CPOM penetrated into the deeper parts later, in summer (Fig.4.) when the maximum storage of CPOM in bed sediments was found. Leichtfried (1988) has reported that the maximum input of organic matter appears after some delay in the bed sediments when floods are decreasing. The total organic carbon (TOC) maximum was found in up to 60 cm of sediment in summer and winter. The source of the summer maximum is bank-erosion bringing already processed soil material into the system. The basis of the winter maximum is mainly leaves shed in autumn. It takes a relatively long time before the leaf litter appears in the bed sediments because of low flood frequency in autumn and winter (Leichtfried, 1991).

The accumulation of benthic FPOM is greater during the spring and summer than other times (Boling *et al.*, 1975; Malmqvist *et al.*, 1978) probably due to laterally transported litter from inundated floodplains during the spring (Mayack *et al.*, 1989) when discharge is high and warm temperatures increase POM processing rates providing a large pool of FPOM (Jones & Smock, 1991). As in these other studies, the maximum FPOM storage in the Morava's river bottom was found in the spring shortly after the discharge peak in March 1990 (Fig. 5). The FPOM amount found in the top layer (0–40 cm) was almost two times higher than in the upper layer. This shows that during the floods, which occur on the

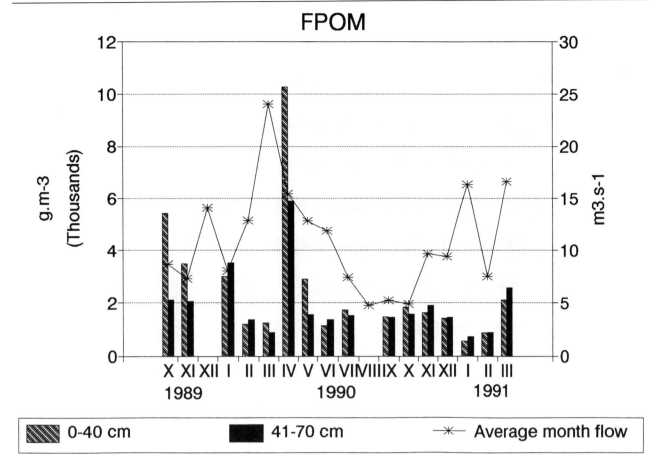

Fig. 5 Seasonal variation of water flow and amount of fine particulate organic matter (FPOM) in river bed sediments.

Morava River each year in January–March (occasionally in April), the top layers up to 40 cm depth are much more influenced by floods than the lower parts of the bottom (Rulík, 1994). The upper layers of the bottom (0–40 cm) are the horizon with intensive circulation of organic matter. The detritus stored in the top several centimetres of sediment is frequently buried and released due to the nearly constant movement of the topmost sediment layer (Metzler & Smock, 1990). Dissolved oxygen concentrations in the top sediment, while lower than on the surface (Poole & Stewart, 1976; Godbout & Hynes, 1982; Strommer & Smock, 1989) are probably not sufficiently low to affect microbial and invertebrate activity and, therefore, the processing rate should be effective here. Deep sediment layers are a long-term deposition zone for organic matter where the much slower processing rates occur probably due to the absence of dissolved oxygen (very often anaerobic conditions). Organic matter storage is scoured only during rare events (Metzler & Smock, 1990).

Although floods cause the loss of organic matter stored deep in the sediment (scouring), higher infiltration rates of surface water into the bed sediments increase the amount of

organic matter especially in deeper horizons (these higher infiltration rates were indicated e.g. by the increase of dissolved oxygen values in bed sediments during floods; Sterba *et al.*, 1992). The transport velocity of water decreases when floods decreased and the sedimentation of transported matter increased. Thus, organic matter appears in the subsurface environment with some delay (see Leichtfried, 1988).

Sediment stability is an important determinant of organic matter retention time within sediment, especially during spates (Metzler & Smock, 1990). Sand has a low erosion velocity and thus is very unstable. Variable sediment distribution results in patterns of organic matter in the deep horizons of the hyporheic zone in the Morava River. The substrate at station B was composed of finer sediment (sand) than at station A (Fig.2). Because of the potentially higher frequency of hyporheic zone disturbance in sandy bottomed streams (Metzler & Smock, 1990) this is probably the reason why almost twice the lower amount of organic matter than that of station A was found. The interstices of station A were more clogged with finer organic material as a consequence of progressive infiltration (downwelling) of surface water into the bed at the upstream end of a riffle (Thibodeaux & Boyle, 1987; White, 1990; Hendricks & White, 1991; White *et al.*, 1992; White, 1993; Boulton, 1993). The lower amount of organic matter at station B is also probably caused by its loss

STATION A

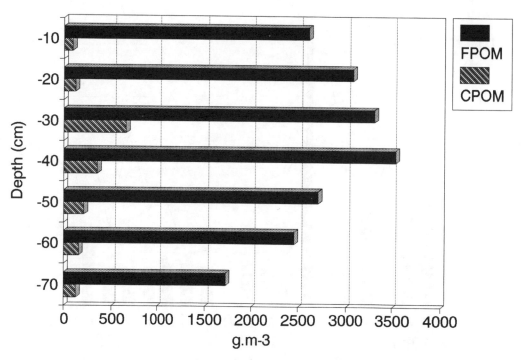

Fig. 6 Average annual values of particulate organic matter in different horizons of river bed sediments at station A.

STATION B

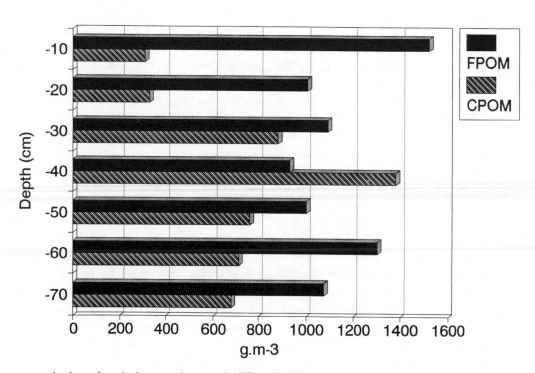

Fig. 7 Average annual values of particulate organic matter in different horizons of river bed sediments at station B.

Table 1. *Average annual values of organic matter and their standard deviations found in different horizons of river bed sediments at both stations*

CPOM Station A

Depth (cm)	Annual average	Standard deviation
−10	94.4196	68.2273
−20	124.083	64.4182
−30	650.758	741.616
−40	353.056	326.638
−50	210.896	144.763
−60	160.103	138.573
−70	129.655	99.5969

FPOM Station A

Depth (cm)	Annual average	Standard deviation
−10	2604.42	1553.70
−20	3076.42	2444.70
−30	3309.02	3623.27
−40	3533.20	4174.53
−50	2706.12	2391.55
−60	2453.89	1019.35
−70	1722.91	967.282

CPOM Station B

Depth (cm)	Annual average	Standard deviation
−10	308.991	612.451
−20	326.779	332.939
−30	881.221	571.997
−40	1381.56	1688.59
−50	762.423	726.866
−60	712.643	33.297
−70	680.921	1055.99

FPOM Station B

Depth (cm)	Annual average	Standard deviation
−10	1519.97	610.182
−20	1004.79	488.081
−30	1090.56	745.108
−40	930.550	565.251
−50	999.140	302.573
−60	1304.58	529.963
−70	1073.79	591.121

due to water upwelling from the bed to the surface water where the organic matter can be released (scoured).

CPOM concentration varied from 402.4 g.AFDW.m^{-3} in the top 0–40 cm of sediment to 505.3 g AFDW.m^{-3} found in the lower part (41–70 cm) of sediment. These results are lower compared to results from other streams (Metzler & Smock, 1990; Smock, 1990). On the contrary, Valett et al. (1990) found comparable concentrations of hyporheic organic matter (0.67 kg.m^{-3}) in cores from 0–40 cm of sediment. FPOM concentrations averaged annually from 2 517.59 g AFDW.m^{-3} at 0–40 cm to 1 926.85 g AFDW.m^{-3} at 41–70 cm of sediment. Metzler & Smock (1990) found an annual mean FPOM concentration of 1.3 kg.m^{-2} and suggested that the total organic matter storage of 4.8 kg.m^{-3} buried in the sediment was far higher than that stored at the channel surface.

Excluding CPOM values at station A the concentration of organic matter did not vary significantly with depth, but results show a general trend to decreasing organic matter with depth. This is in accordance with conclusions published by Williams & Hynes (1974), Godbout & Hynes (1982), and Valett et al. (1990). Decreasing organic matter with increasing depth is a trend opposite to that reported from an Austrian stream by Leichtfried (1985; 1988; 1991).

The variable distribution of organic matter in the bed sediments probably reflects disturbance history both in terms of deposition of OM during floods and time for in situ utilization by heterotrophs (Valett et al., 1990). The buried detritus is maintained by continual burial in the shifting top substrate, with only rare episodic loss of detritus stored deep in the sediment and low processing rates (Metzler & Smock, 1990).

ACKNOWLEDGEMENTS

The author would like to thank Ms Sandra Sweeney for help with the English text and two anonymous reviewers for critical correction of earlier versions.

REFERENCES

Arsuffi, T.L. & Suberkropp, K. (1985). Selective feeding by stream caddis fly (Trichoptera) detritivores on leaves with fungal-colonized patches. Oikos, **45**, 50–58
Bärlocher, F. & Murdoch, J. H. (1989). Hyporheic biofilm – a potential food source for interstitial animals. Hydrobiologia, **184**, 61–67
Boling, R. H., Goodman, E. D., Sickle Van, J. A., Zimmer, J. O., Cummins, K. W., Petersen, R. C. & Reice, S. R. (1975). Toward a model of detritus processing in a woodland stream. Ecology, **56**, 141–151.
Boulton, A.J. (1993). Stream ecology and surface-hyporheic exchange: implication, techniques and limitations. Aust. J. Mar. Freshwater Res., **44**, 553–564.
Bretschko, G. & Leichtfried, M. (1988). Distribution of organic matter and fauna in a second order, alpine gravel stream (Ritrodat-Lunz study area, Austria). Verh. Internat. Verein. Limnol., **23**, 1333–1339.
Bretschko, G. & Moser, H. (1993). Transport and retention of matter in riparian ecotones. Hydrobiologia, **251**, 95–101.

Carling, P. A. & Mc Cahon, C. P. (1987). Natural siltation of Brown trout (*Salmo trutta* L.) spawning gravels during low flow conditions. In *Regulated Streams: Advances in Ecology.* ed. Craig, J. F., Kemper, B. J., pp. 229–244, New York, Plenum Press.

Characklis, W. G. & Marshall, K. C. (eds.) (1990). *Biofilms.* A Wiley – Interscience Publ.. J. Wiley & Sons, INC., New York, Chicester, Brisbane, Toronto, Singapore, 796 pp.

Characklis, W. G. & Wilderer, P. A. (eds.) (1989). *Structure and function of biofilms.* Dahlem Konferenzen 1989. J. Wiley & Sons Ltd, Chicester, New York, Brisbane, Toronto, Singapore, 389 pp.

Crocker, M. T. & Meyer, J. L. (1987). Interstitial dissolved organic carbon in sediments of a southern Appalachian headwater stream. *J. N. Am. Benthol. Soc.*, **6**, 3, 159–167.

Cummins, K. W. (1974). Structure and function of stream ecosystems. *Bioscience*, **24**, 1, 631–641.

Cummins, K. W., Sedell, J. R., Swanson, F. J., Minshall, G. W., Fisher, S. G., Cushing, C. E., Petersen, R. C. & Vannote, R.L. (1983). Organic matter budgets for stream ecosystems: problems in their evaluation In *Stream ecology: applications and testing of general ecological theory.* ed. Barnes, J. R., Minshall, G. W., pp. 299–353, New York, Plenum Press.

Fiebig, D. M. (1992). Fates of dissolved free amino acids in ground water discharged through stream bed sediments. *Hydrobiologia*, **235/236**, 311–319.

Fiebig, D. M. & Lock, M. A. (1991). Immobilization of dissolved organic matter from ground water discharging through the stream bed. *Freshw. Biol.*, **26**, 45–55.

Fisher, S. G. & Likens, G. E. (1973). Energy flow in Bear Brook, New Hampshire: an integrative approach to stream ecosystem metabolism. *Ecological Monographs*, **43**, 421–439.

Fontvieille, D. & Cazelles, B. (1988). Seasonal changes of some physiological and structural parameters in the sediments of an organically polluted stream. *Verh. Internat. Verein. Limnol.*, **23**, 1306–1312.

Fontvieille, D. & Fevotte, G. (1981). DNA content of the sediment in relation to self-purification in streams polluted by organic wastes. *Verh. Internat. Verein. Limnol.*, **21**, 221–226.

Godbout, L. & Hynes, H. B. N. (1982). The three-dimensional distribution of the fauna in a single riffle in a stream in Ontario. *Hydrobiologia*, **97**, 87–96.

Grimm, N. B. & Fisher, S. G. (1984). Exchange between interstitial surface water, Implication for stream metabolism and nutrient cycling. *Hydrobiologia*, **111**, 3, 219–228.

Hendricks, S. P. & White, D. S. (1991). Physicochemical patterns within a hyporheic zone of a northern Michigan river, with comments on surface water patterns. *Can. J. Fish. Aquat. Sci.*, **48**, 9, 1645–1654.

Herbst, G. N. (1980). Effects of burial on wood value and consumption of leaf detritus by aquatic invertebrates in a lowland forest stream. *Oikos*, **35**, 411–424.

Jones, J. B. & Smock, L. A. (1991). Transport and retention of particulate organic matter in two low-gradient headwater streams. *J. N. Am. Benthol. Soc.*, **10**, 2, 115–126.

Leichtfried, M. (1985). Organic matter in gravel streams (Project Ritrodat-Lunz). *Verh. Internat. Verein. Limnol.*, **22**, 2058–2062.

Leichtfried, M. (1988). Bacterial substrates in gravel beds of a second order alpine stream (Project Ritrodat-Lunz, Austria). *Verh. Internat. Verein. Limnol.*, **23**, 1325–1332.

Leichtfried, M. (1991). POM in bed sediments of a gravel stream (Ritrodat-Lunz study area, Austria). *Verh. Internat. Verein. Limnol.*, **24**, 1921–1925.

Lock, M. A. & Williams, D. D. (eds.) (1981). *Perspectives in running water ecology.* New York, Plenum Press, 430 pp.

Macura, L. (1966). *Upravy tokov* (Regulations of rivers). Slovenské vydavatelstvo technickej literatúry, Bratislava, 732 pp.

Malmqvist, B., Nilsson, L.M. & Svensson, B.S. (1978). Dynamics of detritus in a small stream in southern Sweeden and its influence on the distribution of the bottom animal comunities. *Oikos*, **31**, 3–16.

Marshall, K. C. (ed.) (1984). *Microbial adhesion and aggregation.* Dahlem Konferenzen 1984. Berlin, Heidelberg, New York, Tokyo, Springer-Verlag, 426 pp.

Mayack, D. T., Thorp, J. H. & Cothran, M. (1989). Effects of burial and flood plain retention on stream processing of allochthonous litter. *Oikos*, **54**, 378–388.

Metzler, G. M. & Smock, L. A. (1990). Storage and dynamic of sub-surface detritus in sand-bottomed stream. *Can. J. Fish. Aquat. Sci.*, **47**, 588–594.

Moser, H. (1991). Input of organic matter (OM) in a low order stream (Ritrodat-Lunz study area, Austria). *Verh. Internat. Verein. Limnol.*, **24**, 1913–1916.

Munn, N. L. & Meyer, J. L. (1988). Rapid flow through the sediments of a headwater stream in the southern Appalachians. *Freshwater Biology*, **20**, 235–240.

Poole, W. C. & Stewart, K.W. (1976). The vertical distribution of macrobenthos within the substratum of Brazos River, Texas. *Hydrobiologia*, **50**, 151–160.

Rossi, L. & Fano, A. E.(1979). Role of fungi in the trophic niche of the congeneric detritivorous *Asellus* and *A.coxalis* (Isopoda). *Oikos*, **32**, 380–385

Rounick, J. S. & Winterbourn, M. J. (1983). Leaf processing in two contrasting beech forest streams: Effects of physical and biotic factors on litter breakdown. *Arch. Hydrobiol.*, **96**, 4, 448–474.

Rulík, M. (1994). Vertical distribution of coarse particulate organic matter in river bed sediments (Morava River, Czech Republic). *Regulated Rivers*, **9**, 65–69.

Sear, D. (1993). Fine sediment infiltration into gravel spawning beds within a regulated river experiencing floods: Ecological implications for salmonids. *Regulated Rivers*, **8**, 373–390.

Short, R.A. & Ward, J.V. (1980). Leaf litter processing in a regulated rocky mountain stream. *J. Fish. Aquat. Sci.*, **37**, 123–127.

Smock, L. A. (1990). Spatial and temporal variation in organic matter storage in low-gradient, headwater streams. *Arch. Hydrobiol.*, **118**, 2, 169–184.

Strommer, J. L. & Smock, L. A. (1989). Vertical distribution and abundance of invertebrates within the sandy substrate of a low-gradient headwater stream. *Freshwat. Biol.*, **22**, 263–274.

Sterba, O. (1990). Rícní dno a samocistení [River bottom and self-purification]. In: *Sborník referátu ze seminárе "Rícní dno".* CSLS pri CSAV a Geograficky ústav Brno, pp. 69–78, Kuparovice, zárí 1990.

Sterba, O. & Holzer, M. (1977). Fauna der interstitiellen Gewässer der Sandkiessedimente unter der aktiven Strömung der Flüsse [The fauna of interstitial waters of gravel/sand sediments below the active stream]. *Vest. cs. spol. zool.*, **41**, 2, 144–159.

Sterba, O., Uvíra, V., Mathur, P. & Rulík, M. (1992). Variations of the hyporheic zone through a riffle in the R.Morava, Czechoslovakia. *Regulated Rivers*, **7**, 31–43.

Thibodeaux, L. J. & Boyle, J. O. (1987). Bed form generated convective transport in bottom sediment. *Nature*, **325**, 341–343.

Triska, F. J., Kennedy, V. C., Avanzino, R. J., Zellweger, G. W. & Bencala, K. E. (1989). Retention and transport of nutrients in a third-order stream in north-western California: hyporheic processes. *Ecology*, **70**, 6, 1893–1905.

Valett, H. M., Fisher, S. G. & Stanley, E. H. (1990). Physical and chemical characteristics of the hyporheic zone of a Sonoran Desert stream. *J. N. Am. Benthol. Soc.*, **9**, 3, 201–215.

Vervier, P. & Naiman, R. J. (1992). Spatial and temporal fluctuations of dissolved organic carbon in subsurface flow of the Stillaguamish River (Washington, USA). *Arch. Hydrobiol.*, **123**,4, 401–412.

Vervier, P., Dobson, M. & Pinay, G. (1993). Role of interaction zones between surface and ground waters in DOC transport and processing: considerations for river restoration. *Freshwat. Biol.*, **29**, 275–284.

Welton, J. S. (1980). Dynamics of sediment and organic detritus in a small chalk stream. *Arch. Hydrobiol.*, **90**, 162–181.

White, D. S. (1990). Biological relationships to convective flow patterns within stream bed. *Hydrobiologia*, **196**, 149–158.

White, D. S. (1993). Perspectives on defining and delineating hyporheic zones. *J. N. Am. Benthol. Soc.*, **12**, 1, 61–69.

White, D. S., Elzinga, Ch. H. & Hendricks, S. P. (1987). Temperature patterns within the hyporheic zone of a northern Michigan river. *J. N. Am. Benthol. Soc.*, **6**, 2, 85–91.

White, D. S., Hendricks, S. P. & Fortner, S. L. (1992). Groundwater-surface water interactions and the distributions of aquatic macrophytes. In *Proceedings of the First International Conference on Ground Water Ecology.* ed. Stanford, J. A., Simons, J. J., pp. 247–255, American Water Resources Association, Bethesda, Maryland.

Williams, D. D. & Hynes, H. B. N. (1974). The occurrence of benthos deep in the substratum of a stream. *Freshwat. Biol.*, **4**, 233–256.

12 Surface water/groundwater/forest alluvial ecosystems: functioning of interfaces. The case of the Rhine Floodplain in Alsace (France)

M. TRÉMOLIÈRES, R. CARBIENER, I. EGLIN, F. ROBACH, U. ROECK & J.-M. SANCHEZ-PEREZ

Laboratoire de Botanique et Ecologie végétale, CEREG URA 95 CNRS, Institut de Botanique, 28 rue Goethe, F – 67083 Strasbourg cedex, France

ABSTRACT The Holocene Rhine floodplain in Alsace, which represents a large homogeneity of petrographical regions, is used as a model for a comparative study of the interrelation of the different compartments (water-soil-plant) of two alluvial hydrosystems. We show that the species richness of alluvial forest, as well as the diversity of the ecosystems and the geoforms, are the main factors which explain the great efficiency in the functioning of the interfaces, e.g. retention on substrate, uptake and transformation of nutrients into biomass, and hence purification of the groundwater. We studied a canalized river deprived of a floodplain, the rapid transfer of eutrophicants and micropollutants in the canalized river itself and through the channel bed leads to a worsening of both surface and groundwater quality. By contrast a river with a functional floodplain provides large quantities of good quality water to the groundwater table. In this paper, we analyse the processes which occur at the interfaces of the different compartments.

INTRODUCTION

The alluvial hydrosystems are corridors transferring water, sediment, organic matter and organisms (Décamps & Naiman, 1989). The fluxes of matter flow through the longitudinal axis of the river and also the transversal axis to the riparian zones which constitute the land-water interface (Gregory et al., 1991). These zones are characterized by a large diversity of aquatic and terrestrial ecosystems, due to the geomorphological and hydrological dynamics of the large rivers (Amoros et al., 1988). Permanent interaction exists between both ecosystems, thanks to the vector 'water', which allows us to define ecotones, land-water ecotone or water-water ecotone. These ecotones control exchange between two systems by acting as a combination of permeability, elasticity, biodiversity and connectivity processes (Gibert et al., 1990; Vervier et al., 1991). But hydraulic management works considerably affect the interaction processes, for example by suppressing the floodplain and disturbing the exchange proccesses (Carbiener & Trémolières, 1990; Dister, 1994). An additional box in a typical schema of a functional alluvial hydrosystem to other boxes such as soil/water, groundwater/surface water, and labeled 'the hyporheic zone' perhaps ought to be included in the stream-catchment connections, as proposed by Bencala (1993), and defined for example by Valett et al. (1990).

Responses of biocenoses to physical change along the longitudinal gradient of the main channel result in a continuum of biotic adjustments (Vannote et al., 1980). Interdependence of the systems and the dynamic of connectivity as defined by Amoros et al. (1993) also ensure exchange of nutrients and organisms between the different compartments of the fluvial hydrosystem. Composition and structure of biocenoses in the main channel and the lateral arms and also of the riparian forest reflect fluvial disturbance by floods (Nilsson et al., 1993; Gregory et al., 1991). However they can be drastically changed consequent to modifying the exchange and transfer processes.

By monitoring physico-chemical variables and vegetal biocenoses, the purpose of this paper is:

– to define the role of the interfaces (ecotones) – the soil/forest/groundwater system and the river/groundwater system – in the biogeochemical functioning of an alluvial hydrosystem (Fig. 1),

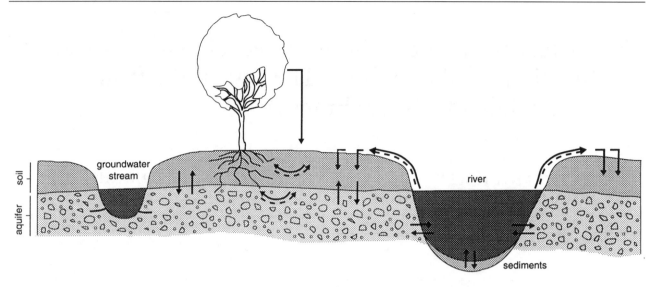

Fig. 1 Interaction and exchange between the different compartments of the alluvial hydrosystem.

– to study the influence of flood waters on the groundwater quality in the case of a river with a functional floodplain mainly covered by alluvial forests and meadows,
– to analyse the changes in the functioning of interfaces resulting from the canalization of the river causing to a reduction or suppression of flood water.

We focus this study on a comparison of the functioning of the two main rivers of the Alsace plain, the Rhine, now deprived of floodplain, and its tributary the Ill, with a functional floodplain. More detailed results were specified in Trémolières *et al.* (1993), and Trémolières *et al.* (1994).

We shall analyse the fluxes of the major limiting mineral nutrients (phosphorus and nitrogen) and of a micropollutant (mercury) which contaminate equally the two rivers. These contaminants can pass from the river to the groundwater directly through the river bed and the river banks (Von Gunten & Kull, 1986; Laszlo, 1989) or through the soil-forest system when transferred by flood waters or by discharge from agricultural uplands (Lowrance *et al.*, 1983; Correll *et al.*, 1992). This last system acts as a nutrient sink. We shall show that the structure of the terrestrial ecosystems (forest successional stages leading to increased specific biodiversity (Adams, 1989; Carbiener & Schnitzler, 1990)) and of aquatic ecosystems (river and side arms) controls the transmission of the mineral elements downstream. The self-purifying processes by fixation, retention in soils, sediment and biomass, and exchange at interfaces will be demonstrated.

STUDY SITE

In the Alsace Rhine rift floodplain, the River Rhine flows for about 200 km over an homogeneous substrate of fluvioglacial calcareous pebbles, which constitutes a large reservoir of groundwater. The recent canalization (1950–1970) deprived the Rhine of the former large floodplain (20 000 ha) except for some islands (Rhinau island for example), isolated by the building of hydroelectric power plants. The Ill, the main tributary of the Rhine in Alsace, has a functional floodplain of about 10 000 ha (Fig. 2). These two rivers define two sectors of contrasting surface geomorphology, substrate and hydrology (Table 1) (Carbiener, 1983; Trémolières *et al.*, 1993, 1994). Despite this human activity, most of the flooded sectors are still forested. In the case of the River Rhine, these forested sectors illustrate the different stages of the successional forest series from a pioneer species-poor, simple structure stage to a mature species-rich stage of high complex structure, according to the duration, the periodicity and the dynamics of floods originated by the Rhine before the canalization (Carbiener *et al.*, 1988; Carbiener & Schnitzler, 1990).

The study site is located in the central part of the Alsace floodplain, corresponding to the old braided and anastomosed sector of the Rhine, where the exchange between surface and groundwater is very active. In this sector, the groundwater, which is close to the surface (−1 m), rises where the soil surface is at the same level as the upper layer of the water table. Therefore groundwater feeds a large network of streams, namely the groundwater-fed streams, the occurrence of which is peculiar to the Alsace Rhine plain. Thus at the springs, the water quality reflects the groundwater quality. The exchange between the river and the groundwater is shown by the transfer of micropollutants (heavy metals such as mercury) and eutrophicants (phosphate, nitrate), which contaminate the groundwater-fed streams.

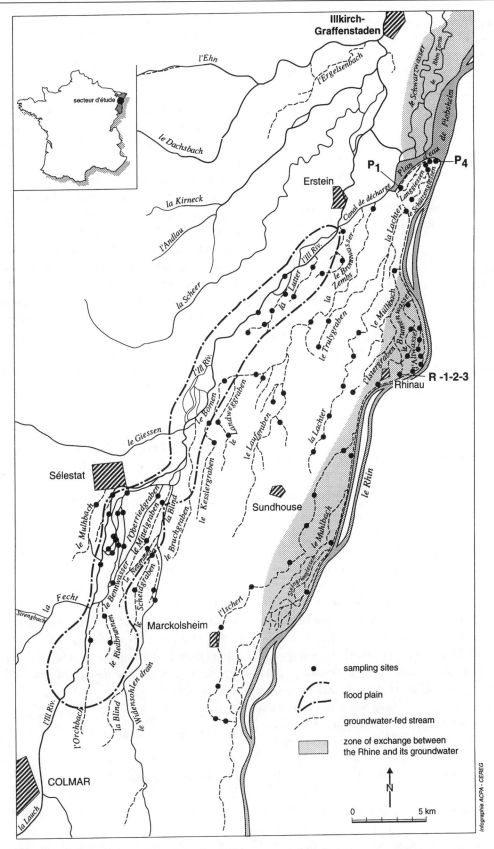

Fig. 2 Geographical situation of the Rhine floodplain in Alsace (France). Localization of the sampling sites.

Table 1. *Characteristics of the hydrology and substrate of the Rhine and Ill sectors*

	Rhine fringe	Ill floodplain
Hydrological regime	summer highwater winter lowwater	summer lowwater winter highwater
Alluvium deposit		
% sand	10	5
% silt	60	35
% clay	30	60
% CaCO3	25	0–5
Soils	poor of colloids basic	rich in colloids neutral to slightly acidic
	not hydromorphic	hydromorphic
Permeability m.s^{-1}	$10^{-1}-10^{-3}$	10^{-4}
Transmissivity m^2.s^{-1}	10^{-1} 10^{-3}	

The Rhinau island, a still flooded Rhine site, is taken as a model to study the influence of the forest on the groundwater quality (Sanchez-Perez *et al.*, 1991 a & b).

METHODOLOGY

The groundwater-fed streams (see sampling sites on Fig. 2) are examined using three methods:

– an analytical method based on the analysis of physico-chemical parameters: pH, temperature, dissolved oxygen, phosphate, nitrate, ammonia and chloride. This last parameter is used as an hydrological tracer of the Rhine-groundwater exchange because of a high level of chloride contamination in the Rhine waters, originating from the potash mines located in the south of Alsace.
– a biological method based on the accumulation of mercury in *Fontinalis antipyretica* Hedw., a very common bryophyte in the groundwater-fed streams. In the 1970s, the Rhine suffered from acute mercury pollution. Mercury was retained in the sediment and is now being released into the overlying water and interstitial water, then transferred to the groundwater. *Fontinalis* is used as a bioindicator of pollution by heavy metals (Mouvet *et al.*, 1986; Gonçalves *et al.*, 1992) and more particularly of mercury transfer (Roeck *et al.*, 1991; Roeck *et al.*, 1993).
– a phytosociological method based on the analysis of aquatic macrophyte communities. The macrophyte communities are correlated to the phosphate and ammonia nitrogen concentrations in the overlying water and thus could be used as indicators of eutrophication related to increase of phosphate and ammonia. Thus we

used the bioindication scale showing the degree of eutrophication with six steps A, B, C, CD, D and E, from oligotrophic to eutrophic level (Table 2), performed by Carbiener & Ortscheit (1987), Carbiener *et al.* (1990).

Analytical procedures were those described in APHA (1985) and specified with the phytosociological surveys in Trémolières *et al.* (1993). The phytosociological surveys consist in a list of plant species to which is assigned a coefficient of abundance (corresponding to a percentage of cover) according to the method of Braun-Blanquet (1964).

The groundwater under the flooded forest was sampled using a series of piezometers, and analysed for the chemical nutrients: phosphate, nitrate, sulphate, carbonate, chloride, ammonium, potassium, sodium, magnesium and calcium. The metal ions are measured by atomic absorption spectroscopy. The ionic balance is used for analytical verification. The maximum error between anions and cations were 4% (Sanchez-Perez *et al.* 1991b). For the chemical analyses, an inter-calibration between the chartered laboratory of Hydrology (Faculty of Pharmacy, Strasbourg) and our laboratory was carried out.

SCALING THE INFLUENCES OF GROUND AND SURFACE WATER EXCHANGES

The role of interfaces in the exchange between the river and the groundwater was analysed on two scales. The smaller scale covers the whole central part of Alsace where we compare the relations between the canalized Rhine and its riparian water table to those of its tributary, the Ill, characterized by a functional floodplain and its corresponding groundwater. The larger scale (corresponding to a stretch of a stream or a parcel of forest) enables us to analyse the factors and processes regulating the transfer of water and solutes from the river to the groundwater, depending on the presence or absence of a floodplain and on the structure of the vegetation.

Observations on a small scale

The canalized Rhine seeps into its groundwater table upstream of the dams and in the river bend, leading to what we call the 'Rhine filtrates' (Fig. 3). Thus the riverside groundwater is contaminated by phosphate, chloride (Robach *et al.*, 1991) and mercury (Roeck *et al.*, 1991) on a fringe of 1–2 km width. Concentrations of contaminants are as high in groundwater as in surface water (Table 3). During the infiltration, processes show reversible reactions due to the formation of a reduced zone as shown by Bourg & Bertin

Table 2. *Synoptic table of communities A-F in the eutrophication sequence related to phosphate and ammonia concentrations, in alkaline running waters. A oligotrophic to F eutrophic community. Degree of species presence expressed by coefficients of Braun-Blanquet (1964): I <20%; II 20–40%; III 41–60%; IV 61–80%; V>80%; between brackets, degree of cover-abundance: +< 5% sparce; 1 <5%; 2 5–25%; 3 25–50%; 4 50–75%; 5>75%. n number of sample per community*

community (n)	A(9)	B(11)	C(15)	D(17)	E(20)	F(16)
Potamogeton coloratus	V(2)	—	—	—	—	—
Batrachospermum moniliforme	II(1)	—	—	—	—	—
Juncus subnodulosus fo subm.	II(+)	—	—	—	—	—
Chara vulgaris	II(+)	—	—	—	—	—
Chara hispida	II(+)	—	I(2)	—	—	—
Ranunculus circinatus	—	—	I(+)	—	—	—
Lamprocystis roseo persicina	II(+)	—	I(3)	I(+)	—	—
Berula erecta (Sium erectum)	V(2)	V(4)	V(2)	III(1)	III(1)	I(+)
Callitriche obtusangula	I(+)	III(+)	V(2)	V(2)	IV(1)	III(+)
Lemna trisulca	—	I(1)	III(1)	III(1)	II(1)	—
Fontinalis antipyretica	—	I(+)	I(1)	II(1)	II(+)	—
Elodea canadensis	—	—	IV(1)	III(1)	II(1)	I(+)
Potamogeton friesii	—	—	II(+)	II(1)	II(1)	II(1)
Sparganium emersum	—	—	II(+)	II(1)	II(+)	IV(1)
Lemna minor	—	—	II(1)	III(1)	V(2)	V(1)
Elodea nuttallii	—	—	I(2)	II(2)	IV(2)	IV(1)
Nasturtium officinale	—	—	—	V(1)	III(+)	II(+)
Groenlandia densa	—	—	—	II(1)	II(1)	—
Zannichellia palustris	—	—	—	II(1)	II(1)	—
Spirodela polyrhiza	—	—	—	I(1)	II(1)	IV(1)
Azolla filliculoïdes	—	—	—	I(1)	I(1)	II(1)
Potamogeton crispus	—	—	—	III(1)	II(1)	I(2)
Myriophyllum verticillatum	—	—	—	I(1)	I(+)	—
Hottonia palustris	—	—	—	I(1)	—	—
Hippuris vulgaris	—	—	—	I(1)	—	—
Potamogeton pectinatus	—	—	—	II(1)	IV(3)	IV(1)
Myriophyllum spicatum	—	—	—	II(1)	IV(1)	V(1)
Potamogeton perfoliatus	—	—	—	I(+)	I(2)	III(1)
Ceratophyllum demersum	—	—	—	I(+)	V(2)	V(1)
Oenanthe fluviatilis	—	—	—	—	I(r)	—
Ranunculus trichophyllus	—	—	—	—	I(+)	—
Potamogeton pusillus	—	—	—	—	I(+)	—
Ranunculus fluitans	—	—	—	—	II(2)	III(+)
Potamogeton lucens	—	—	—	—	—	III(1)
Potamogeton nodosus	—	—	—	—	—	III(2)
Mentha aquatica fo subm	I(r)	I(+)	I(1)	I(+)	I(1)	—
Veronica anagallis aquatica	I(+)	—	I(+)	I(+)	II(+)	—
Myosotis scorpioides	I(r)	—	I(1)	II(+)	II(+)	—
Veronica beccabunga	—	—	—	I(+)	I(+)	I(+)
pH	7.4(0.1)	7.5(0.2)	7.5(0.1)	7.6(0.2)	7.9(0.2)	7.9(0.2)
Conductivity (µS/cm)	608(115)	736(112)	740(99)	657(66)	657(63)	508(52)
Hardness (meq/l)	4.8(1.4)	4.7(0.9)	5(0.7)	3.9(0.4)	3.8(0.5)	3.2(0.3)
N-NH4+ (µg/l)	13.7(7.3)	22.2(13.8)	45.3(27.8)	33.8(31.3)	61.2(40)	255(107)
P-PO43- (µg/l)	7.2(1.7)	13(5.5)	14.9(6.8)	29.4(23.6)	39.9(33)	191.5(116)
N-NO3- (mg/l)	5.5(1.4)	5.1(1.8)	4.7(2.1)	2.9(2.5)	1.6(1.1)	2.5(0.9)

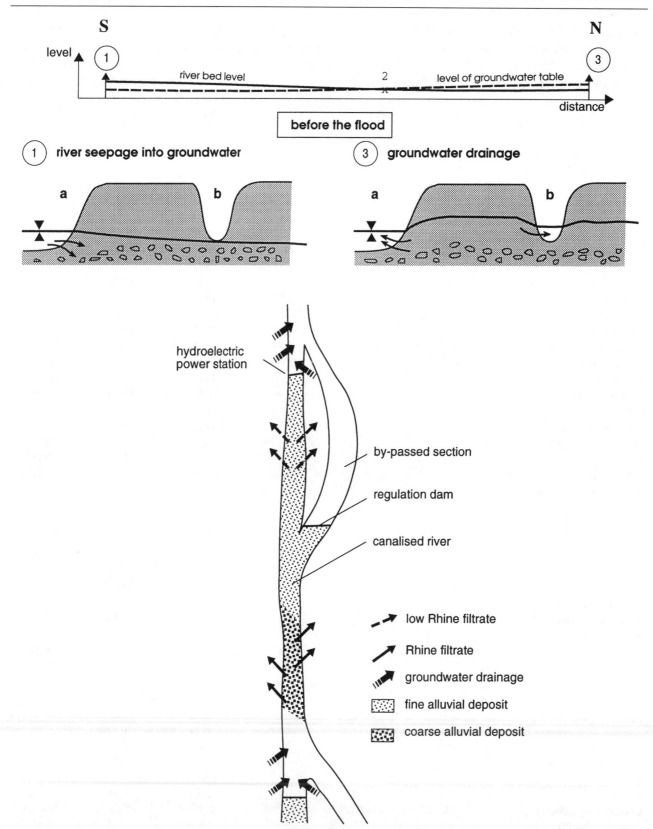

Fig. 3 Model of river-groundwater exchange process related to the topographic filter, proposed for the Rhine floodplain (Trémolières *et al.,* 1993, 1994).

Table 3. *Chemical contents in the Rhine, the River Ill, and their riverside groundwater tables (phosphate, nitrate and ammonia nitrogen and chloride)*

	Canalized Rhine		Ill floodplain	
	Rhine	Riverside groundwater	Ill	Riverside groundwater
P-PO$_4^{3-}$ mg.l^{-1}	40–150	30–100	80–400	1–10
N-NO$_3^-$ mg.l^{-1}	1–2.5	0–2	1–3	3–7
N-NH$_4^+$ mg.l^{-1}	50–500	0–60	100–1000	0–20
Cl$^-$ mg.l^{-1}	30–280	30–230	35–100	30–50

Table 4. *Mercury content in the Bryophyte Fontinalis antipyretica sampled in the Rhine and the groundwater-fed streams (Sh1, S1, R3) compared to the River Ill and its riverside groundwater (Lutter), over the study period (January–November 1990). Hg is expressed in mg.kg^{-1} dry weight*

	January	March	May	July	Sept.	Nov.
Rhine (Rhinau)	0.16	0.22	0.11	0.15	0.28	0.12
Sh1	0.18	0.14	0.19	0.14	0.52	0.21
S1	0.12	0.1	0.1	0.08	0.21	0.17
R3	0.25	0.11	0.17	0.16	0.45	0.62
Ill (Benfeld)	0.41	0.38	0.28	0.38	0.3	0.32
Lutter	0.04	0.05	0.06	0.08	0.13	0.06

(1993) in the case of manganese. Phosphate and mercury were removed from solution by a combination of adsorption and surface oxidation (phosphorus, see for example Manning, 1987; Moutin *et al.*, 1993) or reduction (mercury sulfide) reactions.

As a result of the high level of phosphate and ammonia nitrogen, the vegetation of groundwater-fed streams in the Rhine fringe are lush, characterized by the eutrophic CD, D and E macrophyte communities according to the bioindication scale (Table 2). The bryophyte *Fontinalis* sampled in the groundwater-fed streams has a level of mercury as high as in the Rhine to-day (see sites Sh1 and R3, Table 4). The amounts of nitrate in the groundwater and the groundwater-fed streams are the same as those found in the river, where the level is low (1–2 mg.l^{-1}). This ion acts as a hydrological tracer, like chloride, because of well-oxygenated substrate and thus, in general, absence of denitrification even during the floods (with local exceptions) (Sanchez-Perez *et al.*, 1991b).

In the Ill sector, the water quality variables (phosphate, ammonia, mercury) in the groundwater are at a very low level

despite the high degree of contamination of the river waters (Table 3). In contrast to the riparian Rhine sector, the groundwater-fed streams of the Ill floodplain are characterized by an oligotrophic to mesotrophic vegetation, A, B and C communities. The mercury content of *Fontinalis* is also very low (annual mean of 0.04 mg.kg^{-1} in the Lutter, a groundwater-fed stream located close to the river Ill, Table 4), near the geochemical background values. The nitrate concentrations in the groundwater-fed streams are higher than in the Ill, due to the agricultural use of a large part of the floodplain, but nitrates are very low in groundwater sampled under flooded forest (Correll *et al.*, 1992, Haycock & Burt, 1993).

These results show completely contrasting hydrological functioning depending on whether the river is canalized or not, i.e. with or without a functional floodplain (Trémolières *et al.*, 1993, 1994). In the first case the studied contaminants, phosphate and mercury, are transferred entirely, in the second case they are stopped or slowed down during their transfer.

Observations on a large scale: identification of processes

The increase in the contamination of the riverside groundwater, observed in the case of the canalized river, is directly linked to the disconnection of the forest-soil system from the river and also to the suppression of the diluting water inputs during floods (Carbiener & Trémolières, 1990). Canalization removes the barriers which stop or slow the transfer of nutrients and micropollutants to the groundwater. On the contrary, flooding, even with inputs of eutrophicated or polluted flood waters of the riparian forest or meadow ecosystem, demonstrates the efficiency of filters (system soil-vegetation for example), which preserve the groundwater from contamination by eutrophicants and micropollutants.

PHYSICO-CHEMICAL PROCESSES

The position of the superficial layer of the groundwater relative to that of the overlying (free-flowing) water surface in the river partly explains the infiltration from the river to the groundwater, as verified in the case of the canalized Rhine which is 10 m higher than its plain, or the River Ill, upstream of a line between Colmar and Sélestat (Fig. 3). Moreover the gravelly Rhine substrate which constitutes the whole aquifer of the Alsace Rhine valley, facilitates the transfer of chemicals, nutrients or micropollutants, because of a high permeability (Table 1).

By contrast, the clayey silty alluvial deposits carried by the Ill flood waters constitute thick soils which efficiently retain micropollutants such as mercury and also phosphate uptaken by forest biomass if transferred in the growth season. The strong purifying action by the soil due to

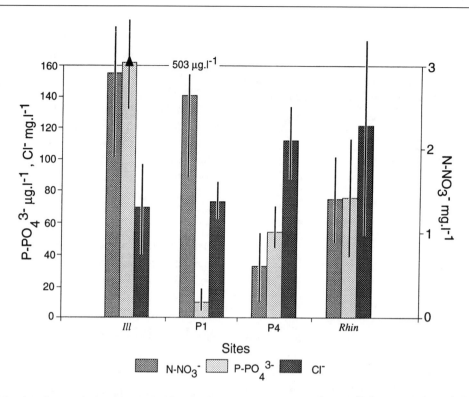

Fig. 4 Mean chloride, phosphate and nitrate concentrations in the drainage channel of Plobsheim (P1, P4), the Rhine and the River Ill. See Fig. 2 for the location of sites: P1 western spring influenced by Ill waters, P4 eastern spring influenced by the Rhine waters.

processes of retention, biodegradation and/or reduction (of nitrate for example) is often invoked in the transfer of micro-pollutants from the river to the adjacent groundwater (Kussmaul & Mühlhausen, 1979; Von Gunten & Kull, 1986; Lehotsky 1986; Laszlo 1989; Ile *et al.*, 1991). In time, problems of increasing intoxication of the soils will occur as they become highly contaminated, and then the risk of transfer of the pollutants to the groundwater will increase. The mercury content of the River Ill was high in the 1970s (0.7 mg.kg^{-1} in *Fontinalis* sampled in the Ill, up to 1.3 mg.kg^{-1} in fish), and has remained almost constant until to-day but at a lower level (about 0.4 mg.kg^{-1} in *Fontinalis*).

An example of a groundwater stream shows the importance of the infiltration/retention process and explains the difference observed between the groundwater quality near the Rhine or near the Ill (Fig. 4). It is an artificial drainage channel, created by the canalization and which is derived from two groundwater-fed sources, located at the former Ill-Rhine confluence (before the canalisation). The western source (P1, see Fig. 2) is influenced by Ill waters and is characterized by low level of chloride and phosphate and a high nitrate level. On the contrary, the eastern source (P4), close to the Rhine, contains high amounts of chloride and phosphate, and low amounts of nitrate, close to those measured in the Rhine (Fig. 4). The low level of phosphate in the

western source shows efficient retention of phosphate on colloids of alluvial deposits carried by the River Ill. In this sector the Ill surface water is at the same level or a little lower than the upper layer of the groundwater, in contrast to the canalized Rhine which at this point is also higher than its plain.

BIOLOGICAL PROCESSES

a) Soil-forest system

In the Rhine islands which remain flooded and are covered by alluvial forests, the groundwater is partly purified. This purification is more efficient under the mature hardwood forests characterized by a high species diversity and a complex structure (Sanchez-Perez *et al.*, 1991b). The depuration efficiency is related to the relations between mineral cycling and specific biodiversity. Thus the successive leaf litter cycles, the vernal geophytes, the summer ivy litter fall (Trémolières *et al.*, 1988) then the classic autumnal litterfall in temperate forests contribute to efficient recycling of organic matter and mineral elements. Moreover, the ligneous and herbaceous species adapted their behaviour to the uptake of nutrients even in excess according to the conditions of mineral availability (see in the case of phosphorus, Weiss *et al.*, 1991; Weiss & Trémolières, 1993). Space time complementarities and differences of the uptake of nutrients by the different ligneous species exist.

In the same way, the groundwater-fed streams in the Rhine fringe have better quality water when they are separated from the river by a large forest, even in the unflooded sectors,

Table 5. *Chemical contents in the alluvial flooded soils compared to inputs by precipitation (6 months of study) and the flood of February 1990. The calculation is based on a depth of 1m of water and 2cm depth of sediment*

	Soil content kg.ha^{-1}	Atmospheric inputs kg.ha^{-1}	Flood inputs kg.ha^{-1}
S	570–1207	12.7	86
Mg	1000–2000	4.6	51
K	620–1443	5.8	34
P	173–408	1.1	27
N	131–160	29.6	8.9

where the forest is rooted in the upper layer of the ground-water table. This is the case of the sources near Rhinau (R1, R2) located at around 500 m from the Rhine (Fig. 2): the ammonia nitrogen and phosphate levels are low (10–20 μg.l^{-1} N$-$NH$_4^+$, P$-$PO$_4^{3-}$) by comparison with the source of Schoenau (Sh1) close to the Rhine (50 μg.l^{-1} N$-$NH$_4^+$, 25 μg.l^{-1} P$-$PO4^{3-}).

Moreover, the forest soils act as an efficient trap for chemicals during flooding (Sanchez-Perez *et al.*, 1993). Thus, in the Rhine soils, we find high storage of phosphate, sulphate, ammonia nitrogen and potassium retained on the colloids, in proportion to inputs through precipitation or flooding (Table 5). This is found despite a light texture (Table 1) and good oxygenation.

In the Ill floodplain (10 000 ha), forest and meadows still cover 7500 ha. This permanent vegetation cover recycles and transforms the mineral nutrients, phosphate and nitrate carried by the flood waters, into biomass. The soils, regularly saturated by flood waters, have anoxic horizons (gley or pseudo-gley) and thus are denitrifying. This process contributes to the decrease of nitrate in the waters which infiltrate, as shown also by Pinay & Décamps (1988), Haycock & Burt (1993). The differences between the phosphate, nitrate and ammonia contents in the flood waters and those in the groundwater are really quite significant.

The destruction of meadows replaced by crops constitutes a risk for the groundwater and the drinking water, because of the related inputs of nitrate (Correll, 1991) and pesticides (Mirgain *et al.*, 1993), whose impact on the forest ecosystem (oak decline in the alluvial forest ?) we do not as yet know.

b) Aquatic vegetation-surface water system
Before the canalization, the groundwaters flowed in the old channels which discharged flood waters during flood episodes and transported fertilizing alluvium to the flooded riparian forest. As the floodplain is suppressed, flood waters only flow in canalized water courses, almost deprived of

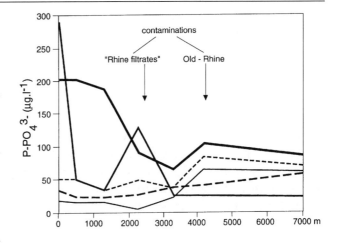

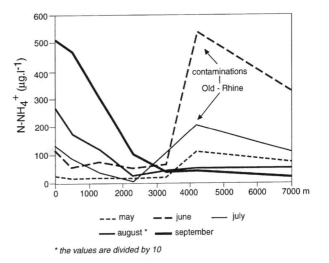

Fig. 5 Variations of phosphate and ammonia concentrations in the Schaftheu, an old lateral arm of the Rhine, according to the distance from the regulation dam, over a growing period. Phosphate expressed in P$-$PO$_4^{3-}$ mg.l^{-1}, ammonia in N$-$NH$_4^+$ mg.l^{-1}.

aquatic vegetation. These artificial streams function as simple systems transferring pollutants downstream. By contrast the old side arms still connected to the main channel provide calm zones conducive to the development of aquatic and semi-aquatic macrophytes and/or phytoplankton which actively use the nutrients (phosphorus and nitrogen). A complex morphology of streams, with successive fords, calm zones and meanders, which increase the time during which the water circulates in the biological filter, amplifies the process of purification. In this type of stream, the aquatic macrophytes are dominant and compete with the phytoplankton. They take part in the purification and clarifying of water by uptake of nutrients, in parallel with the process of retention in sediments. In the example of a side arm still upstream connected to the Rhine (Fig. 5), phosphate and ammonia decrease during the growing season with a high significance ($p<5\%$ for ammonia nitrogen and 12% for phosphorus) on a short distance to the river (around 1000 m) then increase due to the contaminations by the Rhine infiltration (Robach *et al.*, 1991).

CONCLUSION

The riparian flooded zones of rivers are a noticeable biological 'purification reactor', enabling the groundwater table to be fed large quantities of unpolluted waters. To be most efficient, it is necessary to preserve a permanent vegetation cover of forest or meadow, as demonstrated too by Haycock & Burt (1993) or Correll (1991) in the case of discharge from agricultural watersheds. By contrast the disconnection of the floodplain from the polluted canalized river leads to a generalized contamination of the riparian groundwater.

The most efficient way to purify an alluvial hydrosystem is to have the highest specific diversity and richness of ecosystems (case of terminal stage of succession or river with a complex structure), which ensures a better functional diversity, i.e. complementarity between species in uptake of nutrients.

In the Rhine Alsace floodplain, where three nature reserves of alluvial forests were created, the question is to know which type of management to promote for a restoration or conservation of the alluvial systems and as a function of which objectives, such as preserving or improving the groundwater quality, or preserving or restoring the diversity of biocenoses and/or habitats, or limiting the impact of flood on the adjacent areas, downstream of the canalized river. At present, a project on one of the alluvial reserves is to flood again with Rhine waters. In this case, despite the surface water contamination, the risk for groundwater must be limited. However the quality of groundwater-fed streams in this sector in which very clean and pure water flows, could be changed and as a consequence the biocenoses too. Also the options for management of alluvial systems must be based on scientific knowledge. The realization of such a project will allow scientific research on the consequences of reflooding on biodiversity or biogeochemical functioning of the alluvial systems. Such a restored sector could be considered as a natural laboratory for scientific research.

ACKNOWLEDGEMENTS

This work was supported by Sandoz Foundation, PIREN Eau-Alsace and IFARE (Institut franco-allemand de recherche sur l'environnement). We are very grateful for the analytical help of the laboratory of Hydrology (Faculty of Pharmacy, Strasbourg).

REFERENCES

Adams, J. M. (1989). Species diversity and productivity of trees. *Plants today*, 183–187.

Amoros, C., Bravard, J. P., Reygrobellet J. L., Pautou, G. & Roux, A. L. (1988). Les concepts d'hydrosystème et de secteur fonctionnel dans l'analyse des systèmes fluviaux à l'échelle des écocomplexes. *Bull. Ecol.*, **19**, 4, 531–546.

Amoros, C., Gibert, J. & Greewood, M. T. (1993). Interactions entre unités de l'hydrosystème fluvial. In *Hydrosystèmes fluviaux*. ed. C. Amoros & Petts G. E., pp. 169–199, Paris, Masson.

APHA (1985). *Standard methods for the examination of water and wastewater*. 16th ed., American Public Health Association, 1268 pp.

Bencala, K. E. (1993). A perspective on stream-catchment connections. *J. N. Am. Benthol. Soc.*, **12**, 44–47.

Bourg, A. C. M. & Bertin, C. (1993). Biogeochemical processes during the infiltration of river water into an alluvial aquifer. *Environm. Sci. & Technol.*, **27**, 661–668.

Braun-Blanquet, J. (1964). *Pflanzensoziologie*. Springer Verlag Wien, New York.

Carbiener, R. (1983). Le grand Ried central d'Alsace: écologie et évolution d'une zone humide d'origine fluviale rhénane. *Bull. Ecol.*, **14**, 249–277.

Carbiener, R., & Ortscheit, A. (1987). *Wasserpflanzengesellschaften als Hilfe zur Qualitätüberwachung eines der grössten Grundwasservorkommen Europas*. ed. A. Miyawaki, A. Bogenrieder, S. Okuda & White, pp. 283–312, J., Proceed. Inter. Symp. IAVS, Tokyo-Yokohama, 1984, Tokai university press.

Carbiener, R. & Schnitzler, A. (1990). Evolution of major pattern models and processes of alluvial forest in the rift valley (France/Germany). *Vegetatio*, **88**, 115–129.

Carbiener, R., Schnitzler, A. & Walter, J. M. (1988). *Problèmes de dynamique forestière et de définition des stations en milieu alluvial*. Colloques phytosociologiques, "Phytosociologie et foresterie", Nancy 1985, T. XIV 655–686.

Carbiener, R. & Trémolières, M. (1990). The Rhine rift valley groundwater table and Rhine river interactions: evolution of their susceptibility to pollution. *Regulated Rivers, Research and Management*, **5**, 375–389.

Carbiener, R., Trémolières, M., Mercier, J. L.& Ortscheit, A. (1990). Aquatic macrophyte communities as bioindicators of eutrophication in calcareous oligosaprobe stream waters (Upper Rhine plain, Alsace). *Vegetatio*, **86**, 71–88.

Correll, D. L. (1991). Impact on the functioning of landscape boundaries. In *Ecotones The role of landscape boundaries in the management and restoration of changing environments*, ed. Holland M. M., Risser P.G. & Naiman R. J., pp. 90–109, New York, Chapman & Hall.

Correll, D. L., Jordan, T. E. & Weller, D. E. (1992). Nutrient flux in a landscape: effects of coastal land use and terrestrial community mosaic on nutrient transport to coastal waters. *Estuares*, **15**, 431–442.

Décamps, H. & Naiman, R.J. (1989). L'écologie des grands fleuves. *La recherche*, **208**, 310–319.

Dister, E. (1994). *The function, evaluation and relics of near-natural floodplains*. Limnologie aktuell Band/vol2 Kinzelbach (Hg.): Biologie der Donau, Gustav Fisher Verlag, Stuttgart- Jena- New York, 317–329.

Gibert, J., Dole, M. J., Marmonier, P. & Vervier, P. (1990). Surface water-groundwater ecotones. In *The ecology and management of aquatic-terrestrial ecotones*, ed. Naiman R. J. & Décamps, H., pp. 199–225, Man & Biosphere series.

Gonçalves, E. P. R., Boaventura, R. A. R. & Mouvet, C. (1992). Sediments and aquatic mosses as pollution indicators for heavy metals in the Ave river basin (Portugal). *The Science of the total environment*, **114**, 7–24.

Gregory, S. V., Swanson, F. J., McKee, W. A. & Cummins, K. W. (1991). An ecosystem perspective of riparian zones. *BioScience*, **41**, 540–551.

Haycock, N. E. & Burt, T. P. (1993). Role of floodplain sediments in reducing the nitrate concentration of subsurface run-off: a case study in the costwolds, UK. *Hydrological processes*, **7**, 287–295.

Ile, C., Suais, M. F., Durbec, A., Gaillard, B. & Lafont, M. (1991). Caractérisation du pouvoir épurateur des berges d'un fleuve. Application à l'île du Grand Gravier (Rhône). *Hydrogéologie*, **4**, 283–289.

Kussmaul, H. & Mühlhausen, D. (1979). Hydrologische und hydro-chemische Untersuchungen zur Uferfiltration, Teil III: Veränderungen der Wasserbeschaffenheit durch Uferfiltration und Trinkwasseraufbereitung. *Gwf-Wasser/Abwasser*, **120**, 320–329.

Laszlo, F. (1989). Qualität bei der Gewinnung von uferfiltriertem Grundwasser in Ungarn. *Acta Hydrochim. Hydrobiol.*, **17**, 4, 453–463.

Lehotsky, J. (1986). Variations de la qualité des eaux du Danube pendant l'infiltration riveraine. Etude et rapports d'hydrologie, *"Pollution et protection des aquifères"*, ed. UNESCO, 406–419.

Lowrance, R. R., Todd, R. L. & Asmussen, L. E. (1983). Waterborne nutrient budgets for the riparian zone of an agricultural watershed. *Agriculture, Ecosystems and environment*, **10**, 371–384.

Manning, P. G. (1987). Phosphate ion interactions at the sediments-water interface in lake Ontario: relationship to sediment adsorption capacities. *Can. J. Fish. Aquat. Sci.*, **44**, 2204–2211.

Mirgain, I., Schenck, C. & Monteil, H. (1993). Atrazine contamination of groundwaters in eastern France in relation to the hydrogeological properties of the agricultural land. *Environ. Technol.*, **14**, 741–750.

Moutin, T., Picot, B., Ximenes, M.C. & Bontoux, J. (1993). Seasonal variations of P compounds and their concentrations in two costal lagoons (Herault, France). *Hydrobiogia*, **252**, 45–59.

Mouvet, C., Pattee, E. & Cordebar, P. (1986). Utilisation des mousses aquatiques pour l'identification et la localisation précise de sources de pollution métallique multiforme. *Acta Oecol. Oecol. Applic.*, **7**, 77–91.

Nilsson, C., Ekblad A., Dynesius, M., Backe, S., Gardfjell, M., Carlberg B., Hellqvist, S. & Jansson, R. (1994). A comparison of species rich-ness and traits of riparian plants between a main river channel and its tributaries. *J. Ecol.*, **82**, 281–295.

Pinay, G. & Décamps, H. (1988). The role of the riparian wood in regu-lating nitrogen fluxes between the alluvial aquifer and surface water: a conceptual model. *Regulated rivers: Research and Management*, **2**, 507–516.

Robach, F., Eglin, E. & Carbiener, R. (1991). L'hydrosystème rhénan: évolution parallèle de la végétation aquatique et de la qualité de l'eau (Rhinau). *Bull Ecol.*, **22**, 227–241.

Roeck, U., Trémolières, M., Exinger, A. & Carbiener, R. (1991). Utilisation des mousses aquatiques dans une étude sur le transfert du mercure en tant que descripteur du fonctionnement hydrologique (échanges cours d'eau-nappe) en plaine d'Alsace. *Bull. Hydroécologie*, **3**, 241–256.

Roeck, U., Trémolières, M., Exinger, A. & Carbiener, R. (1993). Le transfert du mercure utilisé comme descripteur du fonctionnement hydrologique (échanges cours d'eau-nappe) dans la plaine alluviale du Rhin supérieur. *Annls Limn.*, **29**, 339–353.

Sanchez-Perez, J. M., Trémolières, M. & Carbiener, R. (1991a). Une station d'épuration naturelle des phosphates et nitrates apportés par les eaux de débordement du Rhin: la forêt alluviale à Frêne et Orme. *C.R. Acad.Sc.*, **312**, série III, 395–402.

Sanchez-Perez, J. M., Trémolières, M., Schnitzler, A. & Carbiener, R. (1991b). Evolution de la qualité physico-chimique des eaux de la frange superficielle de la nappe phréatique en fonction du cycle saisonnier et des stades de succession des forêts alluviales rhénanes (*Querco-Ulmetum*). *Acta Oecol.*, **12**, 581–601.

Sanchez-Perez, J.M., Trémolières, M., Schnitzler, A. & Carbiener, R. (1993). Nutrient content in alluvial soils submitted to flooding in the Rhine alluvial deciduous forest. *Acta Oecol.*, **14**, 3, 371–387.

Trémolières, M., Carbiener, R., Exinger, A. & Turlot J. C., (1988). Un exemple d'interaction non compétitive entre espèces ligneuses: le cas du lierre arborescent (*Hedera helix* L.) dans la forêt alluviale. *Acta Oecologica, Oecol. Plant.*, **9**, 187–209.

Trémolières, M., Eglin, I., Roeck, U. & Carbiener, R. (1993). The exchange process between river and groundwater on the central Alsace floodplain (eastern France): I. The case of the canalised river Rhine. *Hydrobiologia*, **254**, 133–148.

Trémolières, M., Roeck, U., Klein, J. P. & Carbiener, R. (1994). The exchange process between river and groundwater on the central Alsace floodplain (eastern France): II. The case of a river with func-tional floodplain. *Hydrobiologia*, **273**, 19–36.

Valett, H. M., Fisher, S. G. & Stanley, E. H. (1990). Physical and chem-ical characteristics of the hyporheic zone of a Sonoran Desert stream. *J. N. Am. Benthol. Soc.*, **9**, 201–215.

Vannote, R. J., Minshall, G. W., Cummins, J. R., Sedel, J. R. & Cushing, C. E. (1980). The river continuum concept. *Can. J. Fish Aquat.*, **37**, 130–137.

Vervier, P., Gibert, J., Marmonier, P. & Dole-Olivier, M. J. (1992). A per-spective on the permeability of the surface freshwater-groundwater ecotone. *J. N. Am. Benthol. Soc.*, **11**, 93–102.

Von Gunten, H. R. & Kull, T. P. (1986). Infiltration of inorganic com-pounds from the Glatt river, Switzerland, into a groundwater aquifer. *Water, air and soil pollution*, **29**, 333–346.

Weiss, D. & Trémolières, M. (1993). Impact des inondations sur la biodisponibilité du phosphore dans deux forêts alluviales de la plaine d'Alsace (France). *C.R. Acad. Sci. Paris*, **316**, Série III, 211–218.

Weiss, D., Trémolières, M. & Carbiener, R. (1991). Biodisponibilité com-parée du phosphore en fonction d'un gradient d'inondablité dans les forêts alluviales (plaine rhénane d'Alsace). *C. R.Acad. Sc. Paris*, **313**, Série III, 245–251.

13 Modelling of hydrological processes in a floodplain wetland

C. BRADLEY* & A.G. BROWN**

* School of Geography, Birmingham University, Edgbaston, Birmingham B15 2TT, England

** Department of Geography, Leicester University, Leicester LE1 7RH, England

ABSTRACT This paper considers the importance of interactions between hydrology and ecology for a floodplain peat bog in Central England. Narborough Bog is covered by wet woodland and a reed-bed dominated by *Phragmites*. There has recently been a reduction in the relative abundance of wetland species due to changing hydrological conditions. The maintenance of near-surface water table elevations is essential to the preservation of the present ecology of this regionally important site, and this is largely dependent upon external factors such as precipitation, evapotranspiration and river levels.

The results of a detailed programme of hydrological monitoring, including the monitoring of water tables, are described and a model of the site hydrology is formulated. This conceptual model is tested using a numerical model (MODFLOW) employing field measured hydraulic parameters. The ability of the model to predict water table responses to varying hydrological inputs is discussed.

The model results are placed within a longer term context which includes the frequency and coverage of overbank events, and the implications of the varying contribution of different components of the water budget for site ecology are described.

INTRODUCTION

Wetlands form ecotones between terrestrial and aquatic ecosystems; they typically possess high levels of species diversity and represent a valuable refugia for rare fauna and flora (Everett, 1989; Wheeler, 1988). There is increasing evidence of their hydrological role in regulating surface and groundwater resources (Carter, 1986), which has strengthened the case for wetland preservation. However, their successful conservation requires quantification of the balance between wetland water inflows and outflows and examination of their relationship to fluvial processes (Mitsch & Gosselink, 1986). Few studies of wetland hydrology have included sufficient field measurements to derive a full water balance, especially in Britain, although the relationship between water tables and vegetation was recognised at an early date by Godwin & Bharucha (1932) at Wicken Fen in Cambridgeshire. Wetlands receive water inflows from a variety of sources which are difficult to quantify, for example floodplain wetlands benefit from periodic overbank floods contributing

water and nutrients (Hammer & Kadlec, 1986). Hydrological modelling represents a potentially invaluable means of studying the hydrology of individual wetlands by incorporating different water fluxes and examining water flow through organic sediments (Gilvear *et al.*, 1993).

The principal processes providing water inflows and outflows in a theoretical floodplain wetland are indicated in Fig. 1. Water inflows include a combination of precipitation, tributary and overland flow from the adjacent catchment, influent seepage and overbank flow from the river. Outflows of water consist of evaporation, evapotranspiration and effluent seepage to the river. Thus the river may provide either recharge or discharge of water depending upon the hydraulic gradient between the floodplain water table and river stage.

Seasonal and diurnal fluctuations in water table depths in a variety of wetlands reflect temporal variations in water storage in response to the hydrometeorological conditions (Koerselman, 1989; Gilman, 1994). The type of wetland vegetation also has an important influence on the hydrological balance affecting evapotranspiration and interception

SLOPE
PROCESSES　　　　　FLUVIAL
PROCESSES　　　　　WETLAND　　　　　ATMOSPHERE

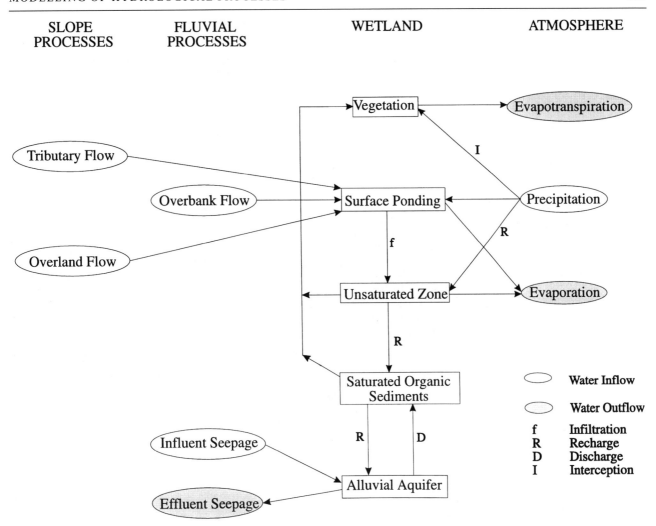

Fig. 1 Conceptual model describing the hydrology of a floodplain wetland system.

storage, however, subsurface seepage to and from a river is harder to quantify. Detailed records of spatial and temporal changes in hydraulic head, measurements of hydraulic conductivity and assessment of overbank floods are required to quantify the significance of river seepage.

STUDY AREA

This paper describes the results of applying a calibrated hydrological model to a specific floodplain wetland, Narborough Bog, in the British Midlands. Narborough Bog is not an example of a unique wetland but is representative of large areas of wetland which have been destroyed in Britain through land drainage, and exemplifies the problems of hydrological management in surviving wetland areas. The site covers an area of 9.5 ha of mixed woodland, meadow and reed-bed scrub, beside the River Soar in Central England. The location of the wetland rela-

tive to the Soar floodplain and adjacent urban areas is shown in Fig. 2, with the direction of surface drainage indicated. The stratigraphy of the central area of the site consists of an accumulation of 1.8 m of herbaceous and wood peat overlying a gravel deposit of 3–4 m thickness. Near the river deposits of silt and clay form a 30 m wide band and overlie a dark wood peat.

Narborough Bog was originally designated as a Site of Special Scientific Interest in 1956 due to the size of the reed-bed which at that time was the largest in the region. Evidence of historical changes in the range of wetland flora is provided by Wade (1919) who described the current reed-bed as a reed swamp 'composed almost purely of *Phragmites communis*' (sic *australis*) with local coverage of *Filipendula ulmaria*, *Phalaris arundinacea* and *Epilobium parviflorum*. In the meadows beside Narborough Bog a rich wet meadow flora was observed; for example *Caltha palustris* was growing in a field which has since been drained to form a sports field. At present the ecology of the reed-bed is undergoing a period of transition. The formerly dominant *Phragmites australis* is now increasingly replaced by *Filipendula ulmaria* and *Epilobium hirsutum*. Photographs taken over the last ten

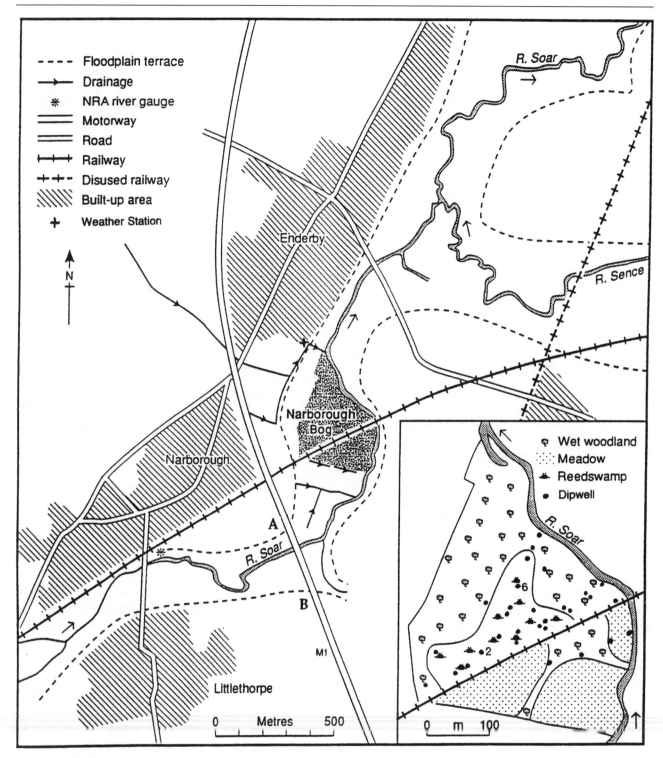

Fig. 2 Map showing the location of Narborough Bog within the Soar floodplain, with internal differences in vegetation communities illustrated in the inset.

glutinosa, *Salix fragilis* and *Salix verminalis*. To the south the meadows still contain a varied wet-meadow flora which includes *Valeriana dioica* and *Dactylorhiza praetermissa*.

years suggest that much of the ecological change in the reed-bed community has occurred comparatively recently.

There have been fewer vegetation changes within the woodland area to the north of the reed-bed since Wade's study. The woodland is an alder-willow association comprised of *Alnus*

HYDROLOGICAL MEASUREMENTS

A programme of water table monitoring was initiated in October 1990 to investigate fluctuations in the water table

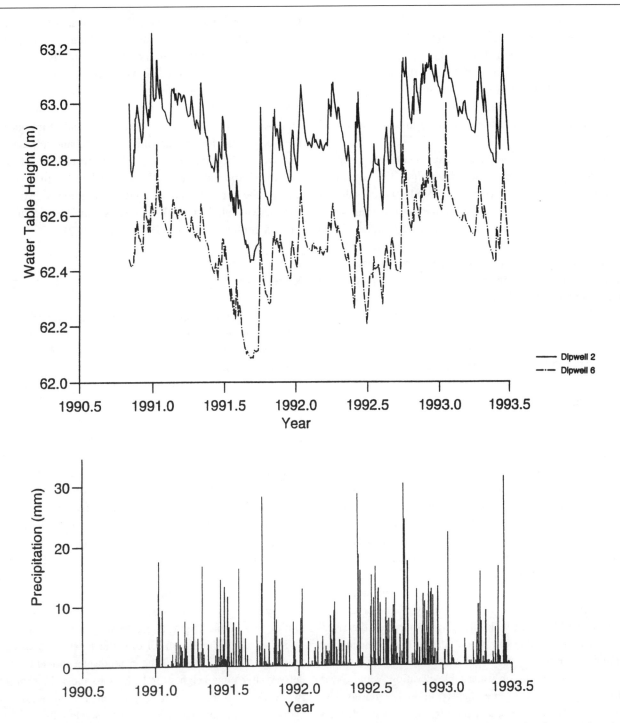

Fig. 3 Time-series plot of water table records for dipwells 2 and 6 and precipitation in 1990–1993.

and consider the significance of seepage to and from the River Soar. A total of thirty dipwells were installed at varying distances from the river as indicated in Fig. 2. The dipwells consisted of plastic pipes with outside diameter 5 cm. The majority were 1.25 m long; three dipwells were 0.5 m long, and were installed to investigate variations in the vertical hydraulic gradient. Holes were drilled in the bottom 25 cm so that water table readings represent an average over this inter-

val. Water table depths were measured using an electrical contact device twice weekly, with daily measurements during major precipitation events or at times of high evapo-transpiration loss. An automatic weather station was also maintained for the duration of the study. Precipitation was measured by a tipping bucket, and daily potential evapotranspiration determined using the Penman formula.

A time-series plot of water table readings for two dipwells (numbers 2 and 6) and daily precipitation for the period September 1990 to June 1993 is given in Fig. 3. Water tables vary over a range of 0.7 m between a maximum height in

January 1991, to a minimum in July 1991. The precipitation distribution appears to be mainly responsible for the pattern of water table fluctuation. Thus the large summer drawdown of the water table in July 1991 reflects the lack of rainfall during a period of high evapotranspiration. The difference in water table height between dipwells 2 and 6, at different distances from the river, illustrates the predominance of effluent seepage.

These routine hydrological measurements were supplemented by experiments to determine the hydraulic conductivity of the herbaceous and wood peat and silt/clay deposits near the river. An infiltration experiment through a peat column with a gypsum crust (Hillel & Gardner, 1969) revealed hydraulic conductivities of $2\rightarrow4$ m day^{-1} for the herbaceous peat. Experiments using the seepage tube method (Luthin & Kirkham, 1949) indicated a hydraulic conductivity of $0.0439\rightarrow0.169$ m day^{-1} for the wood peat, and $7.9\times10^{-3}\rightarrow2.9\times10^{-5}$ for the silt/clay deposits.

HYDROLOGICAL MODELLING

The transient response of water tables to periods of precipitation and evapotranspiration was modelled using the USGS Groundwater Model MODFLOW (McDonald & Harbaugh, 1988). The model was applied to a rectangular area centred on the reed-bed of 360 m$\times$150 m, representing over half the total area of the Reserve. Individual cells were 10 m^2 in area, giving model dimensions of 36 rows and 15 columns. The simplification of stratigraphy required for the model is illustrated in Fig. 4, which is derived from a full stratigraphy obtained by hand augering when installing the dipwells. Three hydrogeological layers were used to represent the gravel and peat deposits. The top layer (Layer One) forms an unconfined aquifer, composed mainly of an herbaceous peat but with a 40 m wide silty-clay area beside the river. Layer Two, consisting of a wood peat with a 20 m silt-clay band near the river, forms a partly or fully saturated unconfined layer depending on water table elevation; while Layer Three was a confined layer entirely composed of gravels. The top and bottom elevations of the three layers were defined from mapping of surface topography and the data obtained by hand augering.

Table 1 summarises the values used to describe internal variations in hydraulic conductivity within each layer of the model. Hydraulic conductivities were obtained from the experimental results described above, and were modified during calibration when model results were compared with field measurements. Specific yield was obtained by examining the water table response of individual dipwells to isolated precipitation events. Gravel transmissivity in Layer Three was derived from published values in Domenico & Schwartz

(1990) and was modified spatially to represent the variable thickness of gravels underlying the wetland. The river was represented by identifying the elevation of the river bed within Layer Three, and using the hydraulic conductivity of fine silt to characterise the permeability of a 40 cm layer of river bed sediments. All exchanges of water between the river and aquifer were assumed to occur vertically through the river bed as the low permeability silt/clay band adjacent to the river restricts lateral seepage. More groundwater flow is also to be expected through the river bed, rather than banks, due to the greater area of contact, and lower hydraulic conductivity of river bank sediments which have a higher silt/clay content (Bathurst, 1988; Sharp, 1977).

RESULTS

The ability of the groundwater model to replicate field water tables is described here for two events, firstly for a precipitation event in December 1991, and secondly a period of sustained evapotranspiration in May 1992. For the rainfall event, initial starting heads were obtained from field measurements on 12 December 1991, and the model response to a total of 7.3 mm of rainfall on 14 December was determined. The results are given in Fig. 5 for Layer Two, which corresponds with the depth interval sampled by the dipwells, with the field measurement points superimposed in plot A. The graphs indicate the stability of the flow net as precipitation infiltrates to the water table, and show the small increase in hydraulic head of c. 0.1 m following rainfall infiltration. The hydraulic gradient remained consistent at 0.35 m/50 m near the river, decreasing to below 0.15 m/50 m with distance. Generally the higher hydraulic gradient adjacent to the river corresponds closely to the location of silt/clay deposits and illustrates their importance in maintaining water table heights at greater distance from the river.

In the second simulation period initial starting heads were used for 15 May 1992, and the groundwater model used to derive water table heights for 17 May to show the result of two days of 7.3 mm evapotranspiration (Fig. 6). Again the hydraulic head configuration remains stable, although the fall in heads is small (c. 0.05 m) despite moderate potential evapotranspiration totals. During this period the hydraulic gradient is greater near the river (0.4–0.5 m/50 m), which demonstrates how the buffering effect of the river marginal sediments is important during a time of falling water tables.

Both model runs provide a realistic indication of the three-dimensional response of hydraulic heads which is supported by field measurements suggesting that the hydrogeology of the site has been represented adequately. The model simulations also recorded a slight vertical hydraulic gradient (c. 0.03 m/1 m), especially within 50 m of the river where the

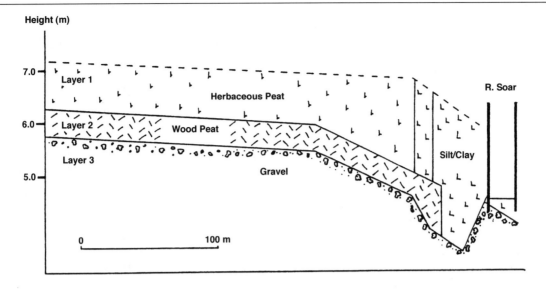

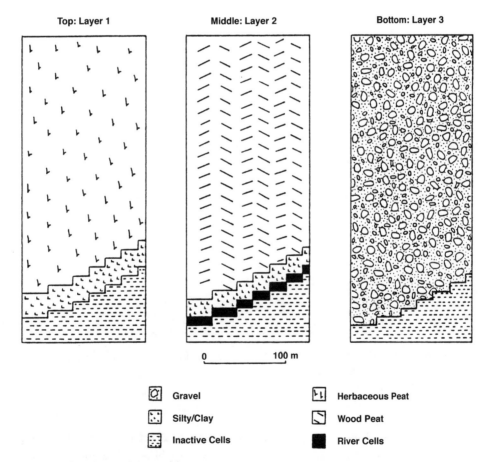

Fig. 4 Hydraulic representation of Narborough Bog, with cross-section (top) and views of Layers one, two and three.

Table 1. *Summary of hydraulic parameters used to characterise the field-site*

Layer one Unconfined aquifer	Bottom elevation (m)			6.5–>3.7
	Thickness (m)			1.2
	Hydraulic Conductivity (m/day)	Herbaceous Peat	Horizontal	1–>2.5
			Vertical	0.2
		Silt/clay	Horizontal	0.1–>0.01
			Vertical	0.01–>0.09
	Yield	Herbaceous Peat		0.18–>0.22
		Silt/clay		0.2
Layer Two Partly saturated/ fully saturated unconfined aquifer	Bottom elevation (m)			5.9–>3.7
	Thickness (m)			0.7
	Hydraulic Conductivity (m/day)	Wood Peat	Horizontal	0.6–>0.4
			Vertical	0.01
		Silt/clay	Horizontal	0.3–>0.01
			Vertical	0.05
	Yield	Wood Peat		0.12–>0.015
		Silt/clay		0.015
Layer Three confined aquifer	Thickness (m)			3–>4
	Transmissivity (m^2/day)	Gravels		10–>700
	Yield			0.05

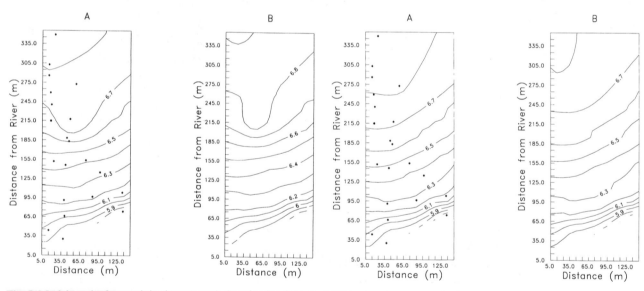

Fig. 5 Model results for precipitation event. A, Starting heads used for 12 December 1991 from field data with measurement points superimposed (*). B, Hydraulic head in Layer Two on 14 December 1991, following 7 mm precipitation.

Fig. 6 Results of evapotranspiration simulation. A, Starting heads on 15 May 1992 with measurement points superimposed (*). B, Hydraulic head in Layer Two on 17 May 1992 following two days of 7.3 mm evapotranspiration.

silt/clay provided an effective confining layer. These results were supported by the field observations of dipwells. The limitations of the hydrological model, however, are its inability to account for seasonal changes in the importance of processes such as precipitation and interception. The relationship between precipitation and water table recharge is determined by factors such as vegetation interception, unsaturated water storage which is proportional to the depth to water table, and preceding moisture conditions. In addition, the subdivision of peat deposits into herbaceous and wood peat classes, and the assumption of homogeneity within each stratigraphic layer, is too simplistic given the characteristic increase in hydraulic conductivity of peat with depth. Furthermore, the results illustrate the difficulty of

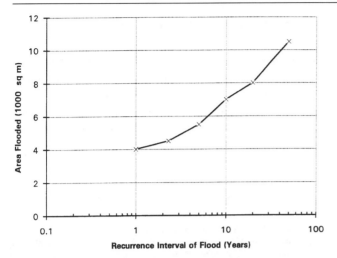

Fig. 7 The relationship of area of Narborough Bog flooded to the return period of flood events.

calibrating a groundwater model for evapotranspiration using estimates of potential evapotranspiration which are essentially radiation driven. Wetland evapotranspiration may be lower than the potential level due to the accumulation of dead vegetation (Gilman, 1994).

OVERBANK FLOODING

The contribution of overbank flooding to water storage of the bog was determined by examining the recurrence interval of floods on the River Soar. The return period for flooding of areas of the bog was estimated using a correlation between recorded stage levels at Narborough over the last three years and the longer record provided by an NRA gauging station upstream at Littlethorpe. The DEM of the bog surface was classed into levels corresponding with flood heights of known return period. This allowed the computation of the cumulative area flooded for return periods between 1 in 1 year and 1 in 50 years, shown in Fig. 7. The concave form of the area graph indicates that the lower bog is sensitive to any change in the frequency distribution of flooding and in particular the more infrequent floods. Even the 1 in 50 years event only inundates about 20% of the bog area. The reason for this is that the majority of the bog surface has accumulated to a level only reached by rare events. This is most probably due to the flow of the small tributary stream, which has been diverted as indicated in Fig. 2, and which could have produced a localised increase in water table height.

The amount of recharge contributed by overbank flooding can be estimated as a function of the infiltration rate and residence time of the flood. This assumes that depressional storage of water in surface pools and subsurface seepage are small. The former is supported by field observations; seepage is harder to estimate but is likely to be small in comparison

to surface infiltration. The rate of infiltration can in the first instance be held constant and the residence time calculated from the historical flow series. Assuming an average cumulative flood residence time inversely proportional to the flood height based upon 48 hours for the 1 in 1 year level to 0.5 hours at the 1 in 50 years height and using a high infiltration capacity of 0.5 cm hr^{-1} then the whole bog would only receive approximately 1200 m³ in an average year (based upon the flood frequency distribution). This represents the equivalent of approximately 230 mm of recharge over the whole bog surface, compared with mean annual precipitation of c. 640 mm. Calculations of evapotranspiration using the Penman formula suggest that potential evapotranspiration may total c. 750 mm, thus indicating the importance of water inflows from overbank flooding. The estimate of this contribution is, however, almost certainly an overestimate due to surface saturation, spatial variation in the infiltration rate (especially within the clay deposits) and return flow to the river.

DISCUSSION

The hydrology of the wetland system at Narborough Bog is dominated by the balance of precipitation and evapotranspiration. For the majority of the time Narborough Bog loses water through effluent seepage to the River Soar. Water flow through the reed-bed consists of a combination of slow infiltration through different hydrogeological layers in response to precipitation, and lateral flow towards the river. Lateral water flow occurs both through the gravel aquifer at the base of the peat deposits, and also near the surface through the phragmites peat. The stability of the system, and its ability to store water, depends upon the significant difference between vertical and horizontal hydraulic conductivities within the phragmites peat and wood peat layers and also upon the 'clogging layer' at the river bed which restricts water flows between the river and the gravel aquifer. Application of a transient groundwater model enables the effects of water storage within the wetland to be determined.

The recent ecological and hydrological changes at Narborough Bog mainly reflect the combination of the annual precipitation distribution, and the diversion of a small stream which flowed onto the reed-bed. Precipitation forms the main hydrological inflow of water to the bog; when a precipitation deficit is coupled with substantial water loss through evapotranspiration the falling water table enables encroachment of dryland plants. Currently overbank flooding is only of local significance, especially as the more frequent events flood only the silt/clay river marginal areas which have a low infiltration rate.

Although the contribution of overbank flows to the water

budget of the site at Narborough is limited, the river remains significant in providing a base level to which water tables in the bog adjust. The area is therefore sensitive to any changes in river regime, and in particular the level and duration of minimum river stage levels. The relationship between the river and local water tables will also vary as a result of management work on the river, in particular dredging of the channel and partial removal of the silt-clay clogging layer would increase the quantities of seepage for both influent and effluent flows. This demonstrates how, in the study of small British wetlands, it is necessary to consider the operation of processes within the entire drainage basin (Newson, 1992).

The interaction of these processes determines the management options which are available to preserve the ecology of the site. Essentially, areal recharge is required at times of precipitation deficit to increase the local height of the water table. The surface elevation of the site suggests that this would appear to be only possible through pumping of river water onto the reed-bed. However, this presents problems of water quality, and would also require an on-going monitoring programme to determine the timing and quantity of pumping necessary. Other strategies involving the control and trapping of overbank flood waters would be of only local significance, possibly prolonging the period of influent seepage of river water.

CONCLUSION

This paper describes the application of a transient groundwater model to a wetland site, and demonstrates that over short time-periods the hydrological response to precipitation and evapotranspiration can be predicated. This is important due to the quantity of internal water storage which limits the possibility of producing a full water balance.

The results of this study indicate the vulnerability of Narborough Bog to changes in water flux. Although it originally developed as a fluvial backswamp, local saturation was maintained by the flow of a small stream through the site. Recent diversion of the stream and some management works on the River Soar are responsible for the hydrological and ecological changes to the site so that Narborough Bog is currently sensitive to any variation in the balance between the water inflows of precipitation and flooding, and losses of evapotranspiration and effluent seepage to the river. Use of a calibrated model should enable the hydrological implication of any changes in water flux to be evaluated.

ACKNOWLEDGEMENTS

We are grateful to the Leicestershire and Rutland Trust for Nature Conservation for permission to work at Narborough Bog, and to English Nature for additional financial assistance. One of us (CB) was supported by a NERC studentship while completing this study. Figure 2 was drawn by Mrs K. Moore.

REFERENCES

Bathurst, J.C. (1988). Flow processes and data provision for channel flow model. In *Modelling Geomorphological Systems* M.G. Anderson (ed), pp. 127–152. Chichester: Wiley.

Carter, V. (1986). An overview of the hydrologic concerns related to wetlands in the United States. *Canadian Journal of Botany*, **64**, 364–374.

Domenico, P.A. & Schwartz, F.W. (1990). *Physical and chemical hydrogeology*. John Wiley: New York, 824 pp.

Everett, M.J. (1989). Reedbeds – a scarce habitat. *RSPB Conservation Review*, **3**, 14–19.

Gilman, K. (1994). *Hydrology and wetland conservation*. Chichester: Wiley, 101 pp.

Gilvear, D.J., Andrews, R. Tellam, J.H., Lloyd, J.W. & Lerner, D.N. (1993). Quantification of the water balance and hydrogeological processes in the vicinity of a small groundwater-fed wetland, East Anglia, UK. *Journal of Hydrology*, **144**, 311–334.

Godwin, H. & Bharucha, F.R. (1932). Studies in the ecology of Wicken Fen -II The Fen water table and its control of plant communities. *Journal of Ecology*, **20**, 157–191.

Hammer, D.E. & Kadlec, R.H. (1986). A model for wetland surface water dynamics. *Water Resources Research*, **22**, 13, 1951–1958.

Hillel, D.I. & Gardner, W.R. (1969). Steady infiltration into crust-topped profiles. *Soil Science*, **107**, 137–142.

Koerselman, W. (1989). Groundwater and surface water hydrology of a small groundwater-fed fen. *Wetlands Ecology and Management*, **1**, 31–43.

Luthin, J.N. & Kirkham, D. (1949). A piezometer method for measuring permeability of soil in situ below a water table. *Soil Science*, **68**, 349–358.

McDonald, M.G. & Harbaugh, A.W. (1988). A modular three-dimensional ground-water flow model. *USGS Technical Water Resource Investigations Book* 6, Chapter A1.

Mitsch, W.J. & Gosselink, J.G. (1986). *Wetlands*. New York: Van Nostrand Reinhold.

Newson, M.G. (1992). Conservation management of peatlands and the drainage threat: hydrology, politics and the ecologist in the U.K. In *Peatland ecosystems and man: an impact assessment*, ed. O.M. Bragg, P.D. Hulme, H.A.P. Ingram & R.A. Robertson. Dept. of Biological Sciences, Dundee University, pp. 94–103.

Sharp, J.M. (1977). Limitations of bank storage model assumptions. *Journal of Hydrology*, **35**, 1/2, 31–47.

Wade, A.E. (1919). The flora of the Aylestone and Narborough Bogs (an ecological study from an original survey). *Transaction of the Leicester Literary and Philosophical Society*, **20**, 20–46.

Wheeler, B.D. (1988). Species richness, species rarity and conservation evaluation of rich-fen vegetation in lowland England and Wales. *Journal of Applied Ecology*, **25**, 331–353.

14 Contribution to the groundwater hydrology of the Amboseli ecosystem, Kenya

A.M.J. MEIJERINK & W. VAN WIJNGAARDEN

International Institute for Aerospace Survey, and Earth Sciences (ITC), Enschede 7500 AA, The Netherlands

ABSTRACT The Amboseli ecosystem consists of the basement plains, the lacustrine saline plains with fresh water swamps and the volcanic slopes of the Kilimanjaro. The area is well known for its large and varied population of wild herbivores, supporting a large tourist industry. The two major fresh spring zones sustaining some 20 km^2 of swamps and an additional 16 km^2 of wet areas chiefly covered by grass, belong to the relatively shallow part of a regional groundwater flow system in the volcanic complex. A 7.5 m rise of the groundwater table since the 1960s and the recent expansion of one of the swamps has caused concern. The rise cannot be explained by the available rainfall data and may be related to more frequent runoff, because of overgrazing, in the catchment feeding episodically the seasonal Lake Amboseli. In addition, tectonic movements may have increased the outflow of the the the deeper diffuse part of the flow system.

The dynamics of the swamps have been studied. Available data suggest that no major changes are likely to affect the swamps in the near future. A good proportion of the annual recharge of the springs, about 14 % of the estimated rainfall, takes place in the Tanzanian part.

INTRODUCTION

The Amboseli ecosystem can be characterized as a semi-arid savanna environment, which shows considerable spatial and temporal variation in resources; climate, soils, vegetation and hydrology. This has an important bearing on the structure and functioning of the ecosystem and has important consequences for the management. The area is well known for its large and varied population of wild and domestic herbivores, upon which a large tourist industry is based. Two major fresh water swamps form an essential part of the entire ecosystem. A rising groundwater table and an increase in one of the swamps, have caused concern for the future management of the park. This paper aims at contributing to the understanding of the hydrology of the Amboseli region.

The ecosystem contains three major landscapes (Fig.1), the basement plains, the lacustrine plains and swamps, and the volcanic footslopes (Touber *et al.*, 1983).

Basement plains

The northern part of the ecosystem consists of gently undulating plains on various gneisses of Precambrian age. The soils developed in these gneisses are in general deep, well drained and acid. The vegetation consists of bushlands or bushed grasslands dominated by *Acacia* spp. and *Commiphora* spp.

The grasslayer is dominated by *Digitaria macroblephora* and *Chloris roxburghiana*. This area is devoid of any permanent water sources. Only during and shortly after the rainy season, the temporary rivers flow and shallow depression (waterholes) are filled with water.

Lacustrine plains and swamps

The central part of the ecosystem, where the Amboseli National Park is situated, consists of a series of lacustrine deposits with a very flat topography. The lowest part forms Lake Amboseli, which is normally dry, except in wet seasons or years with abundant rainfall. The soils of the lake bed are

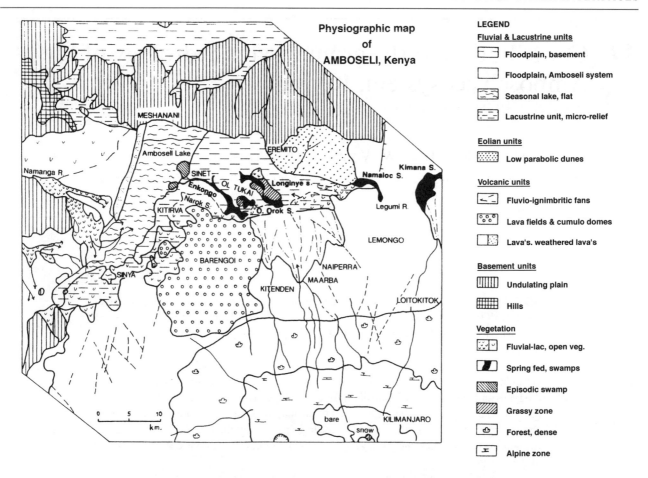

Fig. 1 Physiographic map of the Amboseli region.

poorly drained, clayey and strongly saline. In general this area is devoid of any vegetation.

The slightly higher parts of the lacustrine plains have moderately well to poorly drained alkaline soils, which are often saline at shallow depth. They are covered by a grass-land vegetation dominated by *Sporobolus* spp.

In this area two types of swamps can be distinguished:

(a) spring-fed permanent swamps (Enkongo Narok, Longinye, Namalok) covered by a dense vegetation dominated by *Cyperus* spp., and,

(b) a swamp (Oltukai Orok), which forms a mosaic of semi-permanent small swamps and *Acacia xanthophloea* woodlands.

Drinking water availability for large herbivores is not a problem in this part of the ecosystem because the swamps are perennial.

Volcanic footslopes

This southern part of the ecosystem is situated on the lower slopes of mount Kilimanjaro. It consists of lava flows and partly reworked pyroclastic deposits of mainly Pleistocene

age. In these deposits, deep, well drained and neutral soils have developed, which are however very stony/bouldery. The vegetation ranges from bushed grassland to wooded bush-land and is dominated by various *Acacia* spp. The grasslayer is dominated by the grasses *Pennisetum stramineum* and *Cenchrus ciliaris*. Surface runoff is not strong because of the permeability, but does occur during wet spells.

RAINFALL AND EVAPORATION

The Amboseli area has a semi-arid climate with two rainy periods. The graph of Fig.2 shows the long term record of Makindu (70 km to the north-east of Ol Tukai), Loitokitok (40 km to the east), Namanga (80 km to the west) and two short term records of stations within the Amboseli area itself, Ol Tukai in the centre of the park and Sinya Mine in the west, see the map of Fig. 1.

The data show the typical variation of the yearly rainfall of this type of climate and there are no clear trends in any of the data. The park area itself lies in the rainshadow of the large volcano and has a mean rainfall of about 300 mm. The surrounding areas receive more rainfall, in particular the forest zone at an altitude of 1600 to 2200 m (see map), where the rainfall is estimated at 800 mm annually. At higher alti-

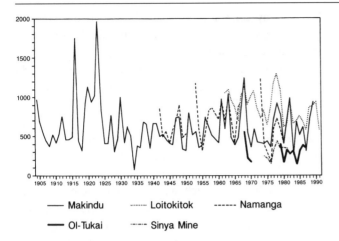

Fig. 2 Annual rainfall of two stations within the park and of three stations in the surroundings with longer records.

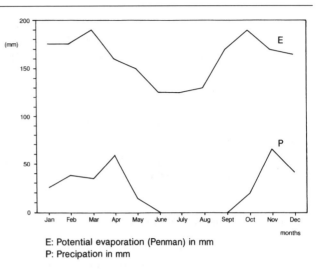

E: Potential evaporation (Penman) in mm
P: Precipitation in mm

Fig. 3 Average (10 years) annual rainfall and potential evapotranspiration of Ol Tukai, Amboseli Park.

tudes the total amount of water available for recharge is extremely difficult to estimate because of fog-drip, which occurs, but of unknown magnitude.

The average annual potential evapotranspiration (Penman) of the park area can be estimated at some 1925 mm using the maps prepared for Kenya by Kalders (1988). The average monthly potential (Penman) evaporation (after Kalders) for Ol Tukai which is located in the centre of the park, at an altitude of 1142 m, is higher than average monthly rainfall (see Fig. 3). The water deficit is reflected by the high salinity of most of the soils in the lacustrine zone.

TEMPORAL ECOSYSTEM DYNAMICS

The area shows both seasonal and year to year variation in resources.

SHORT TERM DYNAMICS

The seasonality of the climate has an important bearing on the short term dynamics of the ecosystem. Firstly, the growing season of the rainwater dependent vegetation is very short, and secondly, surface water is only available during or shortly after the rainy season, except for the permanent swamps. This seasonality of vegetation and water resources has an important effect on the possibilities for large herbivores to use the area (Western, 1973; 1975). The general pattern is that they disperse in the rainy season especially to the basement plains area and to a lesser extent to the volcanic footslopes. In the dry season they concentrate in the lacustrine plains and make intensive use of the vegetation and water of the swamps.

YEARLY DYNAMICS

The typical yearly variation of semi-arid climates affects ecosystems in many ways. In years with normal or above

normal rainfall, vegetation and water resources are abundant, and as a result, large herbivore populations grow strongly, especially livestock. The irregularly occurring dry years, such as 1975 and 1984, may then cause large scale starvation.

There are, however, also long term dynamics in the Amboseli ecosystem that are not clearly related to the variations in yearly rainfall. The generally observed rise in groundwater levels appears to have been responsible for the large scale death of large *Acacia* trees in the 1970s, through an increase of the salt concentration in the soils (Western & van Praet, 1973). However, due to the microrelief in the area, this increase of the groundwater level has also created new sites for the establishment of new generations of *Acacia xanthophloa* trees. The relative high density of elephants (possibly also giraffes and other browsers), has suppressed this regeneration in many places. That this regeneration potential exists, is amply illustrated within protected sites at the research camps and the lodge areas.

As discussed below, the swamp areas change in size from year to year and recent expansion of the downstream part of the Longinye swamp has caused concern. In order to properly understand the long term dynamics, a better insight is needed in the hydrology of the ecosystem.

HYDROLOGY

Occurrences of groundwater in the basement plains are limited to faults, fractures and small parts of the weathered zones and also to the bottom layers of the wide alluvial valleys which are recharged by natural floodspreading.

The basement is an aquiclude for the large regional groundwater flow system of the Kilimanjaro which extends into the lacustrine zone. The depth of the basement below the lacustrine zone and the volcanic zone is poorly known. Lavas and lacustrine deposits have been encountered in drillings, but no basement (Williams, 1972).

The basement configuration is difficult to estimate because of Quaternary tectonics, of which there is evidence. This is not discussed here, except for pointing out that the lacustrine plain can be regarded as a depressional zone due to the loading of the large Kilimanjaro volcano on the basement, whereby the rims are affected by faulting and flexuring.

Regional rise of the groundwater table

All available evidence indicates that within the lacustrine zone the groundwater table has risen during the last 30 years or so.

During a survey by Campbell (1957) the general water table was found to be approximately 5 m below ground level in Lake Amboseli and the Sinya mines in the southwest and 6 to 11 m below the eastern lacustrine zone. The average depth for 11 locations works out to be 8.36 m. Lahi (1967) and Bargman (in Lahi's report), concluded that the elevation in the groundwater table in the Sinya area was attributable to a post-1960 rainfall increase. Western (1973) measured in the period 1968–70 a number of water levels and concluded that the rise had continued since the early 1960s. We found in the Sinya area, Kenyan side, the water level in July 1993, at 0.4 to 0.6 m b.g.l., a further rise of approximately 1.5 to 2 m since Western's observations. The water level in the wells at Ol Tukai rose from 11 m in 1957 to 5 and 4 m b.g.l. in the period 1964–1970 (Western, 1973) and they were at 1.2 to 1.4 m b.g.l. in 1992.

The groundwater levels have thus been rising by approximately 7.5 m since the late 1950s. If an effective porosity of the saline clayey lacustrine deposits is assumed to be 10 %, this groundwater rise over 30 years is equivalent to 25 mm year^{-1} on average. However, the record of the nearest stations with long term rainfall data, Loitokitok, Namanga and Makindu, do not show a corresponding trend of increased rainfall, see Fig. 2.

The July 1992 groundwater level is at many places within the capillary rise zone, and thus a possible further rise will be slowed down or balanced by the high evaporation. Increased salinization can be foreseen, which will affect the lacustrine/fluvial area in the Ol Tukai–Namalok area and will lead to a change in vegetation. What are the causes for this rise and has the rise affected the springs ?

Regional flow system

Water from a regional groundwater flow system within the volcano reaches the lacustrine zone in two ways, by diffuse

flow and by concentrated flow resulting in the powerful springs of Enkongo Narok, Longinye (also called Ol Kenya) and Namalok/Legumi.

The total groundwater catchment area upstream of the Enkongo Narok and Longinye springs, up till the craters, is approximately 750 km^2. The dimensions of the permanent wet areas fed by the springs was 36 km^2 in 1988. By assuming, for those areas, an annual actual evapotranspiration rate of 0.9 of the potential one, a maximum recharge rate of 80 mm is required to maintain such dimensions. The mean annual rainfall of the catchment is estimated to be some 550 mm and thus the recharge is some 14 % of the annual rainfall. This recharge figure is fairly high, but not impossible considering the permeable nature of the area and the low temperatures at the high altitudes.

Two-dimensional groundwaterflow modelling was attempted by using the static water levels and the permeability determinations of the boreholes south of Amboseli (Kitenden, Maarba and Naiperra), see Fig. 1 for the locations. With the many unknowns, a more sophisticated groundwater modelling approach was considered out of place. The model results show that the groundwater discharge area starts roughly at 1450 m altitude extending down into the lacustrine zone. However, apart from a very small spring of less than 2 l s^{-1} at Lemongo, the upper part of the discharge zone in the volcanics is devoid of water.

Thus, it seems that the upwelling water within the lower part of the flow system concentrates in natural subterranean collectors, probably lava tunnels or fracture conduits and emerges at the springs. Some of the major lineaments, shown in Fig.1, could be associated with such conduits. The electrical conductivity (EC) of the Engkoro Narok and Longinye springs is low, 0.15 and 0.16 mS cm^{-1} respectively in July 1992. One would expect higher values if the water had travelled longer distances at greater depths.

Along the south–north section, the groundwater flow system extends well into the lacustrine deposits, but with a low quantity of flow per unit section. The conductivity of the waters in the wells at Ol Tukai ranges between 0.92 and 1.04 mS cm^{-1}.

The Amboseli lake area is partly fed by diffuse groundwater flow from the western Kilimanjaro complex, and by surface runoff from the Namanga catchment. The groundwater levels below the lake are much higher than those to the west of the lake, indicating subsurface flow of – saline – water from the lake into the alluvial deposits of the Namanga. Thus ephemeral surface water and continuous flow of groundwater move in opposite directions in this part of the region.

The springs and swamps

The swamp geography as shown in Fig. 4 is based on classification of spectral data of the Thematic Mapper scene

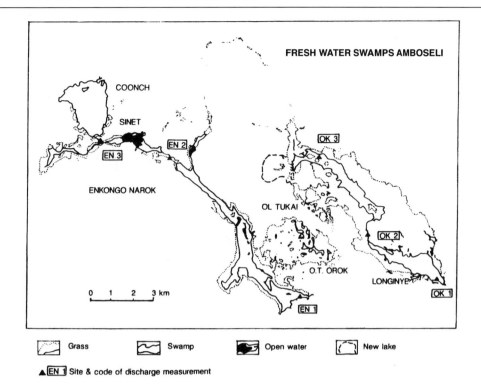

Fig. 4 Map of the swamps within Amboseli National Park, situation of October 1988, except for more recent new lake.

of October first, 1988. The typical characteristics of the fresh water swamps are shown in the schematic section of Fig. 5A. At the peripheries of the swamp, dense mats are formed by sedges (*Cyperus merkii* and *C. laevatigus*), creepers (*Vigra* sp.) and *Cynodon dactylon*. In the more central parts the sedges *Cyperus immensus* and *C. papyrus* are found (Western, 1973).

Water from springs at the upper part of the swamps flows for considerable distances because of the impermeable nature of the compact clays at the bottom, maintaining a swamp vegetation till the evapotranspiration losses outweigh the upstream supply.

To our knowledge, no discharge measurements have been carried out, except for one estimate by Mifflin (1991) at a site of doubtful location. Our discharge measurements took place in July 1992, and are based on current metering with a wading rod, with a close set of verticals. The locations are shown in Fig. 4 and the discharges are listed in Table 1. Within the swamp, below the dense network of hanging roots and stems of the aquatic vegetation, the flow velocity is very low or the water is stagnating. Current metering in the swamps is time consuming because of frequent disturbance by elephants and buffalos and because of the presence of submerged hippo wallowing holes.

The marshy area at Ol Tukai Orok is characterized by many very small marshes with stagnating water and by a tree vegetation dominated by palm trees, *Phoenix reclinata*, see Fig. 5B. Detailed spatial observations of the EC of the waters, which was found to range from 0.45 to over 2.0 mS cm^{-1}, suggest that upwelling groundwater through pipes of higher permeability, possibly old root tunnels, feeds these marshes.

Dynamics of the springs and swamps

According to verbal information from people using the springs, it seems that the springs have a fairly constant discharge throughout the year. Another approximate source of information on the fluctuations of the discharges, are the interpretations of aerial photos, at intervals of a few years since 1950 by Western (unpublished, included in the report by Mifflin, 1991). To that, we have added interpretation of satellite imagery, as indicated in the graph of Fig. 6. The surface areas of the swamps, consisting of water and vegetation growing in water (not the wet to moist grasslands adjoining the swamps proper), have been measured and the results are shown in the graph. Unfortunately, the data available do not allow an indication of seasonal effects. It is possible that the estimates based on the 1974 and 1979 satellite imagery are too high, because it is difficult to separate the swamp from the wet areas. Increased flows of the Enkongo Narok springresult mainly in an extension of the swamp at Coonch, where the water spreads onto the flat seasonal Amboseli lake bed, see Fig. 1. The Longinye swamp has changed appreciably the last few years. It is not known why the upstream part of the swamp has become smaller in size with important reduction of the papyrus. Possibly, vegetal flow obstructions may have broken through, thus reducing

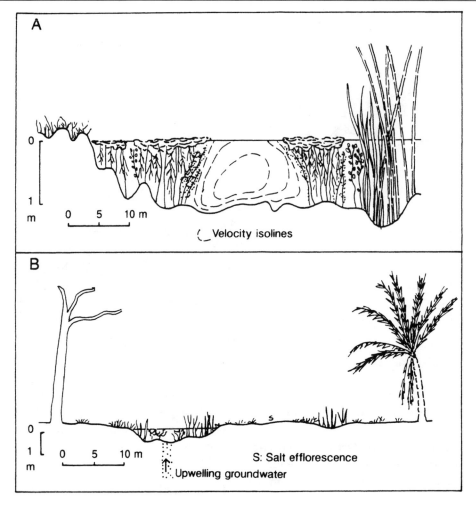

Fig. 5 Schematic section through (A) Longinye/Enkongo Narok swamps fed by springs and (B) Ol Tukai Orok swamp fed by upwelling groundwater.

Table 1. *Discharges and EC of the fresh water swamps, Amboseli National Park*

Spring	Discharge point (see map)	Elev. (m)	Discharge (l/s)	Elec. cond. (S/cm)
Legumi	L1 (upstream)	1220	44	0.24
Namalok swamp	N1 (upstream)	1160	23	0.40
	N2 (middle)	1150	82	0.40
Longinye swamp	OK1 (upstream)	1145	329	0.16
	OK2 (middle)	1140	623	0.18
	OK3 (culverts, near end)	1136	285	0.23
Enkongo Narok swamp	EN1 (upstream)	1140	381	0.15
	EN2 (middle)	1125	410	0.21
	EN3 (Sinet bridge)	1123	217	0.33

the size of the upstream part of the swamp by more rapid drainage. The resulting reduced water loss by evapotranspiration could well account for an increased discharge downstream. Just west of Ol Tukai a new small lake was formed, adding an attraction to the park because of the birds.

The Ol Tukai Orok marshes have not changed much, which may be attributed to more constant groundwater discharge of the lower part of the groundwater flow system.

It is difficult to relate the swamp area to the rainfall from the data available and presented in Figs. 2 and 6. The short term variations in the swamp extension are not contradictory to the supposition that the springs are fed by conduits which draw their water from the groundwater flow system, as explained above. Furthermore, it seems that the rise of the groundwater table (see below) is not reflected by a systematic increase of the swamp areas.

DISCUSSION

The interpretation of the hydrologic data presented so far, raises a few questions of importance for the management of

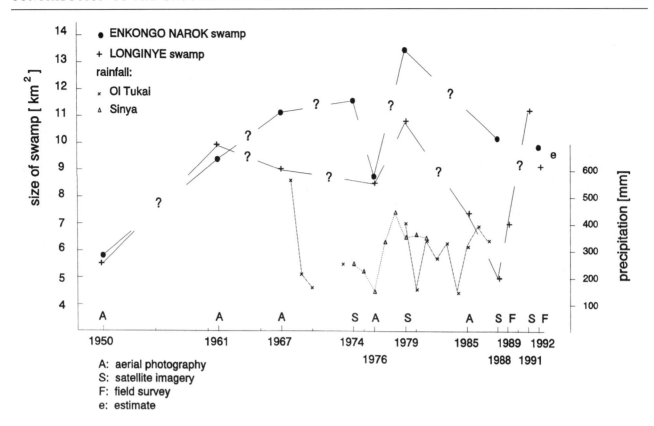

Fig. 6 Size of swamps as interpreted from aerospace imagery at various times. Rainfall of the two stations within Amboseli shown.

A: aerial photography
S: satellite imagery
F: field survey
e: estimate

the park, which are not easily answered. The main questions are as follows.

(a) What will be the future behaviour of the Enkongo Narok and Longinye springs and swamps, which are the basis of the Amboseli park?

(b) What has caused the groundwater level rise and will the levels remain high?

The spring flows are governed by the rainfall on the slopes of the Kilimanjaro and little change in landuse has taken place in this area during the last decades, nor are there plans in that direction. Thus, the same pattern of variation as observed since 1950 may be expected. It seems that appreciable dampening of rainfall variations takes place, even within the upper part of the groundwater flow system and possibly also within the swamps themselves. However, a rainfall excess during wet spells and additional inflow of surface water from the volcano, totalling maximally some 100 to 200 mm, leads to saturation of the entire lacustrine zone, causing widespread inundation, also of the Ol Tukai–Namalok area. This is what happened during the last wet season (spring 1993) and it is in the line of expectation that such may occur more frequently.

With regard to the second question, it is difficult to attribute the groundwater rise of the last 30 years to only the rain-

fall of the lacustrine zones and of the groundwater recharge zone. There may be two additional causes.

(i) Due to overgrazing which has taken place in large parts of the Namanga catchment, more runoff occurring more frequently can be expected to have reached the Amboseli lake. Longer periods of inudation with larger depths will cause a groundwater level rise. Evidence of the overgrazing has been established by Toxopeus et al. (1994, & pers. comm.), who surveyed the vegetation in a large part of the Namanga catchment during 1992/93 and compared his results with those of Touber (1983), made during the late 1970s.

(ii) A period of slightly increased subsidence by loading of the Kilimanjaro volcano could have contributed to the rise in the water table. With a constant volume of the groundwater body, a lowering at the central part by subsidence must be counterbalanced by a rise of the level along the rims. The evidence of the neo-tectonism listed above does not exclude such movements *per se*.

The dynamics of the hydrology affects the touristic infrastructure of the Amboseli Park, but probably not the habitat.

ACKNOWLEDGEMENTS

Mr Charles Irungu has contributed to various parts of the study, particularly to the groundwater modelling. The study was carried out in the framework of theUNESCO–ITC pro-gramme 'Geo–information for sound environmental man-agement' (also IHP M 3.5).

REFERENCES

Campbell, G.D.M. (1957) *Preliminary observations on surface and groundwater conditions – north west slopes of Kilimajaro and Amboseli National reserve*. Report Ministry of Works, Kenya.

Kalders, J. (1988) *Evapotranspiration in Kenya*. Ministry of Water Dev., Kenya, Irrigation section, Nairobi.

Lahi, A.V. (1967). *Hydrology of the Sinya Meerschaum Mine*. Ind. Stud. & Dev. Centre, Dar-es-Salaam, unpubl.

Mifflin, M.D. (1991). *Amboseli National Park hydrology review*. Report, Kenya Wildlife Service.

Touber, L., van der Pouw, B.J. & van Engelen, V.W. (1983). *Soils and vegetation of the Amboseli* – Kibwezi area. Reconnaissance Soil Survey Rep. R6, Kenya Soil Survey, Nairobi.

Toxopeus, A.G., Bakker, X. & Kariuki, A. (1994). *Interactive spatial modelling handbook, User manual Amboseli Case*, Vol.3. Int. Inst. Aerospace Survey and Earth Science/Kenya Wildlife Service/UNESCO.

Western, D. (1973) *The structure, dynamics and changes of the Amboseli ecosystem*. Ph. D. thesis, Univ. of Nairobi.

Western, D. (1975) Water availability and its influence on the structure and dynamics of a savannah large mammal community. *East Afr. Wildlife*, **13**, 265–286.

Western, D. & van Praet, C. (1973). Cyclical changes in the habitat and climate of an East African ecosystem. *Nature*, **241**, 104–106.

Williams, L.A.J. (1972). *Geology of the Amboseli area*. Degree sheet 59 SW Quarter, Geol. Survey of Kenya, Rep.90, Ministry of Natural Resources.

15 The role of hydrology in defining a groundwater ecosystem

J. E. DREHER*, P. POSPISIL** & D. L. DANIELOPOL***

* Breitenfurterstrasse 458, A-1236 Vienna, Austria

** Institute of Zoology, University of Vienna, Althanstrasse 14, A-1090 Vienna, Austria

*** Limnological Institute, Austrian Academy of Sciences, A-5310 Mondsee, Austria

ABSTRACT   Based on two years of observations in a small aquifer in the Danube old arm system (near Vienna), the dynamics of a groundwater ecosystem is analyzed. The results show that differences in the oxygen concentration and the distribution pattern of the meiofauna community are a consequence of seasonal variations of the water level in the old arms. The water level as a boundary condition and also temperature are two important factors that govern the interaction processes between groundwater and surface water. In periods of flood events these interaction processes can achieve greater proportions, giving rise to temporary alterations in the ecosystem. Low water periods inhibit partially the infiltration of surface water into the aquifer.

INTRODUCTION

The delineation of a groundwater (GW) ecosystem within an unconsolidated geological formation is of paramount interest for both basic and applied research (Stanford & Ward, 1992). To perceive the subsurface environment, well defined systems are required. An ecological system consists of both abiotic and biotic components that exchange closely information and/or matter between them (Jordan, 1981). The ecologists dealing with the study of aquatic subsurface environments in porous media had persistent difficulties in defining ecological systems, which should represent more than simple definitions. Danielopol (1980) suggested that beside biological criteria ecologists should also use hydrological ones in order to define an ecological system. Such an approach was pioneered in the 1960s and 1970s by R. Rouch and A. Mangin when they studied the Baget karst system in southern France (see review in Mangin, 1976; Rouch, 1986). Danielopol (1989) showed that for the study of a porous system one of the most appropriate ecological unit should be a well defined aquifer. As hydrogeologists developed appropriate methods to delineate such a system (Castany, 1985), cooperation between ecologists and hydrologists becomes a necessity (Danielopol, 1980; Vanek, 1987; Nachtnebel & Kovar, 1991; Valett et al., 1993). This paper presents an attempt to define a GW ecosystem using hydrological criteria. The aquifer selected is situated in a wetland area close to Vienna within the alluvial sediments of an old arm of the Danube at Lobau (Fig. 1). A preliminary report about the current project was presented in Danielopol et al. (1992). Danielopol & Marmonier (1992) discussed its relevance within the context of present day ecological research in Europe.

MATERIAL AND METHODS

Site description

The former River Danube in the Vienna region was comprised of a closely knit network of branches which covered the broad valley that stretched from the extensive Marchfeld area up to the Austrian/Slovakian boarder (Fig. 1). In turn, geomorphologically the spread of the river gave rise to a typical terraced landscape. The river regulation work carried out since the end of the last century has led to the construction of a number of wiers and dams, and consequently has changed the typical structure of the riverine backwaters. The remnants of the Danube old arms in the region today form a system of GW dominated water bodies which are strongly influenced by the Danube flood regime (Danielopol, 1983;

119

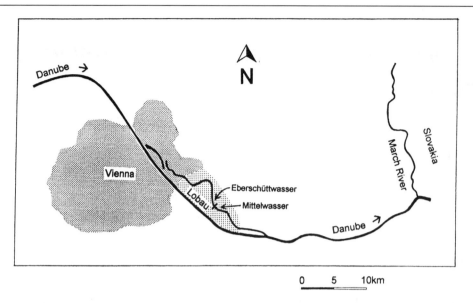

Fig. 1 The location of the Eberschüttwasser (ESW) and Mittelwasser (MW) in the Lobau wetland (48° 9' N, 16° 34' E).

Dreher *et al.*, 1985). These old arms are connected with inflow and outflow channels which are sometimes active.

As a result of alluvial activity in the Danube, thick quaternary gravel layers were formed along its course. However, this activity has led to the formation of highly heterogeneous sediment layers with irregular depositions of an assortment of sediments varying from fine sand to large gravel stones. Fig. 3 shows one characteristic geological profile in the research site reconstructed from a videoscope recording (Niederreiter & Danielopol, 1991). The gravel depositions found at the study site are more than 18 m deep. Changes in the flow regime of the old arm system have promoted sedimentation processes that have resulted in accumulation of fine sediments and gradual clogging of the old river bed. In the water body a sediment layer up to 70 cm thick is found which extends from the reed zone up to the center.

The investigations were carried out in one of these old arms (Eberschüttwasser). The study area is situated close to one of the dams between the Eberschüttwasser (ESW) and the Mittelwasser (MW) (Fig. 2), which is also a relic arm of the Danube. The water level differences between both water bodies (ESW about 30 cm higher than the MW) give rise to an infiltration of surface water, influencing locally the GW stream of the regional Marchfeld aquifer.

Experimental

A battery of 30 piezometers (18 mini-piezometers made of 2.5 cm I.D metal pipes, 10 Plexiglas piezometers of 5 cm I.D, 2 metal wells with 5 cm I.D) was installed over the *c.* 900 m² study area (Fig. 2). The mini-piezometers consisted of 2.5 to 3.5 m long tubes with a 10 cm screen placed at the lower end

(Wells 1 to 26 in Fig. 2). Six of these metal pipes were installed close together (50 cm distance) at different depths to monitor the respective layers, and one Plexiglas well (5 cm I.D) was installed in the reed zone of ESW. The other nine Plexiglas piezometers (D3 to D17) had perforations along the total length of the tubes which enabled multilevel sampling up to 14 m depth. One existing 5 cm screened metal well (T3) was also integrated to the observation network. In all the observation wells the water level as well as temperature profiles were measured at regular weekly intervals. A limnigraph installed at S10 well provided a continuous recording of GW-level and temperature. The water level of the surface waters ESW and MW were measured by two other limnigraphs (PSE and PSM, Fig. 2).

Hydraulic conductivities were measured in all mini-piezometers by the permeability-infiltration test according to constant infiltration rate described by Schuller (1973) and Rouch (1992). However, the determination of the conductivity in the Plexiglas wells was inaccurate due to the flow resistance built across the wall perforations (wide spaces between the slots of the screen). Seepage rate was measured with a simple seepage meter according to Lee (1977) but was restricted to the small shallow shore region. Seepage rate was determined as the change of volume per unit of area and time.

RESULTS AND DISCUSSION

Hydrological outline

The changes of the water level in the ESW as well as of the local aquifer (GW table in connection with ESW) depend mainly on the water level fluctuations of the regional aquifer (Marchfeld). Depending on the Danube water regime, the latter follows the same pattern but with a reduced amplitude

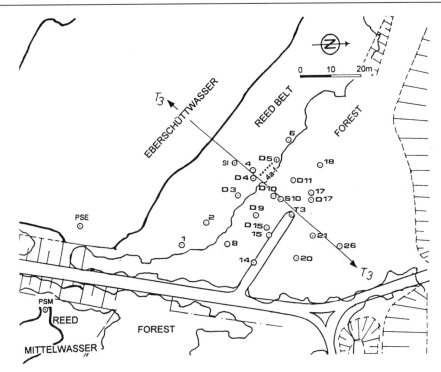

Fig. 2 Experimental area and piezometer network. Mini-piezometers: 1–26; water work well: T3; Plexiglas screened multi-level wells: D3–D17; limnigraphs: S10 (groundwater), PSE and PSM (surface water). The wells referred to in Figs. 3 and 4 are indicated by full circles.

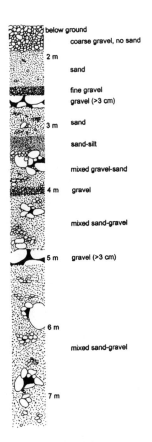

Fig. 3 Cross-section of borehole D15.

and a time lag. The influence of precipitation on the GW table can be neglected (the mean annual precipitation amounts to 560 mm).

Due to the relative narrow distance between the old arm and the main river, the flood events play an important role. During such occurrences the wetland fringe is completely flooded from the downstream water (the MW). The dike between ESW and MW acts as a barrier by reducing the surface water inflow into the ESW. As a result, for a short period of time, the groundwater/surface water system is inverted and exfiltrating from the groundwater into the old arm. When the flood recedes the water level in the ESW becomes higher than the near GW, and the seepage from ESW into GW takes place. Fig. 4 shows the transect T3–T3 (detail in Fig. 2) with the observed maximum flood levels during the study period.

Groundwater maps

Groundwater maps have been constructed based on the measured hydraulic heads of the GW piezometers (Fig. 5). Fig. 5a shows the groundwater map for a representative water level corresponding to a water level below the shore line of the ESW. The main infiltration region can be recognized by the hydraulic gradient along this border. A higher water level at 40 cm is depicted in Fig. 5b (which covers the total reed area) and shows a steeper slope in the hydraulic gradients of the GW table.

Fig. 5c shows the GW map corresponding to an extreme low water level and Fig. 5d represents the isolines to an extreme high water level during the beginning of a flood

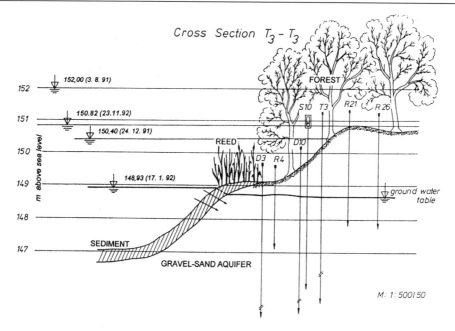

Fig. 4 Cross-section showing flood levels of 1991 and 1992.

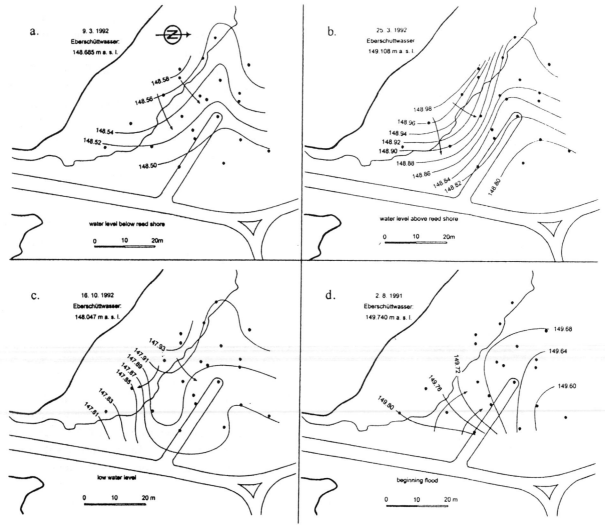

Fig. 5 Groundwater maps showing the GW-flow at different surface water levels (EW).

Table 1. *Estimation of infiltration area from seepage rate and GW flow at two different water tables*

Indicated GW-level	Seepage rate (μm s^{-1})	Average aquifer permeability	Hydraulic gradient (m s^{-1})	GW-flow rate (μms^{-1})	Estimated infiltration area factor
Fig. 5a	1.1	0.8×10^{-3}	0.002	1.6	1
Fig. 5b	ND*	0.8×10^{-3}	0.006	4.8	3

Note:
* ND: not determined

event. In the first case at low water level there is reduced infiltration and the local groundwater flow follows the flow of the regional aquifer (Fig. 5, lower left). In the second case, during the start of the flooding, GW flow changes the direction. This is indicated by the heads in the piezometers 1 and 2 which are higher than the surface water level (Fig. 5d). The hydraulic gradient in the aquifer is comparatively higher.

Fig. 5 shows that infiltration is restricted to the near-shore area (arrows). The GW flow in the north corresponds to the flow of the regional groundwater. A better interpretation of the GW flow in such a system of local and regional aquifer from the measured data is difficult. Winter (1983) and Winter & Pfannkuch (1984–85) discuss similar situations resulting from a theoretical lake-groundwater model. Consequently the quantitative evaluation of GW inflow described below has an estimative character.

Hydraulic conductivities and infiltration rates

The permeabilities of the local aquifer measured in the mini-piezometers show a very irregular distribution of conductivities ranging from 3.6×10^{-6} m.s^{-1} to greater than 0.05 m.s^{-1}. The heterogeneous structure of the profiles (e.g., borehole log, Fig. 3) is reflected in the variations of the measured permeabilities. Nevertheless it can be remarked that low permeability was only found at 3–3.5 m depth in the sandy-silt layer (Fig. 3). Higher conductivities are common in the shallow GW, whereas the highest are found to be at the deeper layers of the aquifer. It can be seen that shallow GW layers are separated from deeper layers by a fine sand layer of variable thickness. The average permeability of the shallow aquifer is estimated to be 0.8×10^{-3} m s^{-1}. The infiltration rates during mean water level (Fig. 5a) measured in the shallow zones of the surface water gave the following results.

(A) The highest infiltration rates (19 to 23 μm s^{-1}) were found to be in the area close to the dike between ESW and MW where the hydraulic gradient is the greatest. This region lies outside the experimental area and will not therefore be considered here. The other parts situated more in the north-western side of the shore area,

which correspond to the infiltration zone of the local aquifer (cross section T3–T3 in Fig. 2), show seepage rates depending on the sediment thickness varying from 0.5 μm s^{-1} (*c.* 20 cm of fine sediment layer) to 1.7 μm s^{-1} (with almost no fine sediments). In the deeper regions of the ESW the fine sediment layers are nearly 70 cm thick which leads to estimated very low seepage rates.

(B) In the above mentioned shore areas where an infiltration pattern can be recognized (Fig. 5a, b, marked with thin arrows) the seepage rate is around 1.1 μm s^{-1} (mean value). When calculated with the average permeability of 0.8×10^{-3} m s^{-1} and the higher hydraulic gradient of 0.006 (Fig. 5b) the rate of GW flow per unit of area amounts to 4.8 μm s^{-1} (Table 1). In order to transport the same water volume, in the latter the infiltration area has to be 4.4 times larger (4.8: 1.1 μm s^{-1}) than the corresponding infiltration area of the lower GW-table (Fig. 5a) under the assumption that the mean seepage flux remains constant. With the same permeability of 0.8×10^{-3} m s^{-1} and the lower gradient of 0.002 (Fig. 5a) the GW flow rate per unit of area drops to 1.6 μm s^{-1}. This flow rate approximately corresponds more to the measured seepage rate of 0.5 to 1.7 μm s^{-1}. In the case considered here, a higher water level of 40 cm leads to the tripling of the flow rate. Therefore it can be concluded that seepage into GW highly depends on the surface water level.

Temperature and oxygen concentration

Groundwater temperature is influenced by climatic changes as well as above ground radiation. The heat transfer into GW depends on the thickness of the unsaturated layer and the water saturation and porosity of the porous media. Groundwater temperature in the study area is therefore strongly influenced by the annual atmospheric temperature changes near the shore areas where the unsaturated layer is the thinnest.

At low surface water level temperature differences between near-shore groundwater and surface water can reach more than 5 °C. At high water level, e.g., during flood events, temperature differences can drop down to zero. A typical example of flood events is shown in Fig. 6. The diagrams show the development of a flood event at well S10 with details of the temperature (Fig. 6a), the hydrographs of GW and surface water (Fig. 6b) and the temperature profiles (Fig. 6c: before, during and after the event). It can be deduced that even if the air temperature decreases down to −10 °C in the period before a flood, the temperature at the GW table changes only a few degrees. When flood begins (Fig. 6b on Dec. 24th), GW temperature increases slightly due to the effect of flow inversion. As a result after the recession of the

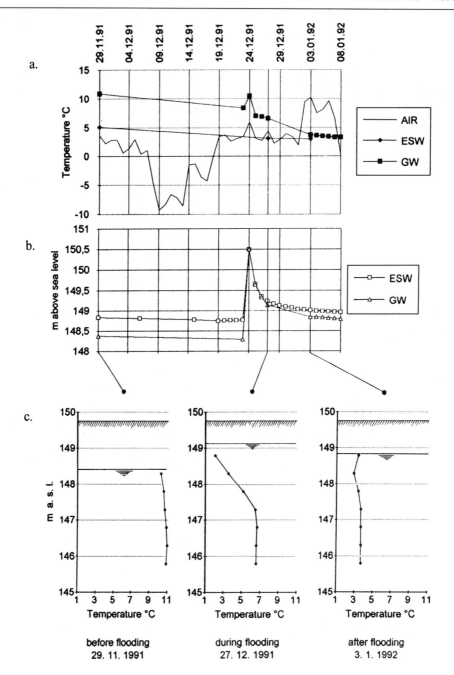

Fig. 6 Development of a flood event: (a) temperature; (b) water level; (c) temperature depth profiles during the flood period.

floods the GW temperature lowers to the same level as the temperature of the surface water.

Fig. 6c shows the temperature profiles in the saturated porous media. Prior to the begining of the flood, the temperature is constant along the profile. Three days after the high water peak, the temperature distribution shows a strong influence from the upper surface layer. The soil being partially frozen, the GW temperature decreases in the upper layers. In the near-shore zone the temperature in the aquifer dropped near to zero. However, there is a temperature recovery in the profile after a week (mean of

3 °C), emphasizing the importance of surface water infiltration.

The oxygen concentration mainly depends on two interconnected factors as illustrated in Fig. 7, the water level fluctuations of the surface water and the GW temperature. Surface water level increases result in an oxygen input into the near-shore GW. This effect can be traced during the floods of December 1991 and November 1992 (Fig. 7a, b). A minor rising of the surface water level in April 1991 seems to have no pronounced effect in the oxygen distribution. This is probably due to very high water temperature leading to a rapid oxygen depletion resulting from increased bacterial metabolism. As oxygen measurements were not continous, short-time peaks were sometimes not recorded.

a.

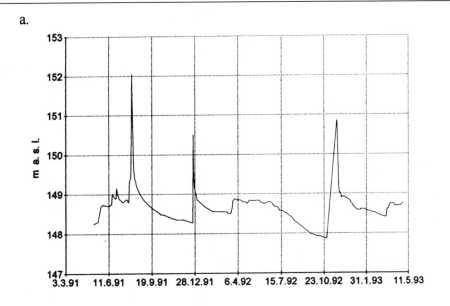

b.

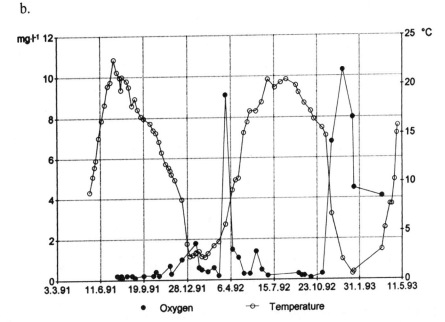

• Oxygen   ⊸ Temperature

Fig. 7 Oxygen distribution related to groundwater level and temperature. (a) Groundwater hydrograph of well S10; (b) corresponding temperature and oxygen concentration at D10.

ECOSYSTEM CONSEQUENCES

The oxygen concentration governed by the hydrological regime of the surface water apparently is the key factor determining the distribution and abundance of the meiofauna. The dominant meiofauna group is the Cyclopoid (Crustacea, Copepoda). Their higher densities coincide with elevated oxygen concentrations in winter and spring (1992 and 1993, Figs. 7b and 8). The high cyclopoid abundance in May 1991 can probably be attributed to more favourable oxygen conditions prior to the investigations. During summer and autumn, when GW becomes hypoxic, the meiofauna numbers decreases drastically. The rapid reestablishment of the cyclopoid assemblage dominated by perennial *Diacyclops* sp. (*D. languidus* and *D. languidoides*) after improving of the oxygen conditions is impressive. Such resilience of meiofauna communities was also observed by Valett *et al.* (1993) in the Sycamore hyporheic system in Arizona. These fluctuations may be due to migration (especially as the cyclopoids are a highly mobile animal group), physical impairment (through anoxia) and propagation, or both.

CONCLUSION

Gibert (1992) mentioned that groundwater ecology has not yet a firm theoretical foundation. She stated that most of the groundwater systems have been described disconnected from

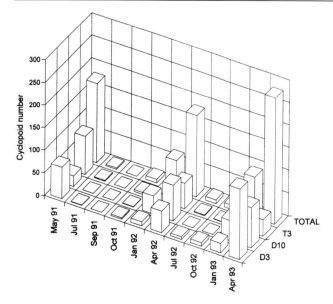

Fig. 8 Cyclopoid abundances (of five liter samples from 2, 4 and 6 m depth) in wells D3, D10 and T3 and totals (seasonal sampling 1991–93).

a global or hierarchical perspective. It is hoped that the data presented here can contribute to a better understanding of the contribution of hydrology in a GW ecosystem. It represents the necessary basic framework for an ecologist to look into various biological phenomena occurring during the exchange processes (between abiotic and biotic parameters) within the aquifer.

ACKNOWLEDGEMENTS

This project was supported by the 'Fonds zur Förderung der Wissenschaftlichen Forschung' of Vienna (P 7881 Bio, attributed to DLD). We thank R. Rouch, A. Mangin and M. Bakalowicz (Moulis) for valuable discussion, and A Gunatilaka (Vienna) for the help during the preparation and for the revision of this note.

REFERENCES

Castany, G. (1985). Liaisons hydrauliques entre les aquifères et les cours d'eau. *Stygologia*, **1**, 1, 1–25.

Danielopol, D. L. (1980). The role of the limnologist in groundwater studies. *Int. Rev. Hydrobiol.*, **65**, 777–791.

Danielopol, D. L. (1983). Der Einfluß organischer Verschmutzung auf das Grundwasser-Ökosystem der Donau im Raum Wien und Niederösterreich. Gewässerökologie. Forschungsberichte Herausgegeben vom Bundesministerium für Gesundheit und Umweltschutz, Vienna.

Danielopol, D. L. (1989). Groundwater fauna associated with riverine aquifers. *J. N. Am. Benthol. Soc.*, **8**, 18 – 35.

Danielopol, D. L., J. Dreher, A. Gunatilaka, M. Kaiser, R. Niederreiter, P. Pospisil, M. Creuzé des Châtelliers & Richter, A. (1992). Ecology of organisms living in a hypoxic groundwater environment at Vienna (Austria): Methodological questions and preliminary results. In *Proceedings of the 1st. Int. Conference on Groundwater Ecology*, ed. Stanford, J. A. & J. J. Simons, pp. 79–90, AWRA, Bethesda, Maryland.

Danielopol, D. L. & Marmonier, P. (1992). Aspects of research on groundwater along the Rhône, Rhine and Danube. *Regulated Rivers*, **7**, 5–16.

Dreher, J.E., Pramberger, F. & Rezabek, H (1985). Faktorenanalyse, eine Möglichkeit zur Ermittlung hydrographisch ähnlicher Bereiche in einem Grunwassergebiet. Mitt. hydrogr. *Dienst Öster.*, **54**, 1–12.

Gibert, J. (1992). Groundwater ecology from the perspective of environmental sustainability. In *Proceedings of the 1st. Int. Conference on Groundwater Ecology*, ed. Stanford, J. A. & J. J. Simons, pp. 3–14, AWRA, Bethesda, Maryland.

Jordan, C. F. (1981). Do ecosystems exist ? *Am. Nat.*, **118**, 284–287.

Lee, D.R. (1977). A device for measuring seepage flux in lakes and estuaries. *Limnol. Oceanogr.*, **22**, 140–147.

Mangin, A. (1976). Les systèmes karstiques et leur méthodologie d'investigation. *Annls. Sci. Univ. Besançon*, **25**, 263–273.

Nachtnebel, H.-P. & Kovar, K. (eds.) (1991). *Hydrological Basis of Ecologically Sound Management of Soil and Groundwater.* IAHS Publ. 202, London, 385pp.

Niederreiter, B. & Danielopol, D.L. (1991). The use of mini-videocameras for the description of groundwater habitats. Mitt. Hydrogr. *Dienst Österreich.*, **65/66**, 85–89.

Rouch, R. (1986). Sur l'écologie des eaux souterraines dans le karst. *Stygologia*, **2**, 352–398.

Rouch, R. (1992). Caractéristiques et conditions hydrodynamiques des écoulements dans les sédiments d'un ruisseau des Pyrénées. Implications écologiques. *Stygologia*, **7**, 13–25.

Schuller, G. (1973). Über Durchlässigkeitsbestimmungen durch hydraulische Bohrlochversuche und ihre Ergebnisse in tertiären Feinsanden (Obere Süßwassermolasse) Süddeutschlands. *Bohrtechnik-Brunnenbau-Rohrleitungsbau*, 24. Jhg., Heft **8**, 291–299.

Stanford, J. A. & Ward, J. V. (1992). Emergent properties of groundwater ecology. Conference conclusions and recommendations for research and management. In *Proceedings of the 1st. Int. Conference on Groundwater Ecology*, ed. Stanford, J. A. & J. J. Simons, pp. 409–416, AWRA, Bethesda, Maryland.

Valett, H. M., C. C. Hakenkamp & Boulton, A. J. (1993). Perspectives on the hyporheic zone: integrating hydrology and biology. Introduction. *J. N. Am. Benthol. Soc.*, **12**, 40–43.

Vanek, V. (1987). The interactions between lake and groundwater and their ecological significance. *Stygologia*, **3**, 1, 17–39.

Winter, T. C. (1983). The interactions of lakes with variably saturated porous media. *Water Resources Research*, **19**, 5, 1203–1218.

Winter, T. C. & Pfannkuch, H. O. (1984/85). Effect of anisotropy and groundwater system geometry on seepage through lakebeds. *Journal of Hydrology*, **75**, 239–253.

16 Typology of water transport and chemical reactions in groundwater/lake ecotones

O. FRÄNZLE & W. KLUGE

Projektzentrum Ökosystemforschung, Christian-Albrechts-Universität zu Kiel, Schauen-burgerstr. 112, 24118 Kiel, Germany

ABSTRACT Exchange reactions are important in the framework of subsurface flow between lakes and adjacent drainage basins. Frequently the groundwater/lake ecotones have the character of both a hydraulic barrier and a hydrochemical buffer. Consequently they exhibit a number of specific geohydraulic and hydrogeochemical features which are grouped into a coherent typology. It forms the basis of a multi-stage modelling approach, involving the professional groundwater model FLONET. The underlying measurements and deductions form part of the long-term German project 'Ecosystem Research in the Bornhöved Lakes District'.

INTRODUCTION

Groundwater/lake ecotones as typically small interfaces control in a very specific way water and related chemical fluxes between lakes and their respective catchment. Therefore reliable material balances require a deeper understanding of the manifold hydrological phenomena related to the transport of solutes with its set of physical, chemical and microbial boundary conditions. Corresponding mathematical models reveal gaps in our knowledge (cf. Naiman & Décamps, 1990; Mitsch & Gosselink, 1992) which must be bridged in order to develop coherent and efficient strategies for the management of lakes in relation to the landuse patterns of their catchments. A necessary first step in this direction is the definition of a process-oriented typology of the water transport and chemical reaction phenomena in groundwater/lake ecotones.

In the present paper this definition is based on the long-term observations, measurements and modelling approaches of the German 'Ecosystem Research in the Bornhöved Lakes District' which is situated some 30 km south of Kiel and covering about 50 km² in terms of interrelated drainage basins (Blume *et al.*, 1992). It is characterized by moraines of the Weichselian Pleniglacial with intercalated outwash sediments including kame deposits, dead-ice depressions with lakes, fens or wetlands, and Late Saalian moraines in the south. Thus the assemblage combines the essential geomorphic features of northern Germany which ensures a very wide applicability of the research results obtained. Owing to the complicated geomorphic history a great variety of natural ecosystems evolved in the course of the last 12 000 years.

Human impact on the highly diversified natural ecosystems of the study area began in the Early Neolithic and increased considerably since the Middle Ages. Consequently the ecosystems have faced long-term and short-term natural and anthropogenic changes which have to be identified, on a comparative basis, in terms of basic processes such as energy, water and nutrient cycles (Fränzle, 1990).

THE HYDROLOGICAL ORIGIN OF THE LAKE WATER

Subsurface exchange of chemical compounds is one component of the water cycle. Its influence on the water and mass balance of the lake can be evaluated with the aid of indices, whereby, when reliable quantitative data are lacking, approximations prove useful:

- all the essential mean in- and outflow terms of the water balance equation,

– the mean proportion of groundwater in the water stored in the lake,

– the mean residence time of the lake water,

– the mean residence time of the groundwater in the partial basins,

– the area of the lake and corresponding values for the surrounding groundwater and surface drainage basins,

– mean values, extrema, frequency distribution of the lake water levels and flood frequency curves.

Values for the residence time of the groundwater in the ecotone area are only meaningful if, with the help of changes in the hydrochemical conditions, the width of the ecotone zone can be clearly distinguished from the adjacent aquifer (cf. Fig. 3). This is of particular interest in the case of wider lake shore wetlands bordering lakes.

With a volume of 10.2×10^6 m³ and a groundwater proportion of about 10%, Lake Belau is a flow-through lake (Dyck & Peschke, 1980). The mean residence time of the lake water amounts to nearly a year. In contrast to this, the mean residence time of the groundwater in the partial groundwater basins varies between 10 and 50 years. The mean water level is relatively constant with fluctuations of $+/- 0.15$ m due to regulation at the outlet.

SYSTEMATIC EVALUATION OF INDIVIDUAL SHORE SECTIONS

In addition to the above macroscale data, microscale evaluations show to which extent single shore sections influence the exchange of water between the individual groundwater sub-basins and the lake. The characteristic features of the water exchange between the lake and the surrounding upland are summarized in Fig. 1 with a differentiation into slope foot, non-flooded wetland, rarely flooded wetland, eulittoral, infralittoral and profundal lake bottom.

Springs or seepage water at the foot of the slope are supplied either permanently or periodically by interflow after precipitation, which is often caused by soil layers of reduced permeability. The water which seeps out can flow off on the surface, gather as ponding water in local depressions or lead to waterlogging in the wetland zone. Hydromorphic soils are characteristic of the relatively flat non-flooded wetland zones. Rising groundwater, badly draining ponding water and capillary water in dry periods exert manifold influences on each other. The wetland zone with short-term flooding or with runoff and erosion generally exhibits marked waterlogging or water saturation. The eulittoral zone is the area of the water level fluctuations of the lake (Wetzel, 1983) while the permanently flooded infralittoral zone exhibits continuous water saturation. Groundwater discharge occurs if the permeability of the sediment is moderate or good. This is most likely to occur at the bottom of the slope as well as in the eulittoral zone extending to the infralittoral. Sedimentation rather than erosion is dominant in the sublittoral and the profundal zones, where there is little advective groundwater exchange.

In the present case the sediment/water interface separates a mixture of unconsolidated particles and interstitial water from the overlying water body. The coupled fluxes of water, solids and solutes moving upward or downward show many specific features whereby hydrophysical actions, chemical reactions and biological activities occur simultaneously.

The major processes responsible for the transport of particles across this boundary layer can be subdivided into the following groups. (1) Sedimentation as laminar accumulation of particles owing to growth of the sediment layer, (2) erosive action or turbulent fluctuations within the sediment/water interface, (3) mixing of sediment and water due to bioturbation, (4) clogging due to fine particles at the sediment surface or within the pores of the originally permeable sediment, (5) internal erosion or piping due to vertical water movement (Fränzle, 1993) caused by a hydrostatic pressure gradient between groundwater and lake.

Clogging and piping have opposite impacts. The hydraulic conductivity is decreased by clogging, resulting in reduction of exfiltration rates from the lake to the groundwater. The internal erosion or piping in low permeable sediments support the preferential exchange between groundwater and lake. Under these circumstances transport effects are more controlled by the hydrophysical and chemical character of the particles than the dissolved substances. All of these overlapping processes are varying with time and space. Both the coupled movement of water and particles and the hydraulic instability of the sediment/water interface largely affect the hydrodynamical and hydrochemical behavior of this important boundary layer. Therefore these questions should be paid more attention in the future.

A combination of all of these processes in one individual shore section is the exception rather than the rule. Thus it is often problematic to clearly classify the water according to its origin due to a continuous mixing in the ecotone area. It should be further noted that the various phenomena differ according to residence time, age and chemical composition (cf. Fig. 4). In comparison to the groundwater flow out of the aquifer, surface runoff, infiltration, interflow close to the surface and the overflow of the lake exhibit considerably more spatial and temporal variability, which must be taken into account when measuring or mapping.

In the past, site observations from the shore zones had focussed on vegetation and morphological features (Friedrich & Lacombe, 1992) whereas the hydrological characteristics of the site have usually been only inadequately

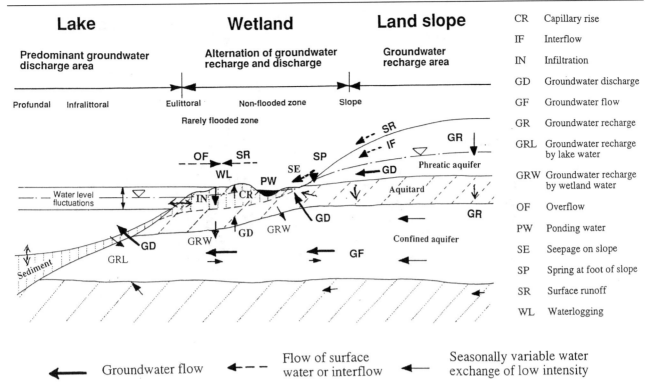

Fig. 1 Characteristic microscale features and zones of water exchange between a lake and its surrounding upland.

considered, although they are essential for elucidating the actual processes occurring during matter exchange in the ecotone. All this information should be obtained by *in situ* observations as well as by using a Geographical Information System (GIS). The application of GIS representations, however, is often problematic because the shoreline does not display a simple geometry and is subject to changes in time. In addition to this, spatial changes of the site and its gradients induce specific dynamics and anisotropy. As a consequence the changes perpendicular to the shoreline generally exceed those parallel to the ecotone by more than a factor of ten.

flow lines requires sufficient geohydraulic knowledge of the regional subsurface flow systems and the position of the groundwater divides. In this connection it should be noted that the surface drainage area and the catchment area of an aquifer can often differ considerably in an alluvial flat. Non-stationary divides, which are dependent on the water level fluctuations in the inland waters as well as on the process of groundwater recharge are not unusual in the hydrology of alluvial plain.

The three-dimensional partial differential equations which describe groundwater exchange are derived from Darcy's law and the continuity equation. The simplified flow regime illustrated in Fig. 2 is based on simulations with the groundwater flow model FLONET (Univ. of Waterloo, Centre for Groundwater Research, Canada). The following types of exchange are distinguished.

TYPES OF HYDRAULIC EXCHANGE FOR GROUNDWATER/LAKE ECOTONES

Aiming at a coherent definition of the different ways in which the exchange of matter can take place between the groundwater and the lake, the following classification is focussed on the exchange of groundwater and based on the relationships between the groundwater basin and the lake at both the catchment and the ecotone scales. The structural differences within the groundwater basin around a lake are the reason for the subdivision of the drainage area into individual geohydraulic flow sections which are operationally uniform. The delimitation of these flow sections along the

(a) Water exchange between aquifer and lake.
  (i) Direct discharge from phreatic or confined aquifers into the limnetic water body without impeding interface sediments (Fig. 2).
  (ii) Free discharge into the hydraulically conductive upper littoral zone, associated with a very limited seepage through the basal lake sediments.
  (iii) Preferentially restricted discharge through littoral and profundal zones due to exceedingly high spatial variability of lake bottom sediments (Fig. 2).
  (iv) Groundwater recharge from lake to aquifer (exfiltration) in places of well permeable lake sediments.

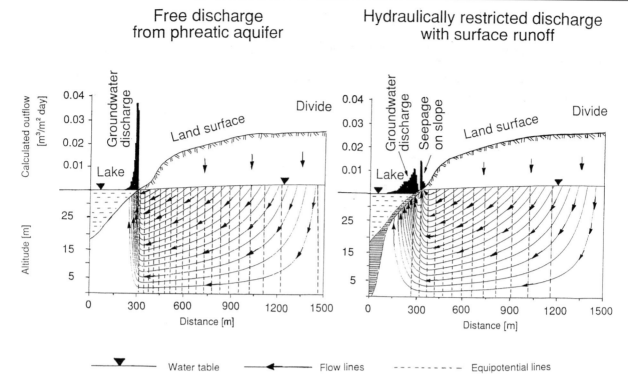

Fig. 2 Two examplary hydrological types of exchange through groundwater/lake ecotones. Note: The vertical exaggeration of the cross sections is 20. It causes distortions of the angles between flow lines and equipotential lines; in reality these lines are perpendicular, because isotropy was assumed.

(v)   Periodic infiltration and exfiltration with non-stationary groundwater divides or great water-level fluctuations in the lake or adjacent aquifer.
(b) Water exchange between wetland and lake.
   (i)   Natural wetland with or without flooding, with or without groundwater underflow.
   (ii) Wetland with water level regulation by ditches or drainage systems.
(c) Hydraulic barriers between lake and surrounding groundwater basin.
   (i)   Subsurface inflow of perched water or local periodic interflow close to the surface.
   (ii) No subsurface water exchange or three-dimensional flow around the impermeable shore barriers.

Two types are shown in Fig. 2; it should be noted, however, that there are no clear-cut distinctions between these types. In subsurface exchange, the water always selects a path that, in the shortest possible distance and with least hydraulic resistance, results in a balance between the spatially varying water levels. Subsurface water exchange is thus based on an energetic principle of minimization. Given sufficient permeability of the shore deposits, it is concentrated in some locations near the waterline and the littoral. In some lakes the erosion or the washing out of the fine sediment particles results in an additional increase of conductivity close to the shoreline. Proceeding from type (a) to type (c) hydraulic resistance increases. The subsurface water avoids those areas with very low permeability, which leads to the lengthening of the flow lines, and the transition to a three dimensional inhomogeneous flow pattern, which makes it very difficult to reconstruct the flow conditions.

A frequently occurring increased hydraulic conductivity in the erosion zone, and a reduced hydraulic conductivity in the marginal shore zone and lake sediment (local effects), and differences in the distance to the groundwater divides (long-range effect) give rise to spatial variability of the exchange rates along the shoreline. A shift of regional groundwater divides is likely to occur if the aquifer between lakes or streams is continuously developed in alluvial flats and the water levels of the receiving waters are headed up or drawn down. In the shore zone, local subsurface divides may even form near the shoreline or in the transitional area between a wider shore wetland and the adjacent aquifer. On the one hand this may be due to greater fluctuations in the lake water level, or to rapid rise in the water level in wetland areas after flooding or storm events in comparison to the adjacent aquifer, on the other.

The reconstruction of the flow lines permits conclusions about the origin of the water, the groundwater recharge areas, and the transported substances: with increasing length of the flow lines and the residence time or age of the groundwater the concentrations of nutrients and xenobiotics decrease with depth below the land surface. In addition to these factors a reduced hydraulic conductivity in the shore zone

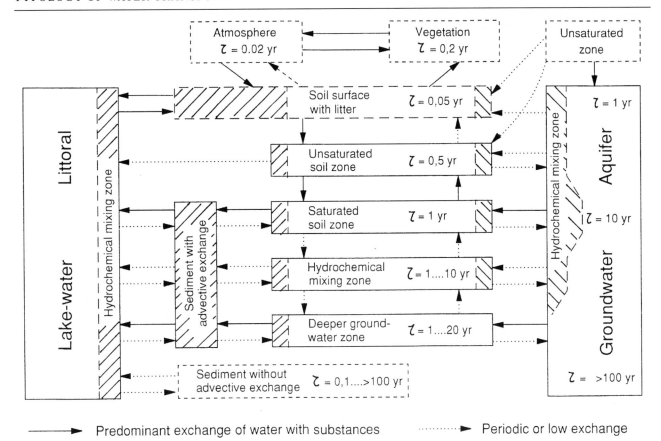

Fig. 3 Interaction between hydrological and hydrochemical influences in mixing zones of groundwater/lake ecotones (age data are only for information).

commonly contributes to occurrence of preferential groundwater flow at local seepage channels. The average time of contact within the ecotone zone will be reduced under such circumstances, and thus decrease the filter and buffer effect of the shore zone for the transported substances. This influence is caused by the microscale heterogeneity of glacial and lacustrine deposits in the lakeshore area (Piotrowski & Kluge, in press). The experimental proof of these phenomena was provided by special measurements of the temperature as well as the specific electrical conductivity of Lake Belau sediments during the last winter season. In some patches with an area of 1 to 100 m² the discharge of groundwater raised the water temperature up to 5 °C and the conductivity was increased by a factor of 2. Analogous results are reported by Vanek (1991) for lakes and coastal waters in Sweden.

In distinctly larger sections of the lake bottom no advective exchange takes place with the groundwater basin. Sedimentation, bioturbation, and diffusion in the sediment/water interface predominate here.

For most lakes inflow from the surrounding basin prevails over subsurface exfiltration (outflow). The types of exchange along the shoreline vary, in particular in lakes which are situated in areas with varied glacial geomorphogeny. These lakes display a surprisingly high spatial variability. The following are predominant in Lake Belau: hydraulically restricted infiltration due to ablation till (Piotrowski & Kluge, in press), natural wetland without flooding, local periodic interflow close to the surface and very limited exfiltration from lake into aquifer.

HYDROCHEMISTRY OF GROUNDWATER/LAKE ECOTONES

In contrast to the groundwater recharge areas, the groundwater discharge zones as the final stage in the subsurface water cycle from recharge areas to lakes exhibit a number of distinctive hydrochemical features (Eriksson, 1985). The following factors determine the hydrochemical conditions in groundwater discharge zones: hydromorphic soils, high biological productivity and the accumulation of particulate and dissolved organic matter, the intercalation of permeable and impermeable sediments, and the occurrence of aerobic transition zones and anaerobic stagnation zones. Focus is here on the hydrolytic and microbiologically catalysed degradation of organic matter, diverse oxidation-reduction reactions, precipitation of calcite, adsorption processes and organic complexation. These processes influence the buffer effect of the shore zone for nutrients and toxic substances (Scheytt & Piotrowski, in press).

The interaction between hydrological and hydrochemical influences is clearly depicted in Fig. 3. The ecotone domain

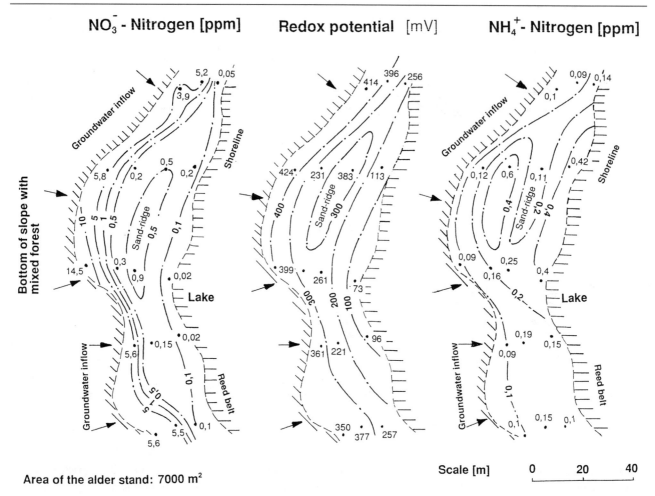

Area of the alder stand: 7000 m²

Fig. 4 Mean nitrate and ammonium concentrations and redox potential in the upper saturated zone of an alder forest on the shore of Lake Belau (from June 1990 to June 1992). Note: Estimated mean values for the advective exchange of the whole aquifer on the basis of observed water heads, simulated flow date, and some groundwater samples out of different depths: groundwater and nitrate inflow to the wetland: 8000 m³ year⁻¹ and 56 kg year⁻¹ as well as groundwater and nitrate inflow to the lake: 9700 m³ year⁻¹ and 25 kg year⁻¹ respectively.

between the aquifer and the lake is vertically subdivided into several compartments. Just as the age of the water increases with depth so does the residence time of the subsurface water in the water cycle. The arrows indicate the direction of water and material exchange. Shaded areas correspond to the hydrochemical mixing zones between the compartments with different hydrochemical conditions. The width of these mixing zones is determined by both periodicity and intensity of water exchange. A more or less stable hydrochemical environment can only develop in the upper saturated zone and the deeper groundwater zone if the wetland is sufficiently wide. In narrow wetlands, groundwater inflow mixes directly with the lake water. Not only the concentration of chemical species but also their lateral and vertical gradients can

exhibit a high spatial variability and temporal periodicity in mixing zones. The causes may be found in the macroscopic dispersion of the advective flow and seasonal variations of the ecotonal water balance.

In the alder forest and pasture wetland around Lake Belau the water, material and energy fluxes between the drainage basin, shore zone and reed-covered littoral are investigated (Blume *et al.*, 1992). Taking the nitrogen cycle as an example, the close relationship between water balance, groundwater flow, deposition, exchange and transformation processes can be demonstrated.

Fig. 4 shows the correlation between nitrate concentration and redox potential as related to the distance from land to shoreline while the concentration of ammonium increases near the shore. No quantitative conclusions about the actual buffer effect of the shore zone on the transport of nitrate from the drainage area can be drawn from the changes in concentration at just one depth. In the example presented here, the concentration of nitrate increases while the concentration of ammonium decreases with depth.

If the groundwater flows in deeper layers through the shore zone, the specific transformations shift into the exchange areas in the lake sediment of the infralittoral or profundal. For water exchange rates above 0.05 m³ (m²day)⁻¹ or at local

Table 1. *Multi-stage model strategy of water and mass exchange through a lake shore*

Stage	Category	Requirements
A	Expert-system assessment	Fundamental knowledge about system structures, functions, dynamics; database, GIS, simple models, few field observations
B	Statistical models (regression, extreme values)	Measurements in space and time
C	Process models (deterministic)	Detailed knowledge about processes and structures
C1	Flow and mass transport models with spatially distributed parameters	Physical theory of flow mechanisms and reactions; detailed knowledge of structures and parameter adaptation
C2	Box-balance and reaction models (stirred-tank reactor)	Conceptual framework, balance equations; measurements for calibration and validation
D	Stochastic models	Long-term measurement programmes in space and time
D1	Stochastic process models (prognosis)	Complete data sets
D2	System-theoretical operator models (black-box, filtering)	Complete data sets with characteristic events
E	Exchangeable modules as intrinsic regulators (e.g., object-oriented models)	Comprehensive knowledge; ecotone typology, ecotone modules; basic data, GIS, simulation language etc.

MULTI-STAGE MODEL CONCEPT

The multi-stage model system which is presented in Table 1 as a comprehensive model concept for a lake/groundwater ecotone is based on the above typology. Within each stage there are on the one hand the hydrological models (water flow or balance), and on the other the coupled water balance, mass transport and reaction models. The more complex the mass transformation model is, the more important is it to have a clear and simple arrangement of structures. Each class of models has a specific realm of application and special requirements as to *a priori* knowledge and data. In the initial expert assessment of the system, all the available information about the hydrological typology of the lake and individual shore sections is consequently used. The estimation of current water and mass balances or simple simulations with commercial models form the core of this evaluation.

Two groups of process models are distinguished: models with spatially distributed parameters (flow and mass transport models) and box balance models. For every finite volume into which the flow pattern is subdivided, geohydraulic and geochemical parameters are necessary. Box models are stirred-tank reactor models for larger compartments which exhibit similar characteristics. Field and laboratory data are required for calibration and validation.

Black-box system models describe the hydrological and hydrochemical transition behavior of ecotones between a lake and the adjacent aquifer. Weather effects and vegetation give rise to individual hydrological zones (cf. Fig. 1). The transition function can be described in the time domain as well as in the frequency domain by means of time series analysis and system hydrology. This procedure permits the application of scale analysis and hierarchical systems theory in ecology (Müller, 1992).

The future goal is to develop interchangeable simulation models for different ecotone types which adequately describe water and mass exchange between drainage areas and lakes as an object-oriented, independent programming regulator (Sekine *et al.*, 1991). These simulation models thus depict the ecotones as black-box modules and are interchangeable like circuit boards. Prognoses and scenario analyses of mass transfer in ecotones could be carried out efficiently by the implementation of these modules in regional groundwater transport models. The realization of this model concept will provide a connection between theoretical knowledge and the results of practical investigations in the framework of different case studies. This system is designed to obtain scale-consistent results for individual groundwater/lake ecotones or for a whole lake, and to optimize experimental programs (Kluge & Heinrich, in press). Furthermore, this model system is intended to fill a gap, hitherto only bridged by

springs, the buffer effect of the groundwater/lake ecotone almost completely disappears with the transfer of the groundwater into the lake.

The high spatial variability and temporal dynamics of the characteristic hydrochemical conditions make measurement as well as modelling difficult. Measurement programs only supply representative data if the number of sampling points is greater than 10 at each depth level, and if the samples are taken over a longer time and preferably fortnightly.

The nitrogen model is such that the transformations of several nitrogen species in the saturated zone can be formulated in the model analogously to the macro-pore concept for aerobic as well as non-aerobic domains between pools of organic matter, dissolved organic nitrogen, dissolved nitrate, dissolved and absorbed ammonium. The two models FEUWA (Kluge *et al.*, in press) and FEUNA form the basis of STOFLU, a more complex model for simulating water and mass exchange between a lake and the adjacent land.

either non-generalizable statements or impracticable model assumptions. User-friendly models of this type are, in connection with a Geographical Information System, a prerequisite for ecotone research, ecotone protection and lake management which adequately account for subsurface exchange with groundwater basins.

ACKNOWLEDGEMENTS

This research is funded by the German Federal Minister of Research and Technology in the framework of the interdisciplinary Project 'Ecosystem Research in the Bornhöved Lakes District' under grant no. BMFT 0339077E.

REFERENCES

Blume, H.-P., Fränzle, O., Kappen, L., Nellen, W., Widmoser, P. & Heydemann, B. (1992). (eds.) *Ökosystemforschung im Bereich der Bornhöveder Seenkette.* Ecosys no. 1, Projektzentrum Ökosystemforschung, Univ. Kiel, Germany.

Dyck, S. & Peschke, G. (1980). *Grundlagen der Hydrologie.* W. Ernst & Sohn, Berlin München, 197–205.

Eriksson, E. (1985). *Principles and Applications of Hydrochemistry.* Chapman and Hall, London-New York, 79–87.

Fränzle, O. (1990). *Ökosystemforschung und Umweltbeobachtung als Grundlagen der Raumplanung.* MAB-Mitteilungen 33, Bonn, 26–39.

Fränzle, O. (1993). *Contaminants in Terrestrial Environments.* Springer-Verlag, Berlin Heidelberg New York, 208–222.

Friedrich, G. & Lacombe, J. (1992). (eds.) *Ökologische Bewertung von Fliessgewässern.* Gustav Fischer Verlag, Stuttgart.

Kluge, W. & Heinrich, U. (in press). Statistische Sicherung geoökologischer Daten, In *Neuere statistische Verfahren und Modellbildung in der Geoökologie,* ed. W. Schröder, L. Vetter & O. Fränzle, Vieweg Verlagsges, Wiesbaden.

Kluge, W., Müller-Buschbaum, P. & Theesen, L. (in press). Parameter aquisition for modeling exchange processes between terrestrial and aquatic ecosystems. ISEM Conf. from 29 Sept.- 2 Oct. 1992 in Kiel, *Ecol. Modelling,* **75/76**, 399–408.

Mitsch, W.J. & Gosselink, J.G. (1992). *Wetlands.* Reinhold Comp., New York, 55–125.

Müller, F. (1992). Hierarchical approaches to ecosystem theory. *Ecol. Modelling,* **63**, 215–242.

Naiman, R.J. & Décamps H. (eds.) (1990). *The Ecology and Management of Aquatic-Terrestrial Ecotones.* Man and Biosphere Series, vol. 4, Parthenon Publ. Group, Casterton Hall, 295–301.

Piotrowski, J.A. & Kluge, W. (1994). Die Uferzone als hydrogeologische Schnittstelle zwischen Aquifer und See: Sedimentfazies und Grundwasserdynamik am Belauer See, Schleswig-Holstein. *Z. dt. Geol. Ges.,* **145**, 131–142.

Scheytt, T. & Piotrowski, J.A. (1995). Ursachen für zeitliche Schwankungen der Grundwasser-Beschaffenheit in einem Seeufer-Bereich. *Wasser und Boden,* **10**, 40–57.

Sekine, M., Nakanishi, H. Ukita, M. & Murakami, S. (1991). A shallow-sea ecological model using an objekt-oriented programming language. *Ecol. Modelling,* **57**, 221–236.

Vanek, V. (1991). *Groundwater flow to lakes and coastal waters – methods and some ecological consequences.* Diss., Lund Univ., Sweden.

Wetzel, R.G. (1983) *Limnology* (2. ed.). Saunders, Philadelphia, 136–137.

17 Development of a water transfer equation for a groundwater/surface water interface and use of it to forecast floods in the Yanghe Reservoir Basin

S. LIU

Department of Hydrology, Institute of Geography, Chinese Academy of Sciences, Beijing 100101, People's Republic of China

ABSTRACT   An equation is developed to describe water transfer on and across a groundwater/surface water interface. The interface is divided into two categories, simple and complex, and conceptualised as having a wedge-shaped profile with depths ranging from zero to a maximum value. The maximum depth is defined as water capacity of the interface. The equation is used to forecast floods of the Yanghe Reservoir Basin in the Hebei Province of China. Because of the scarcity of data describing the basin as well as its irregular shape, simulation using common conceptual models, e.g., the Xin'anjiang Model, is difficult. The equation is derived using principles of fluid dynamics and exhibits a good capability to forecast floods in this case. This approach avoids some confounding concepts, such as hydrograph separation and runoff-formation identification, which, although important, are difficult to incorporate into conceptual models. The relationship between the equation and the conceptual watershed model is explored.

INTRODUCTION

Hydrologists, biologists, and water resources managers have become increasingly interested in groundwater/surface water interfaces as an appreciation of their importance to the integrity of natural systems is revealed, and in response to a renewed appreciation of environmental problems associated with man's manipulation and interference of this interface. Because of the importance of this interactive zone to many of man's needs, it is vital to our common good to develop sustainable, conjunctive management of this resource. Mass transfer is one of the principal functions of groundwater/surface water interfaces (Unesco MAB and IHP Programs, 1992). Mass transfer includes water quantity (simply, water) as well as solute transfer (Liu & Mo, 1993). Models are often used to describe transfer of water, while solute transfer is more often described by convective and diffusive equations. In this paper, based on the interface condition of mass transfer, an equation describing water transfer on and across a groundwater/surface water interface is developed. The solution to the equation is discussed using a case study, and the relationship between the equation and a watershed model is explored. The main goal of this paper is to develop a new method to forecast floods for a particular basin where it would be difficult to use watershed simulation. This new method of using a water transfer equation helps unfold the gray box of the runoff generation. Here, gray box means there is insufficient information about runoff generation but more than what a black box describes.

WATER TRANSFER EQUATIONS

Based on fluid dynamics of porous media, the interface condition of mass transfer (Hassanizadeh & Gray, 1989a) reads:

$$\frac{\partial}{\partial t}(b<\rho>_\alpha^b)+\nabla'\cdot(b<\rho>_\alpha^b \overline{V}^{/b\alpha})-$$

$$\sum_{\beta\neq\alpha}\frac{1}{\overline{O}\,s}\int_{\overline{O}\,A_{\alpha\beta}^b}[\rho\,(W-V)]\cdot n^{\alpha\beta}\,da= \qquad (1)$$

$$[<\rho>_\alpha(\overline{V}^\alpha-W)]\cdot N$$

in which ρ is mass density; V is velocity; i is non advective flux vector; W is the velocity of α-β interface; $[<\rho>_\alpha V^\alpha - W]$ is the jump of $<\rho>_\alpha V^\alpha - W$ across the interface; $<>_\alpha$ is the intrin-

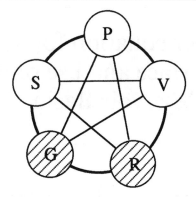

Fig. 1 Four of the ways to form a groundwater/surface water interface.

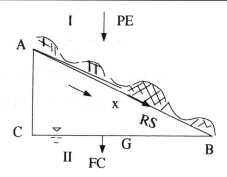

Fig. 2 Abstraction of a groundwater/surface water interface.

sic volume average in α phase; $-\alpha$ is intrinsic mass average in a phase; b denotes the interface. The prime represents the lower dimension spatial vector at the interface. Other phases are lumped as one phase (β phase). $n^{\alpha\beta}$ is unit normal vector to the micro interface $\delta A^{\alpha\beta}$.

There are 16 ways (Liu, 1993) to form a groundwater/surface water interface. In this paper, four of them, i.e., G-R, G-P-R, G-S-P-R and G-P-S-R are considered, as shown in Fig.1. Here G refers to groundwater, R is surface water, V is vegetable water (water in plant), and P is Precipitation (rainfall) and S is soil water. Of the five links (G, P, R, S, V) in Fig.1, vegetable water is neglected. Generally speaking, a groundwater/surface water interface can be conceptualised as a wedge as shown in Fig.2. In rainfall events all the water molecules move into the cross section of the interface (B point in Fig.2) in one dimension (x) along the interface. The depth of the groundwater/surface water interface varies from zero to maximum gradually. The maximum depth defines water capacity of the interface in millimetres. The groundwater/surface water interface is the combination of the simple and complex interfaces (Gray, 1982). We assume α is water phase (ω) and $W=0$, mass transfer between α and β can be omitted and ρ is a constant. Considering the relationship between $<>_\alpha$ and $-\alpha$ (Gray, 1982), the porosity of the side I (the atmosphere), denoted as ε_1, is equal to 1 and that of the side II (groundwater), denoted as ε, equals porosity of the interface. If we then read from the right hand of the identical equation (Hassanizadeh & Gray, 1989b), equation (1) becomes:

$$\frac{\partial}{\partial t}(b\varepsilon\, S_\omega^b <\rho>_\omega^\omega) + \nabla\cdot(b\varepsilon\, S_\omega^b <\rho>_\omega^\omega \overline{V}^{\,lb\omega}) \qquad (2)$$
$$= [S_\omega^1 (<\rho>_\omega^\omega)_1 \overline{V}_1^\omega - \varepsilon\, S_\omega^2 (<\rho>_\omega^\omega)_2 \overline{V}_2^\omega]\cdot N_1$$

in which $S_\omega^{\,b}$, $S_\omega^{\,1}$, $S_\omega^{\,2}$ are the values of α in side I and side II of the interface respectively. N_1 is the normal vector of side I to the interface. It is apparent from the concept that $S_\omega^{\,2}$ means the saturation coefficient. If we make $S_\omega^{\,b}$ equal Θ then

from the meaning of the right hand of the equation (2), we get:

$$\frac{\partial}{\partial t}(b\varepsilon\,\theta) + \frac{\partial}{\partial x}(b\varepsilon\,\theta\, v) = PE' - FC' \qquad (3)$$

in which $V=V'^{b\omega}$, and PE' and FC' are the normal components of mass transfer of side I and side II respectively. PE' means the precipitation excess of a point and FC' is the interaction with groundwater at that point. Equation (3) then represents the water transfer equation of the groundwater/surface water interface. The water balance of the interface can be written from Fig. 2 directly as follows:

$$W_2 - W_1 + RS = PE - FC \qquad (4)$$

where W_1 and W_2 represent soil moisture of the interface at time t_1 and time t_2 respectively. RS refers to runoff, PE means the precipitation excess over the interface, and FC refers to the interaction of the interface with groundwater. PE and FC equal the average of PE' and FC' respectively. When comparing equation (4) with equation (3), the context of the left hand of the equation (3) is clear. The first term of the left hand of the equation (3) refers to temporal variation of water capacity of the interface. It is called Term I, which closely relates the moisture Θ, the depth b and the porosity ε. The second term refers to spatial variation of water transfer, called Term II. With the increasing of ε, Θ, b and velocity V, the water transfer increases. The first term and the second term of the right hand of the equation (3) are called Term III and Term IV respectively. If the angle abc of the interface as shown in Fig.2 is reduced to zero, then b corresponds to zero. From Equation (3) it is seen that PE' then equals FC', and from equation (4) that RS equals zero. This means that there is no water transfer function at the interface. This is an unusual concept to classic hydrology, in which, if the groundwater table rises to the land surface, surface runoff (RS) equals precipitation. However, for this case, in the concept of the above mentioned transfer theory, there is no water transfer. The groundwater is viewed as a reflector mirror and so called runoff is simply the behavior of precipitation at the land surface only. The interface in this case has no water

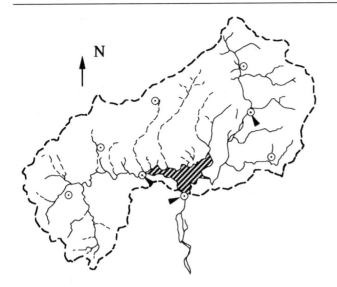

Fig. 3 Sketch of Yanghe Reservoir Basin.

Table 1. *Parameters optimised*

WM (mm)	B	Fc (mm/hr)	IMP	N	NK (hr)	KG (hr)
109.8	0.3	8.96	0.03	2.048	5.75	30

Note:
WM – Storage capacity of the interface;
B – Exponential number of storage capacity distribution curve;
FC – Interaction with groundwater;
IMP – Proportion of impermeable area to the total area;
N – Number of cascade linear reservoirs for surface runoff routing;
NK – Scale parameter of cascade linear reservoir;
KG – Storage coefficient for groundwater routing.

transfer function. The depth of the interface b is defined as the index of water transfer function. The water transfer function increases with increasing value of b.

APPLICATION OF THE WATER TRANSFER EQUATION – A CASE STUDY

The case study involves forecasting floods of the Yanghe Reservoir Basin located at Qinhuangdao, Hebei Province of China (Liu & Liu, 1994). Flood forecasting in this basin is a challenge because of the irregular shape of the basin and because data on the basin are very scarce. As shown on Fig. 3, there are only two river control stations within the basin.

The middle area of the basin, making up almost one third of the whole area, has no control station at all. For the usual hydrology simulation, this lack of data would be a very difficult problem to overcome. However, the water transfer equation discussed above can be used to solve the problem. A description of the rationale follows. To solve the water transfer equation, other equations are often needed, such as momentum transfer and intrinsic equations. Because of inadequate data in our case, this study turns to use catchment simulation to provide the information needed to solve the water transfer equation and to calculate Term I and Term III. On the one hand, based on the fluid dynamics of porous media, we get the mass transfer equation for the groundwater/surface water interface. Using comparative analysis, we obtain the values to substitute into the water transfer equation for the water balance equation of the interface. On the other hand, it is the value of each term of the water transfer equation that explains the meaning of the relevant water balance term. Therefore, the question of how to calculate the relevant water balance terms can be solved by calculating

appropriate terms of the water transfer equation. This paper makes the point that the water transfer equation has some advantages over the watershed model. However, this has nothing to do with the use of catchment simulation as auxiliary information for solving the water mass transfer equation. Without doubt, catchment simulation is a useful tool in hydrology. The following shows that what is assumed to be deduced from the water transfer equation is in fact what is conceptualised from the model. In other words, to some degree the water transfer equation provides support for catchment simulation theory. If one assumes water movement is unidirectional, then Term II represents the hydrograph at the cross-section of the basin. After comparing the genetic algorithm (Wang, 1991) and the Hooke-Jeeves algorithm, the two method were combined to create a powerful and efficient method to estimate parameters for calculating Term II. The eight floods (denoted as 660815, 690902, 700808, 740808, 770802, 810704 and 870826) are used to create optimal object function. Here floods are denoted in year-month-day format. The optimised parameters are shown in Table 1. The result of using other floods for verification is shown in Table 2 and Fig. 4. According to Table 2, the qualified peak of floods is 62.5%, which to a certain extent identifies the capability of using this method to solve the problem.

Although we have data for point precipitation, we must calculate the average precipitation from the concept of Term III and consider how to obtain the average precipitation for the case. The simplest way is to divide the area into subareas according to the weights of hyetal station. In this way, the more subareas divided the less the difference between them. With increasing number of the subareas, they eventually reach points where the differences within the subareas become meaningless. On the other hand, if there are too few subreaches then the representation of the average value of precipitation over the few subareas is of

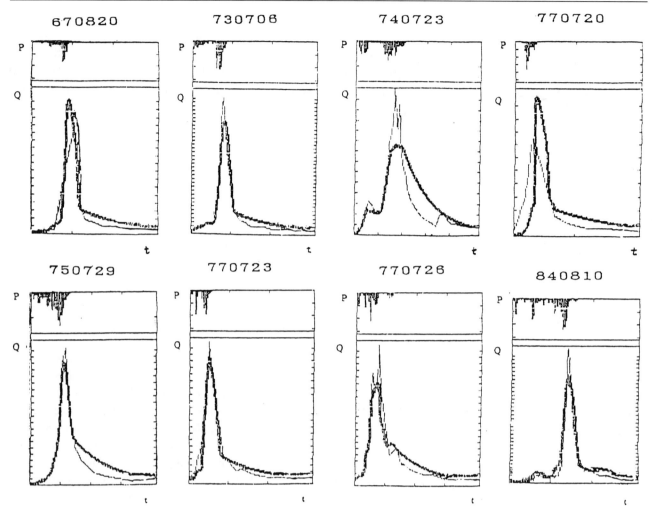

Fig. 4 Comparison between the solution of water transfer equation and Xin'anjiang Model simulation: p is precipitation (mm); Q is flow (m³/s) and t is time (day). Thin line represents measured values and thick line represents calculated values.

Table 2. *The verification result of the hydrograph calculated by interfacial mass transfer equation*

N	$E_{pf}(\%)$	$E_{pt}(\%)$	Q_m (m³/s)	R_{sqr}
670820	−7.28	3.75	1140	0.660
730706	12.28	0.75	2540	0.921
740723	38.66	−1.50	1010	0.714
750729	10.90	1.50	1440	0.919
770720	−36.79	0.59	583	0.220
770723	10.81	−0.25	1480	0.960
770726	25.91	1.25	1640	0.891
840810	19.17	−0.50	2050	0.914

Note:

N – Flood number;

E_{pf} – Error of peak flow;

E_{pt} – Error of flow measured;

Q_m – Peak of flow measured;

R_{sqr} – Residual square error.

questionable accuracy. A scale exists, called the Representative Elementary Length (REL) (Liu, 1993; Liu & Liu, 1994) that is useful in this determination. When the subarea scale is smaller than REL, the effect of the spatial variability within the subareas on the hydrograph is evident, and it is impractical to average the weight of the point data as the average input. When the scale is larger than REL, the representation of the data as an average is questionable. Therefore only REL has been calculated is it possible to correctly determine the input term. Using semivariance analysis, the maximum correlation length, equivalent to REL, is equal to 17.81 km (the area of the basin is 755 km²). Based on this REL, and using step-by-step regression analysis, principle component analysis, grey system theory and the hyetal map, the input term is calculated by weighting the average of the precipitation. This is done at the eight hyetal stations.

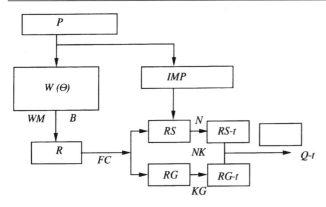

Fig. 5 The flow chart of the solution to water transfer equation (*P* is precipitation, *W(Q)* is soil moisture; *P* is runoff; *RS* is surface runoff; *RG* is groundwater; *-t* represents hydrograph and *Q* is discharge at the outlet).

COMPARISON BETWEEN THE WATER TRANSFER EQUATION AND A WATERSHED MODEL

The Xin'anjiang watershed model (Zhao, 1985) was chosen as an example for comparison. It is difficult for the Xin'anjiang model to yield flood forecasts of the Yanghe Reservoir Basin due to the lack of data and the irregular basin shape, as indicated above. By comparison analysis, it is found that when the unsaturated soil is considered as one layer and the river reach is not divided into subreaches, the simulation result is in agreement with that of the equation solution. This is a very interesting point. When efforts are made to modify a model for better precision the result is to make the model more complex. The comparison made in this paper reaches the opposite conclusion that simpler is better. Our conclusion is reflected in this Frugality Principle. Simply stated, when using the equation to solve flood forecasts, the basics are captured. The forecast resulting from either the water transfer equation or a model can not reflect more information than the data provide. To attempt to have them do more would lead to errors. The flow chart of the equation solution is shown in Fig. 5. From Fig.5 it is seen that the water transfer equation solution avoids confusing concepts often incurred by using the common watershed simulation. These concepts include hydrograph separation and runoff formation identification. Moreover, all the assumptions to deduce the water transfer equation are the same as conceptualised in the model.

CONCLUSIONS

A water transfer equation for movement across the groundwater/surface water interface is developed (equation 3). The

equation describes the properties of the interface and links groundwater and surface water. By comparing the water transfer equation with water balancing equation, the terms of the water transfer equation are explored. The water transfer term provides a new perspective to some surface runoff concepts of classic hydrology. The thickness of the interface is defined as the index of water transfer function. The water transfer equation is used to forecast floods of the Yanghe Reservoir Basin in the Hebei Province of China. Use of the equation has advantages over using common conceptual catchment models, especially when data are lacking and the basin is of irregular shape. The water transfer equation avoids the difficulty of using hydrograph separation and runoff formation identification involved in common conceptual models of catchment simulation.

ACKNOWLEDGEMENTS

The author gratefully acknowledges the tutoring of Professor Changming Liu and discussions with Dr Xingguo Mo during the development and writing of the thesis for the Ph.D degree from the Chinese Academy of Sciences. This paper is a part of that dissertation. Thanks are also due to Dr S.M. Hassanizadeh and Dr Janine Gibert for their review and comments, and John Simons for kind help in editing. This work was partially supported by the Chinese Natural Science Foundation and the Foundation for Diverting the Qinglong River to Qinhuangdao City.

REFERENCES

Gray, W.G. (1982). Derivation of vertically averaged equations describing multi-phase flow in porous media. *Water Resour. Res.*, **18**, 1705–1712.
Hassanizadeh, S.M. & Gray, W.G. (1989 a). Boundary and interface conditioning porous media. *Water Resour. Res.*, **25**, 8, 1706–1715.
Hassanizadeh, S.M. & Gray, W.G. (1989 b). Derivation of condition describing transport across zones of reduced dynamics within multi-phase system. *Water Resour. Res.*, **25**, 529–539.
Liu, S. (1993). (in Chinese) Water transfer on across groundwater/surface water interface. Dissertation for Ph. D to Chinese Academy of Sciences. Institute of Geography, Beijing.
Liu, S. & Liu, C. (1994). (in Chinese) The concept of Representative Elementary Length and it application to flood forecast of Yanghe Reservoir Basin. *Yangtze River*, **45**, 8, 23–25.
Liu, S. & Mo, X. (1993). Water exchange processes at land surface. In Exchange process at the land surface for a range of space and time scale, ed. Bolle-H. J. *et al.*, pp. 249–252, Proc. of IAMAP-IAHS'93 Congress. IAHS Publ. no 212.
Unesco MAB & IHP Programs (1992). Ecotones news, No.2, April.
Wang, Q.J. (1991). The genetic algorithm and its application to calibrated conceptual rain-runoff models.*Water Resour. Res.*, **27**, 9, 2467–2471.
Zhao, R. (1985). (in Chinese) Hydrological simulation in Catchments—Xin'anjiang Model and Shan'bei Model. *Water Resources Press.*, Beijing.

18 Uses and limitations of ground penetrating RADAR in two riparian systems

G. C. POOLE*, R. J. NAIMAN*, J. PASTOR** & J. A. STANFORD***

* Center for Streamside Studies, AR-10, University of Washington, Seattle, Washington 98195, USA

** Natural Resources Research Institute, University of Minnesota, Duluth, Minnesota 55811, USA

***Flathead Lake Biological Station, The University of Montana, Polson, Montana 59860, USA

ABSTRACT Ground penetrating RADAR was used in an attempt to map sediment accumulation in active and abandoned beaver (*Castor canadensis*) ponds in northern Minnesota and to map buried paleochannels of the Flathead River floodplain in Montana. We attempted to map ice thickness, water depth, sediment depth, depth to parent material (bedrock or clay), thickness of soil horizons, organic deposits (peat), frost penetration, and depth to the water table in the beaver ponds. Ground penetrating RADAR successfully located some of the subsurface interfaces between these layers but water saturation and the high clay content of the soils interfered with the ground penetrating RADAR signal while the physical complexity of the subsurface hampered data interpretation. In Montana, paleo-channels and water tables were located, but the stony nature of the substrate prevented immediate excavation for verification. In both Montana and Minnesota, success depended strongly on physical characteristics of the sites and specific interfaces. Generally, our efforts were only successful where the physical subsurface interfaces had abrupt, well defined boundaries, and where clay content was low.

INTRODUCTION

Ecological applications of ground penetrating RADAR (GPR) have included classification of soils and remote examination of their structure (Doolittle, 1982; Ulriksen, 1982). Specifically, GPR has been used to examine depths of organic soils (Ulriksen, 1982; Baraniak, 1983; Shih & Doolittle, 1984), map depth to bedrock (Olson & Doolittle, 1985; Doolittle et al., 1988; Collins, Doolittle & Rourke, 1989), measure depth to the water table (Houck, 1982; Sellman, Arcone & Delany, 1983; Hubbard, Asmussen & Perkins, 1990), examine soil characteristics (Hubbard et al., 1990; Truman et al., 1988; Collins et al., 1986; Olson & Doolittle, 1985), improve soil surveys (Doolittle, 1987), study soil micro-variability (Collins & Doolittle, 1987), and measure depth to permafrost (Doolittle, Hardisky & Gross, 1990). Few studies, however, have reported the limitations of GPR. Ulriksen (1982) and Shih & Doolittle (1984) discussed how increases in soil moisture, dissolved salts, and clay content result in increased signal attenuation and therefore decreased signal penetration. Doolittle (1982) and Shih &

Doolittle (1984) also briefly discussed how the character of the subsurface interface (including interface abruptness) affects success with GPR.

In spite of these limitations, and given the successful adaptation of GPR to soil science and biology, we believe there is considerable potential for the application of GPR in ecology — especially with respect to mapping and understanding sediment deposits and areas of hyporheic flow in riparian systems. In riparian systems dominated by the presence of beaver ponds and in floodplains of gravel-bed rivers, the interaction between groundwater and surface water is heavily dependent upon the physical characteristics of sediments and the manner in which these sediments are deposited. By using GPR to examine sediments, we hope to evaluate GPR as a potential tool for gaining a better understanding of the hydrology of hyporheic floodplain gravels and the role of wetland soils in nutrient and carbon flux in active and abandoned beaver ponds. These two aquatic ecosystems present a range of conditions against which we can test the usefulness and limitations of GPR technology for precisely locating and identifying subsurface interfaces in riparian areas.

Beavers build dams and flood land thereby altering bio-geochemical cycles and accumulations of sediments. Since 1927, beavers have converted 13% of the 294 km^2 Kabetogama Peninsula (Voyageurs National Park, MN) to meadows and ponds. These ecosystems are dynamic and affect boreal forest drainage networks for decades, or even centuries (Naiman, Johnston & Kelley, 1988). However, the total accumulation of materials in the ponds, and their physical transformation after the ponds drain and meadows are formed, remain largely unexplored. The mass and character of this material, variations in the water table, depth of frost penetration, and the distribution of saturated zones and organic deposits determine subsequent vegetative succession, system productivity, and ecological diversity.

The hyporheic zone of gravel-bed rivers is the subsurface volume of the river floodplain within which river water actively circulates (Triska *et al.*, 1989). In the Kalispell Valley floodplain of the Flathead River, Montana, river water penetrates gravel substrata as much as 3 km laterally from the river channel. Groundwater flow rate is dependent on the porosity of the gravel and the slope of the floodplain. Flow rates exceeding 0.1 cm/s were measured in zones of very porous substrata characterized as paleochannels (abandoned, buried stream channels which contain fluvially sorted deposits of rounded cobbles and gravel which essentially pipe ground waters through the floodplain substrata at rates much faster than within the adjacent substrata). The hyporheic zone of the Flathead River supports a food web comprised of over 80 taxa of metazoans, including large *Plecoptera*. Understanding this food web requires resolution of the cross-sectional and longitudinal structure of floodplain substrata, especially the very porous paleochannels (Stanford & Ward, 1988, 1993).

How GPR works

A GPR system sends pulses of electromagnetic waves into the ground and records the strength of reflected signals along with the elapsed time since the signal was generated (Ulriksen, 1982). Reflections are strong when the signal passes across an abrupt boundary between two substances of differing dielectric constants. GPR maps changes in the physical characteristics of the subsurface by detecting changes in the dielectric constant of adjoining layers of different density, water content, or chemical composition. However, GPR does not directly record specific information about subsurface interfaces (i.e., measures of density or water content). It locates the point at which these properties change. Simply put, GPR does not generally describe subsurface interfaces; it locates them.

GPR is most effective given adequate knowledge of the subsurface environment. GPR can significantly reduce the number of conventional subsurface samples (e.g., samples taken by digging, drilling, or other means of exposing the subsurface) needed to map subsurface features. Once a subsurface layer is identified, only occasional samples are needed to verify accuracy. Rather than sampling discrete points along a transect with conventional methods, GPR can be used to obtain a continuous subsurface map of the transect.

STUDY SITES

Minnesota

The study site is located on the Kabetogama Peninsula of Voyageurs National Park near the United States/Canada border approximately 60 km east of International Falls, MN. The bottoms of beaver ponds generally are composed of a layer of fine organic matter on top of shallow sediments that overlie glacio-lacustrine clays or granitic bedrock. The richly organic clayey soils are generally shallow (<30 cm) in meadows, but peat formations exceeding 3 m in depth may be present in older, larger meadows or ponds. In spite of cold temperatures (-40 °C or colder), surveys were conducted in the winter because summer vehicular access is impeded by wet and muddy conditions. In winter, frost intrudes into the upper soil layers in the meadows and ice forms on the ponds. During the winter of 1990–91, frost intrusion was generally 15 to 50 cm, while ice thickness reached 60 cm.

Montana

The Flathead River is a sixth order river that drains a large portion of the Glacier National Park and Bob Marshall Wilderness complex. Key biological and physical attributes of this large river system are described in detail by Stanford & Hauer (1992) and Stanford & Ward (1988). We studied two broad floodplains of the Flathead River — the Kalispell Valley on the mainstem and the Nyack Valley on the middle fork near West Glacier, MT. At both sites the thalweg of the river meanders across an extensive floodplain complex composed of low terraces bisected by active flood channels and higher terraces on which paleochannels are traced as surficial depressions meandering across the terrace surface. These paleochannels also are delineated by a water table near the surface and the presence of hydrophytes such as alder (*Alnus* spp.), dogwood (*Cornus* spp.), and black cottonwood (*Populus balsamifera* (Torr. & Gray) Brayshaw). On the downstream ends of the floodplains, springbrooks often upwell from paleochannels. Varying depths (0 m–3 m) of fluvial silts and clays overlie the cobbles and gravels which are deposited 10–20 m thick over an impermeable clay layer (Tertiary age) or bedrock (Precambrian mudstones). Water

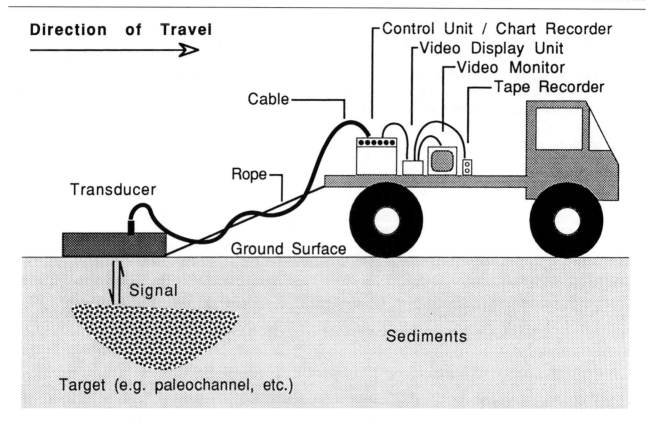

Fig. 1 Schematic of a GPR system mounted on a vehicle. The transducer is towed along the ground surface. Signals are generated and collected by the transducer and carried to the control unit through a cable. The control unit filters the signal, prints a strip chart, and transmits the filtered signal to a video display unit. The video display unit outputs to a video monitor for real-time viewing as well as an audio cassette recorder for data storage.

table elevations on the floodplains vary from 4–8 m below surface (based on data from monitoring wells) and the average slope of the floodplains is about 3%.

METHODS

System description and operation

We used a Subsurface Interface RADAR System 3 manufactured by Geophysical Survey Systems, Inc. (GSSI).* The system included two transducers (120 MHz and 300 MHz), and a video display unit. Additional accessories included a portable cassette recorder for data storage, and a video monitor to view data while recording. The entire system was converted to operate on AC power because cold temperatures and weight considerations prohibited the use of batter-

* Brand names used in the description of this system are for informational purposes only and do not constitute endorsements by the authors.

ies. A 300 watt generator supplied power to the system (Fig. 1).

A basic GPR system is composed of two components – a transducer that sends RADAR impulses of a predetermined wavelength and receives the reflected signal, and the control unit, which interprets and filters the signals from transducer and records the data. The transducer is dragged across the ground surface along a predefined transect, sending and receiving impulses several times per second. Each impulse is one sample. The number of samples per unit length depends on the sampling rate and the speed at which the transducer is moving. The transect must be relatively free of obstructions such as logs or rocks that would lift the transducer off the ground. Maintaining contact between the transducer and the ground surface is essential while collecting data.

Systems mounted on vehicles are practical because the weight of the system, supplies, and accessories may exceed 100 kg. Further, the GPR system can be powered by the vehicle's electrical system. Batteries can be used, but because they draw down quickly in cold temperatures and can not always be recharged easily, the advantages of using the vehicle's alternator or a generator are obvious. Finally, the vehicle protects the system from the weather. The system cannot be operated in heavy rain because of exposed electrical connectors. However, operating the system from a vehicle necessitates that the transect be clear and flat enough for the vehicle to travel along at a constant speed.

In Minnesota, the equipment was mounted on a rescue

sled pulled by a snowmobile. The video display unit, video monitor, and tape recorder were mounted inside an insulated waterproof box to protect the electronics from moisture. The box and the control unit were placed in the sled, and the transducer towed behind. The generator was mounted on a rack on the rear of the snowmobile.

Transects, spaced 20 m apart, were established using a surveyor's transit and tripod. The transects formed a grid, running both perpendicular and parallel to the general direction of water flow through each study site. Information about ice/frost thickness, water depth, sediment/soil horizon thickness, depth of peat deposits, and depth to 'parent material' (either glacio-lacustrine clays or granitic bedrock) was collected using a Dutch auger ~10 cm in diameter at each transect intersection. Subsurface features were then profiled by towing the GPR system behind the snowmobile along the transect. GPR data stored on cassette tape were later loaded onto a personal computer for analysis using the RADAN software package developed by GSSI. Information collected with the auger was used to verify the GPR data.

In Montana, the system was operated from a four wheel drive vehicle. The generator and control unit were placed in the back of the vehicle and the transducer was towed from a rope connected to the trailer hitch. Two transects were established on the Kalispell Valley floodplain. The lower transect began at the river's edge and continued perpendicular to the river for 250 m. The upper transect was approximately 1 km from the river's edge on one of the higher terraces and continued parallel to the river for approximately 100 m. On the Nyack floodplain, one transect 200 m long was surveyed parallel to the river alongside a seasonal side channel. Information about depth to the water table and location of hyporheic channels was derived from local wells and compared with the GPR measurements.

RESULTS

Water/Sediment Interface (Minnesota). GPR surveys taken through fresh water often provide excellent profiles of the water/sediment boundary (Ulriksen, 1982). However, the pond bottoms in Minnesota consisted of extremely fine organic and inorganic sediments that remain partially suspended in the water column resulting in a diffuse water/sediment boundary. The boundary was not distinct enough to cause a reflection noticeable through the background noise of the GPR data (Fig. 2).

Depth to Parent Material (Minnesota). In flooded ponds, we were able to map the depth to parent material (sediment/clay boundary; Fig. 2.) along approximately 60% of the transects. The water and organic sediments in the ponds were easily penetrated by the RADAR and the sharp boundary

between the sediments and parent material often provided a strong return signal. However, in wet meadows we were unsuccessful. GPR signals generally will not penetrate clays (Doolittle, 1982; Hubbard *et al.*, 1990). The meadow soils are high in clay content (50%–90% clay) and the thickness of the organic soils were sometimes less than the signal wavelength so reflections were not detected.

Thickness of Peat Deposits (Minnesota). We successfully mapped the thickness of peat deposits in ponds because their site characteristics are optimal for the use of GPR. Subsurface features are simple (composed of a layer of *Sphagnum* moss covering the peat with the peat extending to the parent material), the boundary between the peat and clay is distinct, and the thickness of the layers exceeded the signal wavelength. While the *Sphagnum*/ peat boundary is not sharp enough to detect with the GPR, the *Sphagnum* layer is relatively uniform in thickness (80–90 cm). We therefore assumed the depth to clay, less 85 cm, was the approximate thickness of the peat.

However, when peat depth exceeded 3.5 m, the RADAR signal was not powerful enough to reach the clay layer and return. The peat was saturated with water and water attenuates GPR signals rapidly (Ulriksen, 1982). A lower frequency transducer would have penetrated further, but cost was prohibitive for obtaining more transducers.

Frost Intrusion and Soil Depth (Minnesota). Frost intrusion varied considerably throughout each meadow, ranging from 15 to 50 cm. Soil thickness (soil surface to parent material) also varied substantially, from about 10 to 50 cm. Further, the meandering stream left occasional deposits of sand and small gravel scattered throughout the meadow. When the frostline, the soil/parent material interface, sand deposits, logs, stones, and other anomalies were detected by the GPR system, the high spatial variability and juxtaposition of these features resulted in a complex subsurface and the features were generally indistinguishable.

Water Table (Minnesota & Montana). Locating the water table also can be hampered by a diffuse interface. In finer soils, such as those in Minnesota, the capillary action of the soil may cause a diffuse transition zone between saturated and unsaturated soils. In sandy or gravelly locations such as in Montana, the boundary is typically much sharper and more easily detected using GPR (Fig. 3), although verification of depth from nearby wells was still necessary to be sure the weak reflection recorded was indeed the water table. Houck (1982) and Sellman *et al.* (1983) also report success in locating the water table. Olson & Doolittle (1985), however, had more difficulty and also cite fine soils as the probable cause.

Paleochannels and Hyporheic Flow (Montana). Attempts to locate zones of rapid hyporheic flow within the Flathead River floodplains were successful. Fig. 4 shows known paleo-

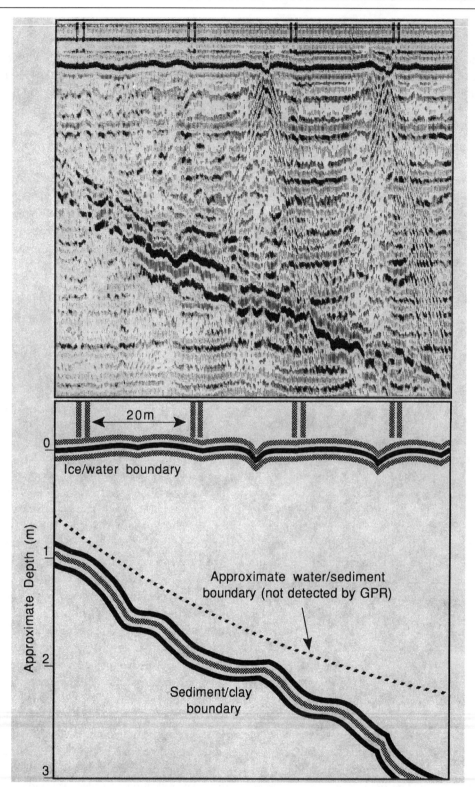

Fig. 2 Portion of GPR profile on a flooded beaver pond in Minnesota showing ice/water boundary and sediment/clay boundary (original data above, schematic below). Notice that the water/sediment boundary known to exist was not detected by the GPR system.

channels located on the upper transect of the Kalispell Valley. A 'U' or 'V' shaped feature with reduced signal penetration below seems to be the characteristic signature of paleochannels. The increased water content of paleochannels appears to attenuate the GPR signal more rapidly than the finer, less porous deposits known to surround them.

Figs. 5 and 6 show probable paleochannels which were

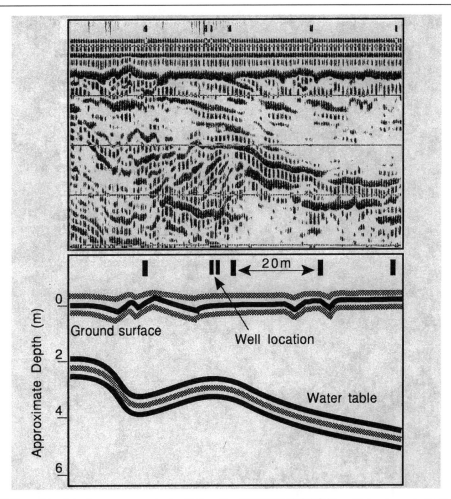

Fig. 3 Portion of GPR profile of the lower transect in the Kalispell Valley (starting at river edge, running perpendicular to river) showing the water table (original data above, schematic below). The water table is actually flat. The apparent curve in the water table is caused by a small hill on the ground surface which temporarily increased the distance between the water table and the transducer.

unknown prior to our GPR survey. Fig. 5 is a section of the lower transect in the Nyack Valley. Notice the similarity to the paleochannels in Fig. 4: an apparent buried channel with decreased signal penetration beneath. Fig. 6, a portion of the transect on the Nyack floodplain, shows what appears to be a paleochannel buried beneath several meters of more recent deposition. Decreased signal penetration below this feature is not as dramatic as in the previous examples. Perhaps the porosity of this channel is not as high as in the previous examples and the paleochannel may therefore have more restricted hyporheic flow. Alternatively, the reduced penetration may be less evident because the signal was already nearing its maximum penetration when it encountered this feature. Unfortunately, at the time of the survey, we lacked the drilling equipment needed for verification of the exact nature of the features in Figs. 5 and 6.

DISCUSSION

Ground penetrating RADAR has proven effective in many applications in the ecological and soil sciences. However, in our experience, the success of the system depends strongly on the site and subsurface characteristics. The following paragraphs describe constraints which are essential to the success of any GPR study and which should be considered when deciding whether or not to employ GPR.

Locating vs Identifying Subsurface Features. GPR does not generally identify subsurface features, it locates subsurface features. The only way to positively identify a subsurface feature is to expose it. Once a feature has been identified using another means, GPR may be useful for mapping the extent of that feature by correlation with evidence from boreholes. Alternatively, if a feature is known to occur within a discrete area, GPR may be able to pinpoint its location and thereby eliminate the need to unearth the entire area to find it. If the feature has a unique shape, such as with paleochannels, reasonable certainty can be obtained from the GPR data alone but absolute certainty still requires exposing the feature.

Feature Characteristics. GPR is best suited to locating features with sharp boundaries. If the boundary is gradual or

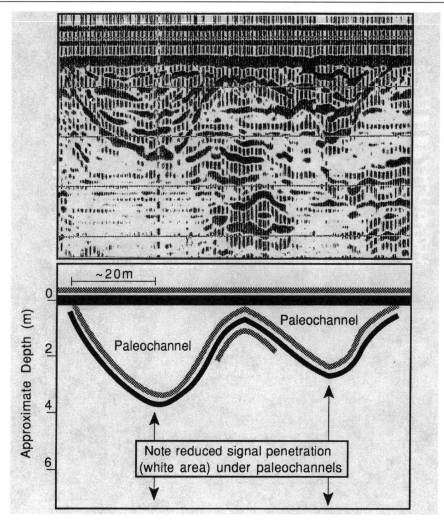

Fig. 4 Portion of GPR profile of the upper transect in the Kalispell Valley (parallel to active river channel, 1 km distant) showing the subsurface features of the hyporheic zone (original data above, schematic below). Zones of high porosity (referred to as a paleochannels) are indicated in the profile.

not well defined, GPR may not detect it. Each of the features we successfully detected (the water table and paleochannels in Montana and the peat/clay boundary in Minnesota) was well defined. Those features without distinct boundaries (pond bottoms and water table in Minnesota) were not detected.

Site Characteristics. Soils with high clay content are not generally penetrated by the GPR system. The boundary is generally easy to detect if, as with the peat deposits in Minnesota, the subsurface feature lies directly above clay. However, the chance of locating a feature is reduced if it lies below a clay layer or below soil high in clay content, as was the case with frost intrusion and parent material in the Minnesota meadows. Deposits high in metals and salts may also interfere with penetration (Shih & Doolittle, 1984; H. M. Valett, University of New Mexico, personal communica-

tion). GPR will not penetrate as far through fresh water or wet soils as through dry soils. Fresh water does not block the signal, but more powerful transducers may be necessary to obtain the desired penetration.

Transducers. Different transducers are designed to collect different types of data. High frequency transducers collect detailed, high precision information near the surface (< 2 m). They can detect smaller or poorly defined subsurface features which low frequency transducers can not. Low frequency transducers collect less detailed, lower resolution data, but have the ability to penetrate further into the subsurface (maximum of roughly 30 m depending on site characteristics). Matching the transducer to the type of information being sought is essential. In our study, a more powerful transducer would have allowed us to map peat depths beyond 3.5 m.

Subsurface Complexity. GPR is most suited to mapping areas with relatively simple subsurface structure. As the complexity of the subsurface increases, more and more additional information about the subsurface is needed to properly interpret the output, thereby reducing one of the primary benefits of GPR – labor reduction. In parts of the

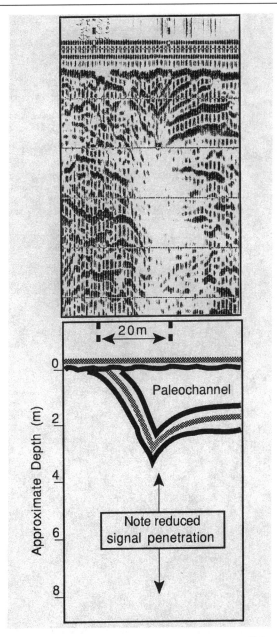

Fig. 5 Portion of GPR profile in the Nyack Valley (parallel to active river channel) showing a probable paleochannel which was discovered during the GPR survey (original data above, schematic below).

Minnesota meadows where clays did not block the signal, the subsurface was complex enough to render most data uninterpretable. However, in Montana, where the subsurface was not complex, little supplemental information was needed to interpret the images.

CONCLUSION

GPR has the potential to be a useful tool for exploration and mapping ecologically significant subsurface structures. The

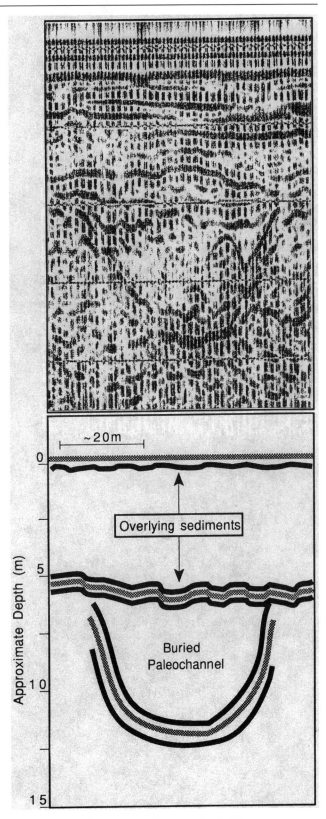

Fig. 6 Portion of GPR profile from the Nyack Valley (perpendicular to active river channel) suggesting the presence of a buried paleochannel (original data above, schematic below).

ability to locate these structures should help increase our understanding of groundwater systems and their interactions with surface waters. There are, however, important limitations of GPR – especially when working in riparian systems where soils tend to be saturated with water. GPR techniques should have a good chance of success if, as with our Montana study, each limitation is carefully considered and no incompatibility is apparent. However, as each of these considerations deviates from the optimum, like in our Minnesota site, the chance of success deteriorates significantly. We hope that, having more clearly defined the limitations and capabilities of GPR, others will be better able to determine whether or not it is likely to be useful in their investigations.

ACKNOWLEDGEMENTS

The field assistance of Cal Harth, Paul Nordeen, and Kurt Fogleburgh is gratefully acknowledged. Carol Johnston provided information about soils on the Kabetogama Peninsula. Peter Gogan at Voyageurs National Park and Tom Fenner at Geophysical Survey Systems, Inc. were extraordinarily helpful throughout this study. Maury Valett provided insight as well coordinating a loan of equipment from Stuart Fisher. Support for this project was provided by the National Science Foundation.

REFERENCES

Baraniak, D.W. (1983). Exploration for surface peat deposits using ground penetrating RADAR. In *Proceedings of the International Symposium on Peat Utilization*, ed. C.H. Fuchman & S.A. Spigarelli, pp. 105–21. Bemidji, MN: Bemidji State University, October 10–13, 1983.

Collins, M.E., G.W. Schellentrager, J.A. Doolittle, & Shih, S.F. (1986). Using ground-penetrating radar to study changes in soil map unit composition in selected histosols. *Soil Sc. Soc. Amer. J.*, **50**, 408–12.

Collins, M.E., & Doolittle, J.A. (1987). Using ground-penetrating radar to study soil microvariability. *Soil Sc. Soc. Amer. J.*, **51**, 491–93.

Collins, M.E., J.A. Doolittle, & Rourke, R.V. (1989). Mapping depth to bedrock on a glaciated landscape with ground-penetrating radar. *Soil Sc. Soc. Amer. J.*, **53**, 1, 806–12.

Doolittle, J.A. (1982). Characterizing soil map units with the ground-penetrating radar. *Soil Surv. Horizons*, **23**, 3–10.

Doolittle, J.A. (1987). *Using ground-penetrating radar to increase the quality and efficiency of soil surveys*. Soil Survey Techniques: Soil Science Society of America Special Publication #20. Madison, WI: Soil Science Society of America.

Doolittle, J.A., R.A. Rebertus, G.B. Jordan, E.I. Swenson, & Taylor, W.H. (1988). Improving soil-landscape models by systematic sampling with ground-penetrating radar. *Soil Surv. Horizons*, **29**, 46–54.

Doolittle, J.A, M.A. Hardisky, & Gross, M.F. (1990). *A ground-penetrating radar study of active layer thicknesses in areas of moist sedge and wet sedge tundra near Bethel, Alaska, U.S.A. Arctic and Alpine Research*, **22**, 175–82.

Houck, R.T. (1982). *Measuring moisture content profiles using ground-probing radar*. Technical Report, pp. 1–10. Springfield, VA: XADAR Corp.

Hubbard, R.K, L.E. Asmussen, & Perkins, H.F. (1990). Use of ground-penetrating radar on upland costal plain soils. *J.Soil Wat. Conse.*, **45**, 399–405.

Naiman, R.J., C.A. Johnston, & Kelley, J.C. (1988). Alteration of North American streams by beaver. *BioScience*, **38**, 753–62.

Olson, C.G. & Doolittle, J.A. (1985). Geophysical techniques for reconnaissance investigations of soils and surficial deposits in mountainous terrain. *Soil Sc. Soc. Amer. J.*, **49**, 1490–98.

Sellman, P.V., S.A. Arcone, & Delany, A.J. (1983). *Radar profiling of buried reflectors and the ground water table*. Report 83–11: pp. 1–10. Hanover, NH: Cold Regions Research and Engineering Laboratory.

Shih, S.F., & Doolittle, J.A. (1984). Using radar to investigate organic soil thickness in the Florida Everglades. *Soil Sc. Soc. Amer. J.*, **48**, 651–6.

Stanford, J.A. & Hauer, R. (1992). Mitigating the impacts of stream and lake regulation in the Flathead River catchment, Montana, USA: an ecosystem perspective. *Aquatic Conservation*, **2**, 35–63.

Stanford, J.A, & Ward, J.V. (1988). The hyporheic habitat of river ecosystems. *Nature*, **335**, 64–6.

Stanford, J.A. & Ward, J.V. (1993). An ecosystem perspective of alluvial rivers: connectivity and the hyporheic corridor. *J. N.Am. Benthol.Soc.*, **12**, 48–60.

Triska, F.J., V.C. Kennedy, R.J. Avanzino, G.W. Zellweger, & Bencala, K.E. (1989). Retention and transport of nutrients in a third order stream in Northwestern California: hyporheic processes. *Ecology*, **70**, 1893–1905.

Truman, C.C., H.F. Perkins, L.E. Asmussen, & Allison, H.D. (1988). Using ground-penetrating radar to investigate variability in selected soil properties. *Jour.Soil Wat.Cons.*, **43**, 341–5.

Ulriksen, C.P.F. (1982). *Application of impulse RADAR to civil engineering*. North Salem, NH: Geophysical Survey Systems, Inc.

III

Malfunction of groundwater/surface water interfaces: causes and methods of evaluation

19 Heterogeneity of groundwater-surface water ecotones

V. VANEK

Université Lyon 1, URA CNRS 1974, Ecologie des Eaux Douces et des Grands Fleuves, Hydrobiologie et Ecologie Souterraines, 43 Bd du 11 novembre 1918, 69622 Villeurbanne cedex, France

Present address: VBB VIAK Consulting Engineers, Geijersgatan 8, S-216 18 Malmö, Sweden

ABSTRACT Groundwater-surface water ecotones control hydraulic exchange between groundwater and surface water and, together with land-water ecotones, serve as a temporary or permanent sink of catchment-derived inorganic and organic matter. These factors and the chemistry of water entering an ecotone, determine the ecotone's effect on water quality. One of the problems when studying ecotones is their high heterogeneity in space and time. The heterogeneity is caused by several thousand years of human impact, which has been exacerbated during the last century, and by local geology, surface water effects and various ecotone-related processes (capillary-fringe effect, evapotranspiration, anti-clogging, clogging). The heterogeneity is measurable and at least partly predictable, and should be considered when preparing sampling strategies and evaluating data. Ecotones often function as filters of nutrients and particulate matter, thus improving the quality of water passing through them. The adsorption or buffering capacity of the ecotone, however, may be exhausted which may result in the enrichment of water by organic matter, dissolved metals etc. This 'malfunctioning' occurs often naturally as a result of decomposition of accumulated organic matter, but may also be due to the excessive leaching of fertilizers or other substances from upstream areas. Losses from catchments may be diminished by increasing the area and improving the function of the ecotones. The ecotone-oriented measures, however, must be combined with modified land use of the whole catchments if we want to use the landscape in a sustainable way.

INTRODUCTION

Groundwater-surface water ecotones mainly occur in two types of environments, beneath surface water bodies and along the shores where they overgo gradually into land-inland water ecotones. Both these environmental types serve often as a temporary or permanent sink of inorganic and organic matter. This matter may either be derived from the catchment or produced in the ecotones, using catchment-derived nutrients. Groundwater-surface water ecotones are generally characterized by hydraulic exchange between surface water and groundwater, and by locally elevated concentrations of organic matter or various catchment-derived substances. Differences in chemistry between groundwater and surface water in combination with these two characteristics result in high biogeochemical activity, expressed as horizontal gradients in oxygen, temperature,

light etc. This activity influences the quality of water passing through the ecotone.

Usually, water flow through surface water beds and shores is highly variable both in space and time (Vervier *et al.*, 1992). In many places it may be close to zero and insignificant for the total water budget, but even then it may still be high enough to impact water chemistry and biology. The hydraulic exchange is seldom obvious without special measurements and sampling programs, and only recently has it been recognized as an important factor for global cycles of mass and energy. Biologically-oriented ecotone studies often are directed towards biogeochemical processes within the ecotone itself. However, the amount of hydraulic exchange and the original quality of water entering the ecotone should also be considered to understand the functioning and ecological role of these systems (Gibert *et al.*, 1990). This paper treats various sources of spatial and temporary heterogeneity of

151

groundwater-surface water ecotones, with respect to both their hydraulics and biogeochemistry, and discusses the importance of these factors for ecotone functioning.

In a few cases, the role of ecotones as filtering and buffering systems improving water quality has been recognized and used, such as in artificial groundwater recharge schemes or when establishing protection areas around valuable surface water resources. Only recently, it has been realized that the growing environmental problems of both global and local character are not compatible with sustainable development of our society. In the discussions of waste water and solid waste handling, the terms 'end-of-pipe' and 'close-to-source' approach have been used. 'End-of-pipe' means to collect and treat the wastes in big, centralized treatment plants or landfills. 'Close-to-source' means to reduce production of wastes and encourage their reuse in small systems that are linked closely (in space, economically . . .) to the waste producers (Niemczynowicz, 1991). Even if the 'end-of-pipe' approach is still preferred by many technical designers and decision makers, its limitations are becoming more obvious. Instead, the more ecological 'close-to-source' approach has to be adopted, and it is believed that better understanding and use of ecotones may play an important role in this future development.

ECOTONE HYDRAULICS

Basic theory

Some of the precipitation falling on land surfaces infiltrates the ground and continues as subsurface groundwater flow. Groundwater sooner or later discharges into surface water, and on its way passes groundwater-surface water ecotones. There are, however, several exceptions from this rule which may cause the ecotones to be by-passed (Ward, 1975; Chow *et al.*, 1988).

a) Saturated overland flow, which is runoff generated by rain falling on water-saturated soil surface. Soil may be saturated by rain, by discharging groundwater or by surface water brought in during flood periods. The first case occurs when rainfall intensity exceeds infiltration capacity of soil surface. This type of runoff is called Hortonian overland flow, and is a typical feature of arid and semiarid areas, and of urban regions or other regions heavily altered by human activity.

b) Subsurface unsaturated flow which develops in highly anisotropic or heterogeneous soils, or in areas with steep slopes. Here, the infiltrated water may not reach groundwater but runs laterally within the unsaturated zone. Eventually this water may form ephemeral, perched water tables.

c) Return flow of groundwater that seeps out in land-based discharge areas and runs overland. During the rain event, this kind of runoff mixes with saturated overland flow.

d) Evaporation or evapotranspiration of shallow groundwater.

e) Finally, groundwater may be collected artificially via wells, ditches or drainage pipes.

Thus, the importance of ecotones in regulating water-related fluxes in the landscape may be suppressed, simply because they do not come in contact with all types of water flows.

Between every open water body and a surrounding geological material there will be some exchange of water. This is given by the fact that all geological materials are to some extent permeable to water, and that water levels in the two systems (open water body and water-saturated matrix) are seldom the same, being governed by different mechanisms. In open water bodies, water level is primarily determined by gravitational forces and the elevation and hydraulic characteristics of the river bed, lake outlet etc. In geological material, water table elevation is determined by an equilibrium between the rate of water input and the resistance of geological material to the gravitational flow transporting water away. Theflow in a porous medium is expressed by Darcy's Law which states that (Freeze & Cherry, 1979)

$$Q = K\,i\,A \qquad (1)$$

In our case (flow across the ecotone), Q=water flow across the sediment-bottom boundary (also called seepage rate or 'Darcy velocity'), K=permeability (hydraulic conductivity), i=hydraulic gradient in the direction of water flow and A=cross-sectional area of flow. In the equation, K is a characteristic of the geological material, A is given by the geometry of the system and i is a driving force causing water to flow.

Water always flows in the direction of decreasing hydraulic gradient. If the groundwater table is higher than the adjacent surface water level, groundwater will seep out through the bottom sediments. If it is lower, surface water will infiltrate groundwater. The same applies in a vertical sense – if a piezometric level in deeper groundwater is higher than in shallow groundwater, water will flow upwards and vice versa (Fig. 1). Some of commonly used names describing the situation of groundwater inflow are a groundwater discharge area, a discharge lake or a gaining stream. In the case when surface water infiltrates groundwater we speak about groundwater recharge, recharge lake or loosing stream. Commonly we find both recharge and discharge conditions in the same water body. In such a case we speak about flow-through lakes or streams with gaining and loosing reaches.

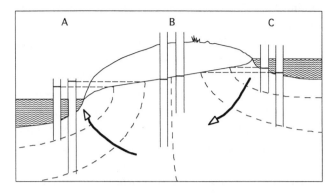

Fig. 1 Direction of groundwater flow (arrows) and equipotential lines (hatched) in a groundwater system connecting two surface water bodies. Water levels in piezometer nests A, B and C indicate downward, horizontal and upward groundwater flow, respectively.

McBride & Pfannkuch (1975) have shown that in lakes, the absolute (either negative or positive) groundwater velocity across the bottom-sediment boundary (seepage rate) usually is highest near the shore and decreases exponentially with increasing distance from shore. This basic pattern (Fig. 2A) applies to all water bodies surrounded by homogeneous porous media, and is determined by the geometry of the system. Sometimes it may be accentuated by the occurrence of low-permeable sediments further offshore, for example by soft sediments in the central parts of lakes.

External factors causing the spatial and temporal variability of ecotone hydraulics

The basic seepage pattern (Fig. 2A) is only found in homogeneous geological formations that sometimes occur in sand and gravel deposits. In most cases, however, the pattern is more or less modified, either in space or time. The actual deviation from the basic seepage pattern can often be explained as an effect of some of the factors described below. Spatial and temporal heterogeneity of ecotone hydraulics usually lead to heterogeneity of biogeochemical properties and water quality, problems which I will treat separately.

FACTORS RELATED TO GEOLOGY OF THE ADJACENT AQUIFER

Most natural aquifers are heterogeneous and anisotropic, and it is easy to understand that the hydraulics of an ecotone bordering such an aquifer will differ from an ecotone which is adjacent to a homogeneous, isotropic sand deposit. Some common examples of geologically-caused seepage patterns are given in Fig. 2B-D. For example, if sediments consist of layers of different permeability, groundwater flow may preferably follow high permeability layers. This is the case where low-permeable, horizontal layers of clay, peat etc. are underlain by a water-bearing material of higher permeabil-

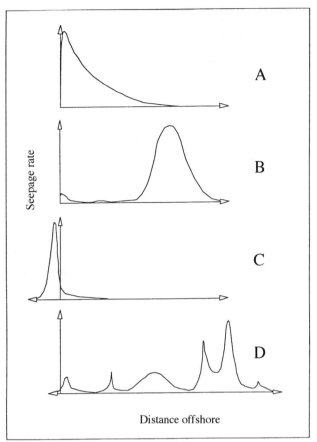

Fig. 2 Some seepage patterns, i.e. relationships between seepage rate and distance from shore, found in lakes and coastal areas (further explanations in text).

ity. If such a low permeable layer is located near the shore, it will deflect groundwater discharge from deeper layers further away from the shore (Fig. 2B). The low-permeable layers may be many meters thick and extend several kilometers offshore, as the aquitards of large artesian aquifers (Kohout, 1965). The layers may, however, also be thin (<10 cm) and still alter seepage patterns (Guyonnet, 1991).

Another pattern (Fig. 2C) is typical for areas where a very shallow water table intercepts the ground surface before reaching a surface water body. Such conditions may occur in low-permeable glacial till areas where permeability is high near the surface but declines rapidly with the increasing soil depth. In this case, most of the flow is concentrated into macropores, such as tree hole channels, rodent holes, discontinuities near rocks or stones etc., and the flow often creates a distinct seepage horizon just above the average lake or river level. A similar pattern is common in marine coastal areas where fresh groundwater tends to float on denser sea water. Temporal fluctuations often associated with this seepage pattern are similar to those of capillary fringe effects, and will be mentioned later.

Highly irregular seepage patterns (Fig. 2D) are typical for

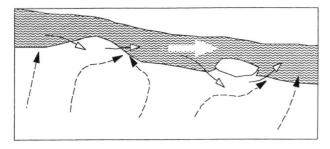

Fig. 3 Flow pattern of surface water underflow (full arrows) and groundwater discharge (broken arrows) in streambeds, highly dependent on streambed shape and geomorphology.

ecotones bordering karstified or fissure-rock aquifers where most of the flow occurs in subaquatic springs, pockmarks (crater-like bottom structures) or other highly active, limited seepage zones.

Seepage patterns determined by local geological conditions are often fixed in place, and react in a predictable manner to hydrological changes. In certain areas, however, rapid changes may occur. For example, irregular seepage patterns of calcium rich sediments (Fig. 2D) are often unstable in time due to the clogging of macropores and creation of dissolution channels.

When comparing ecotones adjacent to lakes with those of rivers and marine coastal zones, we find many similarities but also some differences. One major difference is that geological history of ecotones close to lakes (particularly small ones) is almost entirely dependent on local geological conditions while along rivers and coastal areas there may be a mixture of local as well as more distant geological influences.

SURFACE-WATER RELATED FACTORS
Bottom-sediment morphology and other bottom characteristics are highly dependent on surface water, namely on its movements and its content of suspended matter. Water movements may be due to gravitation (running water currents), wave and wind action, tidal effects or human activities (e.g. ship navigation), and cause the sorting and redistribution of bottom sediments according to main sedimentological characteristics, i.e. particle size, shape and specific weight. The sorting effect is more or less predictable (Håkanson & Jansson, 1983) and results in a quasi-stable pattern of bottom areas characterized mainly by the erosion, transportation or sedimentation processes. Particles are usually largest in erosional bottoms, medium in size in transportation areas and smallest in sedimentation bottoms. If water currents over the erosion and transportation-type bottoms are strong enough they will induce the creation of bottom structures such as ripples, sand banks, gravel bars or sequences of riffles and pools.

Particle size is one of the most important factors deter-

mining the sediment permeability. Therefore, one may expect that permeability will be highest within the erosion areas and decrease towards the sedimentation areas. Usually, erosional bottoms are concentrated close to the shore, in areas where for geometrical reasons, seepage rates are high. Thus, the processes of erosion, transportation and sedimentation do not greatly alter the basic seepage pattern (Fig. 2A).

Similarly, as in a porous medium, surface water will only flow if there is a difference in pressure between two points. If surface water currents occur close to the permeable bottom or river bank, pressure differences will be transmitted into the porous material and cause water to flow (usually much slower) inside the material and across the bottom-sediment boundary. Probably the best described currents of this type are those occurring in riffles and pools in small streams, where surface water enters the stream bed at the upstream end of a riffle, continues underground for some distance as a so-called underflow, and reenters the stream (usually together with some deeper groundwater) at the end of the riffle (Fig. 3; Hendricks & White, 1991). Similar patterns are common in larger rivers where surface water circulates through the gravel or sand bars or through the stream banks, e.g. in meandering streams (Thibodeaux & Boyle, 1987, Creuzé des Châtelliers, 1991). In large lakes and marine coastal areas, pressure changes associated with the tides and the wave movements cause interstitial water to move through the sediment. This circulation, sometimes called 'subtidal pump', seems to occur mainly on a small scale of individual ripples, down to a water depth of 100–200 m (Riedl et al., 1972, Shum, 1992).

A somewhat similar mechanism occurs when the surface water level suddenly rises above the adjacent water table. Groundwater usually adjusts to this change, but reacts more slowly which gives surface water time to spill over and fill unsaturated zones above the water table. This water eventually returns when the surface water level has dropped. This so-called bank storage effect is known from rivers and regulated dams where it influences propagation of flood waves (Pinder & Sauer, 1971), but also from marine beaches ('tidal pump' of Riedl 1971, seawater filtration by beaches, McLachlan, 1989) where it leads to mixing of saline water with inflowing fresh groundwater.

Surface-water related heterogeneity of the ecotones may exhibit marked variability in time, particularly in rivers with highly fluctuating discharge or in high-energy beaches. In small lakes, low-energy beaches or groundwater-fed smaller streams, this kind of ecotone heterogeneity usually is less pronounced and easier to predict.

HUMAN IMPACTS
Indirect human impacts on groundwater-surface water ecotones are mainly coupled to modifications of land use, which in turn change hydrological cycle and fluxes of solutes and

particulate matter in the landscape. These modifications began several thousand years ago with deforestation and creation of pastures. Replacement of deep-rooted vegetation by shallow-rooted vegetation usually leads to a decrease in evapotranspiration and increase in total runoff. In semiarid and arid lands, high rainfall intensities often are combined with low infiltration capacities of the soil, particularly in deforested areas where surficial soil layers desiccate into low permeable crusts. Under such conditions, deforestation leads to serious erosional problems due to increased overland runoff, and at the same time to a lower groundwater table. In most humid regions, the infiltration capacity of the soil exceeds rainfall intensities. Here, deforestation leads to a rise in the groundwater table. The increased runoff accelerates weathering and leaching of salts and nutrients from soils, and increases their transport out of the catchment (Ruprecht & Schofield, 1991; Lepart & Debussche, 1992). Higher total runoff increases also the magnitude and velocity of floods, which leads to erosion and morphological modifications of river channels and floodplains.

Even if deforestation may have variable effects on soil erosion, water table elevation etc., it always reduces water retention capacity of the catchment. Human impacts, which follow deforestation, accelerate erosion and transport of particulate matter out of the catchment through increased grazing pressure, the introduction of arable farming and irrigation. In semiarid regions and other sensitive areas, these activities may easily lead to complete breakdown of soil structure and desertification of waste areas (Ward, 1975; Zaletaev, 1996). Even in less sensitive, humid areas, most catchments are leaking salts and nutrients and losing great amounts of soil, often irreversibly. All these changes increase gradually the stress on groundwater-surface water ecotones (and all other ecotones in the landscape), both in the form of increasing, more variable hydraulic load and increasing transport of dissolved and particulate matter.

During the first period of land-use modifications, land-water and groundwater-surface water ecotones themselves often were left aside as low productive or unmanageable areas. This situation changed profoundly some 100 or 150 years ago when pressure on all natural resources increased as a result of increasing human population and improved mechanization. Since then, many ecotones were either completely destroyed (drainage of lakes and wetlands, straightening and regulation of rivers for flood protection or navigation, culverting of small agricultural streams), or heavily modified (conversion of riparian zones into arable land, lowering of groundwater table for agricultural needs or as a secondary effect of stream regulation coupled to increased erosion of stream bottoms, by-passing of ecotones via drainage pipes or ditches etc.). At the same time, stress on the remaining ecotones increased and diversified (new pollutants, fertilizers

etc.). Even if most remaining ecotones adapted remarkably well to these dramatic changes, there is a growing number of examples where the assimilative or buffering capacity of the ecotone is endangered or exhausted.

Ecotone-related processes

In addition to the above described external processes, ecotone hydraulics may be influenced by several internal mechanisms depending on processes within the ecotones themselves. The division into external and internal processes, however, is rather artificial as most of the 'internal' processes occur also outside the ecotones. Ecotone-related processes may be divided into several categories.

LOCALIZED WATER INPUTS OR OUTPUTS

These factors are not only typical for groundwater-surface water ecotones but occur also in other groundwater ecotones such as wetlands and areas with shallow water tables. In soils consisting mostly of fine to medium sand, there is an extensive (several tens of centimetres thick) capillary fringe, an almost saturated sediment layer just above the water table. If the water table is shallow, only a small amount of precipitation is needed to fill the remaining soil pores and quickly raise the water table to the ground surface (Gillham, 1984). This may create local groundwater mounds, reverse the flow within the ecotone or result in an overland runoff of precipitation water mixed with shallow groundwater.

The capillary fringe effects are an example of temporal, precipitation-dependent changes of ecotone hydraulics, which typically last only for a short period of several hours or days. In humid areas with shallow or low-permeable soils, variable precipitation or some human impacts (e.g. deforestation) may lead to the creation and dissipation of temporal swamps or ephemeral streams on a whole-catchment basis. In such a case, the total area of the catchment actually contributing flow to the stream ('variable source area', Fig. 4) and therefore also the extent of groundwater-surface water ecotones changes in time. These changes often occur on a temporal scale of days to months, sometimes several years.

Phreatophytic vegetation which has roots extending down to the groundwater zone is a common feature of most ecotones. Evapotranspiration of this vegetation may cause local lowering of the water table and modify groundwater flow velocities and directions, either on a diurnal or a seasonal basis. For example, willow rings surrounding small prairie lakes (potholes) in Canada are known to control their hydrology (Meyboom, 1966). Another example occurs in small streams or irrigation channels which during summer often exhibit marked diurnal fluctuations, both in flow and in electrical conductivity (due to the evaporative concentration of salts; Calles, 1982).

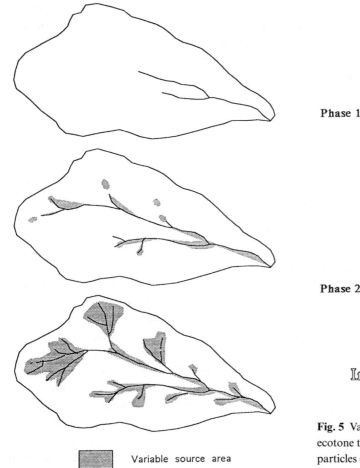

Fig. 4 The area contributing directly to the stream flow (variable source area) may change its size in time, particularly in a catchment with shallow or low-permeable soils. At the same time changes in the relative proportion of various runoff forms occur (subsurface runoff, saturated overland flow, return groundwater flow etc.) (simplified from Chow *et al.*, 1988).

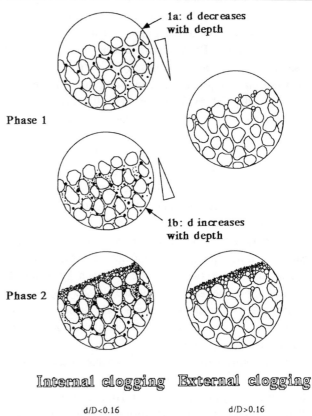

Fig. 5 Various phases of clogging a groundwater-surface water ecotone that has an original mean particle size D, by smaller particles (particle size d) brought in by surface water flow. Initially (Phase 1), mechanical clogging is the main process. Later (Phase 2), the average particle size around sediment-water interface (external clogging) or within the whole sediment layer (internal clogging) decreases. Then, clogging usually accelerates due to various physico-chemical or biological clogging processes (simplified from Sinoquet, 1987).

MECHANICAL FILTERING, CLOGGING AND ANTI-CLOGGING

Filtering of small particles is a process occurring in all porous media. In the ecotones. surface water often is a permanent source of both allochthonous and autochthonous particles of all sizes. Depending on particle size, these particles may either be trapped on the surface of the ecotone (external clogging, cf. 'Surface-water related factors') or brought into the porous matrix (Fig. 5). Mechanical clogging by fine particles brought in from the outside is likely to be more important in surface-water infiltration areas, as groundwater usually contains much less suspended, readily movable particles.

Flow velocity within the ecotone or at least its parts usually is much slower than in the open water, but at the same time higher than the average regional groundwater flow velocity. Moreover, unconsolidated types of fine sediments

which often cover surface-water bottoms (gyttja, marl etc.) are easily flushed away or fluidized by groundwater flow (Fig. 6). Then, groundwater flow may create limited areas of very soft, fluidized sediments or crater-like bottom-sediment structures, pockmarks. These structures seem typical for areas where groundwater flow for geological reasons is concentrated into limited bottom areas (fissured rocks, karstified regions, bottom areas which are incompletely covered by low-permeability layers of clay or peat) (Vanek, 1992).

BIOCHEMICAL PROCESSES OF MIXING, CLOGGING AND ANTI-CLOGGING

Even if bioturbation occurs both in the groundwater and the surface water zone, it probably is most important in the uppermost 10–30 cm thick sediment layer where various benthic organisms actively pump water through the sediment (insect larvae, mollusca) and mix the sediments through their burrowing activity. Another example of bioturbation impor-

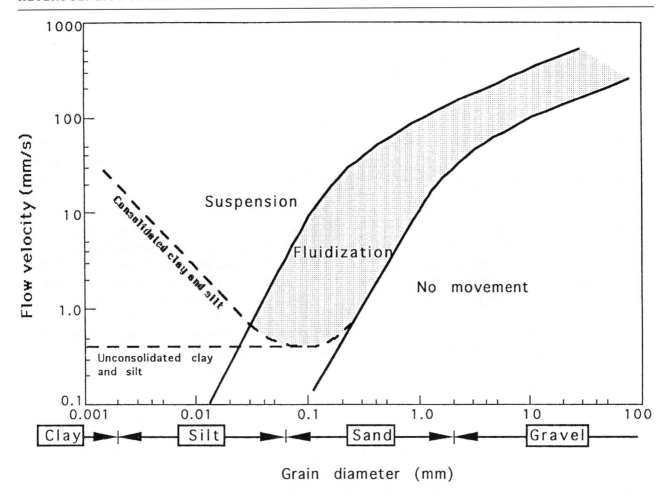

Fig. 6 Empirically derived effect of upward water flow velocity on the behaviour of particles of variable size (Lowe, 1975).

tant in ecotones containing low permeable soils is the transport of water and gases (both passive and active) through or along plant roots and stems.

Most microorganisms in the groundwater zone are probably attached to particles (Ghiorse & Wilson, 1988). Bacterial numbers are surprisingly high also in pristine aquifers, but the organisms often are inactive in a dormancy stage (Fliermans & Balkwill, 1989). If there are enough nutrients or organic matter, some of the particle-bound microorganisms create thick biofilms that are rich in polysaccharides and may easily clog sediment interstices. Experiences from slow sand filters of waterworks and from natural ecotones (Husmann, 1978; Danielopol, 1984) show that some meio- and macroorganisms (protozoans, tubificids, chironomids) may partly or completely counteract this kind of clogging by increasing sediment permeability, both indirectly (sediment ingestion and formation of pellets) and directly (grazing of bacterial biofilms).

Groundwater entering surface water often is low in oxygen and high in dissolved iron or manganese. Groundwater-discharge areas of this water type are usually highly visible due

to the presence of red or black encrustations of metal oxides which may cover large bottom areas (lake ores) and cause groundwater flow to divert to the borders of these areas and lower their permeability (Krabbenhoft, 1984). Formerly, these encrustations or nodules were collected as an important local source of iron, and it has been observed that new encrustations developed at the same place within 10–20 years (Naumann, 1922).

Gas bubbles present one of the most effective ways of lowering permeability of bottom sediments. The gas may be created under deeply anaerobic conditions (methane, hydrogen, hydrogen sulphide), under slight anaerobiosis (nitrogen and other products of denitrification), or as an effect of pressure release when groundwater approaches the surface (e.g. degassing of the deeper groundwater which sometimes is saturated or over-saturated with various gases). Carbon dioxide usually does not cause any serious clogging due to its high solubility in water.

Under some circumstances, e.g. continuous gas production, increasing temperature or sudden drop of surface water level, gas bubbles mix the sediments and alter their permeability as they escape into the atmosphere (gas ebullition). In marine areas large, sudden gas escapes from the sediments may form pockmarks similar to those created by groundwater flow, but often much larger (up to 300 m in diameter and 30 m depth, Hovland & Judd, 1988).

Carbon dioxide released under decomposition of organic matter and as a product of respiration dissolves calcium carbonate and other carbonates according to the equation

$$CO_2 + H_2O + CaCO_3 \rightarrow Ca^{2+} + 2HCO_3^-$$

If a lake or river bottom is covered by a calcite-rich sediment it is likely that groundwater will form solution channels and escape preferably as point discharges. In contact with the atmosphere, the excessive carbon dioxide escapes. The above equilibrium moves to the left and calcite precipitates as thin encrustations, travertines or calcareous tuffs.

Molecular diffusion is important in all cases where waters of different quality meet and where there is insufficient mixing by other means (groundwater flow, bioturbation etc.). In groundwater ecotones, diffusion and other slow advective processes (due to the sediment compaction, thermal convection etc.) are only important when average groundwater flow velocity is less than a few centimetres per year (Vanek, 1993).

TEMPORAL VARIABILITY

Most of the above mentioned internal factors increase not only spatial but also temporal heterogeneity of ecotones. For example, diurnal or seasonal fluctuations may be expected in all processes of biological character (evapotranspiration, bioturbation, bacterial growth, denitrification). Gas clogging of the sediments, calcite dissolution and other not entirely biological processes are also temperature-sensitive. In general, temporal fluctuations are damped in ecotones or their parts which are fed by deeper groundwater. This water exhibits stable temperatures throughout the year and varies often less in quality and discharge than either shallow groundwater or surface water.

In lakes, the temperature difference between lake water and the inflowing groundwater is often sufficient to influence water density and mixing. In summer, the colder groundwater will tend to sink to the deepest parts of the lake. During winter, the warmer groundwater will float to the lake surface. Eventually, water and sediment freezing near the shores may temporarily clog the near-shore seepage zones and force seepage flow to deeper sediment layers. Only when groundwater influence near the shores is strong enough to counteract the cooling, will near-shore seepage zones remain ice-free.

Finally, water temperature influences viscosity of water. Usually, the viscosity effect is included in the coefficient of permeability K (Eq. 1), and may be neglected in groundwater studies where temperature fluctuations are small. In surface-water fed ecotones, however, the yearly temperature amplitude of 20 °C is not uncommon. Lower viscosity during the summer causes up to 30% higher seepage rates than found in the winter.

ECOTONE BIOGEOCHEMISTRY

Typically, ecotones are characterized by gradients in light, temperature, pH, redox, oxygen, organic carbon, sediment permeability, water flow velocity etc. Of these, gradients in organic matter and oxygen are usually the most important determinants of water quality. The filtering and retention ability of ecotones (both groundwater-surface water ecotones and the closely associated surface water-land ecotones) cause the entrappment of organic carbon and nutrients. Mineral dissolution (except the dissolution of calcite in some ecotone types) and other slow processes typical for the groundwater zone usually have insufficient time to act in the limited ecotone zones. On the other hand, redox-related processes, which often are slow or insignificant in unpolluted groundwater, may in ecotones exhibit high activity. These processes involve cycling of nitrogen, iron, manganese, phosphorus, organic carbon and other solutes, and are theoretically rather well understood. Their relationship to each other and to the environmental variables within the ecotones, however, is known only poorly.

One of the problems when studying biogeochemical processes within ecotones is the previously discussed heterogeneity of water flows. Up to a certain flow velocity, biogeochemical activity is likely to increase together with the increasing fluxes of nutrients etc. to biologically active surfaces as well as metabolites etc. away from surfaces. Also, most biogeochemical processes occur at higher rates with increasing temperatures. If flow velocity is suboptimal (too low or too high) and/or temperature is too low, biogeochemical activity may be limited due to a lower local supply of organic matter, poorly developed redox gradients, too low temperature etc., or simply because the transient time of water within the ecotone is too short. Consequently, the highest biogeochemical activity within the ecotone may be associated with the areas or periods of slow to medium seepage velocities. Within erosional bottom areas or in highly active seepage zones such as pockmarks, springs or sediments influenced by artificial groundwater recharge, activity may be lower. Another effect of spatial and temporal heterogeneity is that various processes that act independently under simplified laboratory conditions (nitrification and denitrification as a typical example) act simultaneously in the ecotones and are impossible to separate.

Ecotone heterogeneity should be considered when preparing sampling programs or, at least, when evaluating data. Otherwise, our data may be representative only for a small part of the system, or the origin of the sampled water may not be determined (Fig. 7). Spatial and temporal heterogeneity are best described by repeated surveys of chemistry or hydraulic gradients of shallow porewater, bottom-near

Winter

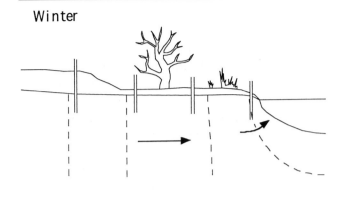

Summer

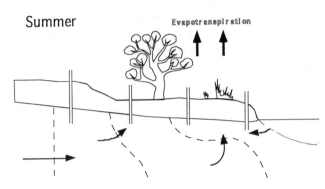

Fig. 7 An example of a groundwater-surface water ecotone with strong temporal fluctuations in groundwater-flow conditions. Variability in this case is caused by evapotranspiration within the ecotone in combination with seasonal fluctuations of shallow groundwater flow and surface water level. Deeper groundwater flow is more constant. Here, water quality changes may depend more on variable proportions of shallow groundwater, deeper groundwater and infiltrated surface water than on biogeochemical processes within the ecotone.

surface water or near-shore groundwater. Some of the methods that allow quick sampling of many points are bottom-contact electrical-conductivity sediment probes, shallow-porewater suction samplers, portable wellpoint samplers or, in some situations, remote sensing techniques (Gandino & Tonelli, 1983; Vanek & Lee, 1991; Vanek, 1993).

Assimilative capacity of the ecotones

A total load that an ecotone is able to assimilate or transform is given by the difference between the input and the output. In groundwater-surface water ecotones, water-related transport usually is a major process even if other processes such as the trapping/filtration and successive release/erosion of particulate matter and the release or fixation of gases (O_2, N_2, NO_x, CO_2) may also be important. The assimilative capacity of an ecotone may roughly be expressed as a long-time difference in water concentrations (IN minus OUT) times the average water flow through the ecotone. In this context, it is important to recognize which time scale should be considered.

Water flow velocity through ecotones depends highly on sediment permeability and may range from hours or days in highly permeable sediments to many years in ecotones containing soft sediments or other low-permeable materials. Highly permeable ecotones are likely to react quickly to possible changes in inputs. Their total assimilative or sorption capacity is likely to be lower when compared to ecotones containing clays or other material of high specific surface and high organic content, but the total volume of water passing through the ecotone is much higher. Typical examples of this ecotone type are highly permeable river beds of variable topography which may effectively control the content of nutrients and other reactive solutes in river water (Sterba et al., 1992; Broshears et al., 1993). Low-permeability ecotones react slowly but may store more nutrients or other matter. It has been shown that even slow groundwater flow through soft sediment layers (few centimeters per year) may determine exchange of solutes across sediment-water interfaces, an important factor influencing water quality of lakes and other slowly moving waters (Vanek, 1993).

In European and some other countries with a long history of highly developed agriculture and forestry, ecotones have been modified over several thousand years. Most important changes, however, occurred during the last century and included three main factors: (1) reduction of total ecotone area, (2) increasing inputs, mainly from the agricultural land, and (3) increasing exploitation of ecotones themselves. The combination of these factors resulted in the increased input of soil from catchments to ecotones which often have reduced the permeability of ecotones, and caused a gradual shift from hydrologically controlled ecotones to aerobically and anaerobically controlled ecotones (Vervier et al., 1992).

Water quality changes and ecotone malfunctioning

What kind of water quality changes can be expected in an ecotone? This depends largely on the quality of water entering the ecotone, the velocity of flow, and the content of organic matter within the ecotone. For example, if the input consists of nitrate-rich groundwater, denitrification will be the main biogeochemical process. Nitrate will be converted into gaseous nitrogen (or, sometimes, nitrous oxides) and some of the particulate organic matter will be oxidized to carbon dioxide. Usually, however, oxidation will not be 100% and there will be some increase in concentration of DOC and ammonia. If there is a well developed oxygenated layer near the sediment-water interface, these can again be removed by nitrification-denitrification and aerobic respiration.

In ecotones receiving anaerobic groundwater which is rich

in iron or manganese, most activity (metal precipitation) will be concentrated in the superficial oxygenated layer. Only under specific conditions, e.g. if groundwater is also polluted by organics or rich in sulphate, the development of sulphate-reductive or methane fermentation zones may be expected.

Mechanical filtering will be an important process in ecotones receiving surface water or shallow subsurface runoff that carry many particles. Usually, this water also contains enough dissolved oxygen to support nitrification of ammonia and aerobic degradation of organic matter, which for most organics is the quickest and most effective way of degradation.

Particularly complex may be biogeochemical processes within the ecotones where several water sources mix, such as in permeable river bottoms where the underflowing, well oxygenated river water meets the upwelling groundwater, often poor in oxygen, in tidally-influenced beaches or in other highly dynamic systems.

In groundwater recharge zones, surface water will be mechanically and biochemically filtered as it infiltrates the ground. If the water passes through an anaerobic zone it can easily become enriched by organics, dissolved metals etc. This is why artificial groundwater recharge systems work well only in highly permeable ecotones containing limited amounts of organic matter or metallic oxides. In other cases, it is usually better to by-pass the ecotone and to lead surface water into special infiltration basins or ditches which, if needed, can be drained and cleaned. Usually, there is also an unsaturated zone between these structures and the water table, which further improves oxygenation of the infiltrating water and aerobic degradation of organic matter. Without doubt, these artificial systems are more effective than most natural ecotones (with the exception of denitrification or few other processes depending on low oxygen conditions), but the loads used in these water treatment systems usually are much higher than is common in natural ecotones (Asano, 1985).

Thus, we may conclude that water quality changes within an ecotone will be most profound if water passes first through an anaerobic and then through an aerobic layer. Ecotones containing mainly anaerobic sediments or lacking aerobic layers on the effluent side will have negative effects on water quality. Exceptions to this are denitrification ability and the ability to degrade some organics (e.g. some pesticides). Usually, the oxygen content in an ecotone is closely coupled to sediment permeability, particularly to permeability of sediment-water interfaces. If this permeability is low (e.g. due to the clogging), only a microzone of oxygen-saturated sediment may exist near the surface. If also deeper layers of the ecotone are anaerobic, there will be no significant water quality improvement.

The remaining question is, how to look at ecotones that contain little oxygen on the effluent side, and become enriched in organics, various nutrients, metals or metabolites? Are they malfunctioning? In my opinion, they are malfunctioning only if the change from aerobiosis to anaerobiosis is caused by pollution or other human impact. In many cases enrichment is a natural process where organic matter, which has been derived from the catchment and deposited within the ecotone, is decomposed, even if not as completely as under aerobic conditions. This can hardly be called a malfunctioning. Serious ecotone malfunction occurs if water flow through the ecotone is hindered or stopped, e.g. by eroded material (siltation), some other clogging process, or if the ecotone is by-passed via drainage pipes or ditches.

In general, the number and the area of groundwater-surface water ecotones (and most other types of ecotones as well) have declined drastically during the last century. At the same time, the hydraulic stress and load of particulate and dissolved matter have increased. This has resulted in the fact that many ecotones are no longer able to assimilate or buffer the inputs. Ways to improve include enlargement of the ecotones (or intensification of some of their functions) and reduction of inputs from the catchment. These two approaches have to be used simultaneously, but reduction of fluxes from the catchment should always be our primary goal. Today, most catchments are leaking nutrients, soil particles and various chemicals. Even if ecotones to some extent may be used to stop these (often irreversible) losses, they cannot solve whole-catchment problems such as deterioration of soil and groundwater quality. This can only be done by improved agricultural practices, by restoring old, previously removed ecotones, by applying various anti-erosion measures, and other actions increasing water retention capacity and resulting in 'nutrient tight catchments'.

REFERENCES

Asano, T., ed. (1985). *Artificial recharge of groundwater*. Boston: Butterworth Publishers.

Broshears, R.E., Bencala, K.E., Kimball, B.A. & McKnight, D.M. (1993). Tracer-dilution experiments and solute-transport simulations for a mountain stream, Saint Kevin Gulch, Colorado. *U.S. Geological Survey Water-Resources Investigations Report*, 92–4081.

Calles, U.M. (1982). Diurnal variation of electrical conductivity of water in a small stream. *Nordic hydrology*, **13**, 157–164.

Chow, V.T., Maidment, D.R. & Mays, L.W. (1988). *Applied hydrology.*, New York: McGraw Hill.

Creuzé des Châtelliers, M. (1991). *Dynamique de répartition des biocœnoses interstitielles du Rhône en relation avec des caractéristiques géomorphologiques*. PhD Thesis. Lyon: Université Claude Bernard Lyon I.

Danielopol, D. (1984). Ecological investigations in the alluvial sediments of the Danube in the Vienna area – a phreatobiological project. *Verh. int. Verein. theor. angew. Limnol.*, **22**, 1755–1761.

Fliermans, C.B. & Balkwill, D.L. (1989). Microbial life in the deep terrestrial subsurface. *Bioscience*, **39**, 370–377.

Freeze, R.A. & Cherry, J.A. (1979). *Groundwater*. Englewood Cliffs, N.Y.: Prentice-Hall.

Gandino, A. & Tonelli, A.M. (1983). Recent remote sensing technique in fresh water submarine springs monitoring: qualitative and quantitative approach. Verslagen en Mededelingen Commissie voor hydrologisch Onderzoek TNO 31, 301–310.

Ghiorse, W.C. & Wilson, J.T. (1988). Microbial ecology of the terrestrial subsurface. *Advances in Applied Microbiology*, **33**, 107–172.

Gibert, J., Dole, M.J., Marmonier, P. & Vervier, Ph. (1990). Surface water – groundwater cotones. In *The ecology and management of aquatic – terrestrial ecotones*, ed. R.J. Naiman & H. Décamps, pp. 199–225. Parthenon Publishers.

Gillham, R.W. (1984). The capillary fringe and its effect on water-table response. *Journal of Hydrology*, **67**, 307–324.

Guyonnet, D.A. (1991). Model analysis of effects of small-scale sediment variations on groundwater discharge into lakes. *Limnology & Oceanography*, **36**, 341–343.

Håkanson, L. & Jansson, M. (1983). *Principles of lake sedimentology*. Berlin: Springer-Verlag.

Hendricks, S.P. & White, D.S. (1991). Physicochemical patterns within a hyporheic zone of a Northern Michigan river, with comments on surface water patterns. *Canadian Journal of Fisheries and Aquatic Sciences*, **48**, 1645–1654.

Hovland, M. & Judd, A.G. (1988). *Seabed pockmarks and seepages. Impact on geology, biology and the marine environment*. London: Graham & Trotman.

Husmann, S. (1978). Die Bedeutung der Grundwasserfauna für biologische Reinigungsvorgünge im Interstitial von Lockergesteinen. *GWF Wasser Abwasser*, **119**, 293–302.

Kohout, F.A. (1965). A hypothesis concerning cyclic flow of salt water related to geothermal heating in the Floridian Aquifer. *Transactions of the New York Academy of Sciences, Series*, **2**, 28, 249–271.

Krabbenhoft, D.P. (1984). *Hydrologic and geochemical controls of freshwater ferromanganese deposit formation at Trout Lake*, Vilas County, Wisconsin. MSc Thesis. Madison: Univ. of Wisconsin.

Lepart, J. & Debussche, M. (1992). Human impacts on landscape patterning: Mediterranean examples. In *Landscape boundaries. Consequences for biotic diversity and ecological flows*, ed. A.J. Hansen & F. di Castri, pp. 76–106. Springer-Verlag Ecol. Studies Vol. 92.

Lowe, D.R. (1975). Water escape structures in coarse-grained sediments. *Sedimentology*, **22**, 157–204.

McBride, M.S. & Pfannkuch, H.O. (1975). The distribution of seepage within lakebeds. *Journal of Research of the U.S. geological Survey*, **3**, 505–512.

McLachlan, A. (1989). Water filtration by dissipative beaches. *Limnology & Oceanography*, **34**, 774–780.

Meyboom, P. (1966). Unsteady groundwater flow near a willow ring in hummocky moraine. *Journal of Hydrology*, **4**, 38–62.

Naumann, E. (1922). *Södra och mellersta Sveriges sjö- och myrmalmer. Deras bildningshistoria, utbredning och praktiska betydelse*. Sveriges Geologiska Undersökning, **C 297**, 1–194.

Niemczynowicz, J. (1991). Environmental impact of urban areas – the need for paradigm change. *Water International*, **16**, 83–95.

Pinder, G.F. & Sauer, S.P. (1971). Numerical simulation of flood-wave modification due to bank storage effects. *Water Resources Research*, **7**, 63–70.

Riedl, R.J. (1971). How much seawater passes through sandy beaches? *Int. Rev. ges. Hydrobiol.*, **56**, 923–946.

Riedl, R.J., Huang, N. & Machan, R. (1972). The subtidal pump; a mechanism of interstitial water exchange by wave action. *Marine Biology*, **13**, 210–221.

Ruprecht, J.K. & Schofield, N.J. (1991). Effects of partial deforestation on hydrology and salinity in high salt storage landscapes. Extensive block clearing. *Journal of Hydrology*, **129**, 19–38.

Shum, K.T. (1992). Wave-induced advective transport below a rippled water-sediment interface. *Journal of Geophysical Research*, **97**, 789–808.

Sinoquet, C. (1987). *Impacts d'une ballastière en eau sur la qualité des eaux souterraines – Cas de deux ballastières alsaciennes et modélisation mathématique appliquée aux échanges hydrochimiques et hydrothermiques*. Ph. D Thesis. Strasbourg: Univ. Louis Pasteur.

Sterba, O., Uvíra, V., Mathur, P. & Rulík, M. (1992). Variations of the hyporheic zone through a riffle in the River Morava, Czechoslovakia. *Regulated Rivers*, 7, 31–43.

Thibodeaux, L.J. & Boyle, J.D. (1987). Bedform-generated convective transport in bottom sediments. *Nature*, **325**, 341–343.

Vanek, V. (1992). The role of groundwater in the hydrology of some Bavarian lakes. *Archiv Hydrobiologie/Supplement*, **90**, 63–84.

Vanek, V. (1993). Groundwater regime of a tidally influenced coastal pond. *Journal of Hydrology*, **151**, 317–342.

Vanek, V. & Lee, D.R. (1991). Mapping submarine groundwater discharge areas – an example from Laholm Bay, southwest Sweden. *Limnology & Oceanography*, **36**, 1250–1262.

Vervier, Ph., Gibert, J., Marmonier, P. & Dole-Olivier, M.-J. (1992). A perspective on the permeability of the surface freshwater – groundwater ecotone. *Journal of the North American Benthological Society*, **11**, 93–102.

Ward, R.C. (1975). *Principles of hydrology*, 2nd ed. London: McGraw-Hill.

Zaletaev, V.S. (1996). Ecotones and problems of their management in irrigation regions. This issue.

20 Failure of agricultural riparian buffers to protect surface waters from groundwater nitrate contamination

D. L. CORRELL, T. E. JORDAN & D. E. WELLER

Smithsonian Environmental Research Center, P.O. Box 28, Edgewater, Maryland 21037, USA

ABSTRACT   For two years we studied the flux of nitrogen moving in shallow groundwater from double row-cropped uplands through a flood plain and into a second order stream in Maryland. Two floodplain sites were compared: one forested and the other vegetated by grass. At both sites, the soil layer through which the groundwater moved was very sandy. The nitrate concentrations leaving the crop fields were 20–30 mg N l^{-1} and averaged 25 mg N l^{-1}. Nitrate concentrations declined about 32% on average from the field edge to 48 m into the forest and this decrease was about 44% on average in the grassed buffer. These decreases were greater in the winter than in the summer. Nitrate to chloride ratios declined about 43% across the riparian forest transect. Declines in nitrate concentration were not accompanied by off-setting increases in dissolved organic N or ammonium. Soil E_h averaged 191 mV and 263 mV at 33 m and 48 m into the forest, respectively. While nitrate removal rates were the highest of three study sites we have investigated in the Maryland Coastal Plain, nitrate concentrations entering the stream channel were still high (12–18 mg N l^{-1}). The flux of nitrate in groundwater from the farm fields at this site clearly exceeded the nitrate removal capacity of these riparian buffers.

INTRODUCTION

Coastal receiving waters are often overenriched with nutrients, especially in cases where the drainage basins are intensively farmed or support large populations of humans (Beaulac & Reckow, 1982; Turner & Rabalais, 1991). This is clearly the case for the Chesapeake Bay on the eastern seaboard of the United States, where diffuse drainage basin discharges contribute approximately two-thirds of the nitrogen, one-quarter of the phosphorus, and all of the silicate inputs to the Bay (Correll, 1987). In this estuary, over-enrichment with both nitrogen and phosphorus (Gallegos *et al.*, 1992; Jordan *et al.*, 1991a,b; Malone *et al.*, 1988) contribute to excessive plankton blooms and extensive reaches of hypoxic waters (Officer *et al.*, 1984). Much of the nitrogen inputs to Chesapeake Bay are nitrate from croplands, which infiltrates through the soils to groundwater, then percolates to surface water streams before entering the Bay.

Our past and continuing work has demonstrated the importance of landscape structure, particularly the configuration of riparian buffers, in controlling nutrient discharges from agricultural watersheds of the coastal plain (Correll & Weller, 1989; Correll, 1991; Correll *et al.*, 1992; Jordan *et al.*, 1993; Peterjohn & Correll, 1984, 1986). In these studies over 80% of the nitrate entering the riparian forest in shallow groundwater drainages from croplands was removed at all times of year. Other studies have reported similar findings in the coastal plain (e.g. Lowrance *et al.*, 1984; Gilliam & Skaggs, 1988) and in some other systems (Haycock & Pinay, 1993; Labroue & Pinay, 1986; Pinay & Labroue, 1986; Pinay & Décamps, 1988; Schnabel, 1986).

The present study extends our research on coastal plain riparian buffers to a new site which receives groundwater with nitrate concentrations two to three times higher than the previous two sites (Peterjohn & Correll, 1984; Jordan *et al.*, 1993). This site also has subsoils with high hydraulic conductivity due to high gravel and sand content. Thus, conditions at this site were favorable to observe saturation of this riparian buffer's nitrate removing potential.

SITE DESCRIPTION

The study site (39° 2' N, 75° 57' W) was a flood plain along a second-order stream draining approximately 4.5 km² of land. This stream is a tributary near the lower end of German Branch, a fifth-order stream draining 52 km² of land. German Branch in turn is a tributary of the Choptank River, which is the largest river on the eastern shore of Chesapeake Bay and drains directly into the Bay. Approximately 90% of the uplands at the study site were used for row crop production. The fields were double-cropped and spray irrigated during dry periods. Two areas within the flood plain were studied. One was vegetated with a stand of mixed species of deciduous hardwood trees. The other was vegetated with mown grass. The surface sediments in this region of the coastal plain are part of the Pensauken formation, which forms the Columbia aquifer. Throughout the region, this aquifer is perched on the less permeable sediments of the Chesapeake Group (Bachman & Wilson, 1984), which are highly clayey with hydraulic conductivity of less than 1×10^{-4} cm hour^{-1} (Jordan et al., 1993).

METHODS

We installed transects of groundwater wells at both the forested and grassed areas (Fig. 1). The wells were installed with bucket augers and were lined with polyvinyl chloride pipes, which were perforated from just below the soil surface to the bottom of the well. The wells extended from 1 to 4 m below the soil surface with the bottoms of the wells at approximately the same elevation as the stream bed. Wells were arranged at various distances from the crop fields with sets of three replicates spaced 10 m apart laterally (Fig. 1). We sampled the wells about once a month from May 30, 1991, till May 25, 1993, by first pumping the wells dry, then sampling the water that immediately refilled the wells. We filtered the samples through 0.45 µm pore-size membrane filters. We measured dissolved Kjeldhal N, nitrate, ammonium, phosphate, total P, organic C, chloride, pH, and conductivity (Correll & Weller, 1989; Jordan et al., 1993). Triplicate analyses were routinely performed on about 10% of the samples to provide a check on analytical precision. At the forested area we also measured E_h below the water table with platinum electrodes (Faulkner et al., 1991) placed near wells 32 and 42 (Fig. 1).

RESULTS AND DISCUSSION

Both the forested and grassed riparian buffer areas (Fig. 1) seemed to remove nitrate from groundwater. Nitrate

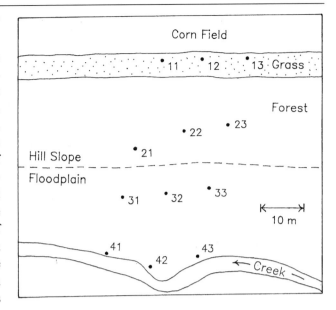

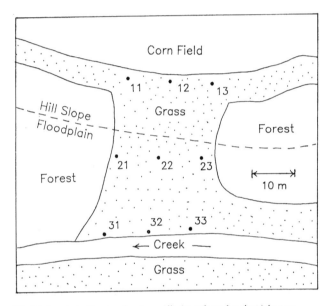

Fig. 1 Layout of groundwater wells (numbered points) in two riparian buffers (39° 2' N, 75° 57' W) on the flood plain of a second-order stream, tributary to the German Branch of the Choptank River in Maryland, United States. In the forested buffer platinum electrodes were located near wells 32 and 42 for monitoring soil E_h.

concentrations declined significantly with distance from the crop fields (Figs. 2 & 3). At the edge of the fields, NO_3^- concentrations averaged 25 mg N l^{-1} at both buffer areas. In the forested buffer, the NO_3^- concentration declined to 17 mg N l^{-1} at 48 m from the field. In the grassed buffer, the NO_3^- concentration declined to 14 mg N l^{-1} over a distance of 37 m. Concentrations of dissolved organic N and dissolved ammonium (Figs. 2 & 3) did not increase significantly, so the NO_3^- was not converted to dissolved reduced forms of nitrogen. In the forested buffer, dissolved ammonium

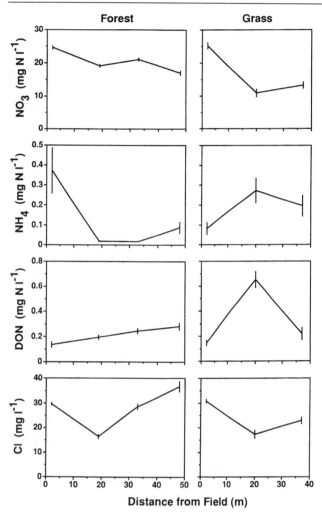

Fig. 2 Concentrations in groundwater versus distance from upland crop fields along transects through riparian buffers. Values are means of monthly samples over a two year period from May 30, 1991, to May 25, 1993. Brackets are standard errors of means.

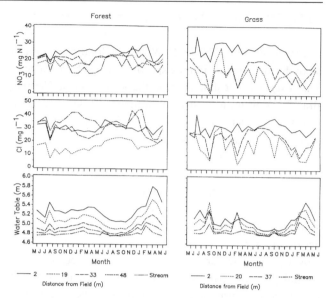

Fig. 3 Nitrate and chloride concentrations and water table elevations versus date at different distances from the crop field along transects through riparian buffers. Sampling was from May 30, 1991, to May 25, 1993. Stream water elevations are also shown with water table elevations. Elevations are relative to an arbitrary reference point.

decreased 0.3 mg N l^{-1} while dissolved organic N increased by 0.12 mg N l^{-1}. In the grassed buffer, dissolved ammonium increased by 0.1 mg N l^{-1} while dissolved organic N increased by less than 0.1 mg N l^{-1}.

Lateral flow toward the stream channel is indicated by the slope of the water table (Fig. 3). However, despite the regional presence of a clayey Chesapeake Group layer, it is possible that some groundwater from more distant source areas could have entered the riparian areas from sandy layers between the sampling depth and the clayey layer. Concentrations of Cl^- can indicate dilution by another water source or concentration by evapotranspiration. At the forested site, Cl^- decreased from about 30 mg l^{-1} at the edge of the field to 16 mg l^{-1} 19 m from the field, then increased to 36 mg l^{-1} 48 m from the field (Fig. 2). The low Cl^- at 19 m suggests possible mixing with another water source. Transient dilution of both Cl^- and NO_3^- by rain and infiltration of overland flows is suggested when low concentrations

coincide with peaks in water table elevation following rain (Fig. 3). The increase in Cl^- between 19 and 48 m from the field in the forested area may be due to evapotranspiration. If so, NO_3^- concentrations should also increase, unless NO_3^- was being removed in the buffer.

Nitrate removal seemed to be less effective in the early summer than in the winter, judging from the steeper concentration gradients in winter (Fig. 3). Nitrate may be removed by denitrification, which can occur when E_h is below 300 mV (Correll & Weller, 1989). Between November and March soil E_h in the forest buffer averaged 191 mV (standard deviation=25) near well 32 and 263 mV (sd=21) near well 42.

In the grassed riparian area, Cl^- concentration decreased 42% from 30 mg l^{-1} near the field to 18 mg l^{-1} 20 m away. Over the same distance, NO_3^- dropped 56% from 25 to 11 mg N l^{-1}. Some of the NO_3^- decline may result from dilution rather than removal of NO_3^- in the riparian zone. Chloride and NO_3^- concentrations responded to transient increases in water table elevation as at the forested site (Fig. 3). However, the tranient increases in water table elevation also resulted in reversals of water table slope with the water table at 20 m and 37 m from the field higher than at the edge of the field. This suggests a faster rate of recharge of groundwater at 20–37 m from the field than at the edge of the field. Such recharge would be a possible source of dilution effects on Cl^- and NO_3^-.

The water table slopes usually indicated flow towards the

stream, but the slopes were not as steep in the grassed area as in the forested area (Fig. 3). This suggests a higher rate of groundwater flow through the forest buffer than through the grassed buffer. Thus, the mass of NO_3^- removed could be higher in the forest than in the grass despite the greater change in NO_3^- concentration within the grassed buffer.

Our results clearly indicate that conditions at this groundwater/surfacewater ecotone combined to exceed its buffering capacity. The contibuting conditons included high nitrate concentrations (mean of 25 mg N l^{-1}), and high velocity of groundwater movement due to a coarse-grained substrate and fairly high water table slopes. Groundwater entering the stream channel had an average nitrate concentration of 14 or 17 mg N l^{-1} in the two areas. While this is much higher than we found in two other Coastal Plain riparian sites in Maryland (less than 1 mg N l^{-1}; Correll & Weller, 1989; Jordan et al., 1993), the decrease in nitrate concentration within the riparian buffer at this site was greater than at the other two sites.

ACKNOWLEDGEMENTS

This research was supported by a grant from the National Science Foundation (BSR-89–05219) and a grant from the Governor's Research Council of Maryland.

REFERENCES

Bachman, L. J. & Wilson, J. M. (1984). *The Columbia Aquifer of the eastern shore of Maryland*. Report 40, Maryland Geological Survey, Baltimore, Maryland, United States.

Beaulac, M. N. & Reckhow, K. H. (1982). An examination of land use-nutrient export relationships. *Water Resources Bulletin*, **18**, 1013–22.

Correll, D. L. (1987). Nutrients in Chesapeake Bay. In *Contaminant Problems and Management of Living Chesapeake Bay Resources*, ed. S.K. Majumdar, L.W. Hall Jr. & H.M. Austin, pp. 298–320. Philadelphia: Pennsylvania Academy of Science.

Correll, D. L. (1991). Human impact on the functioning of landscape boundaries. In *The Role of Landscape Boundaries in the Management and Restoration of Changing Environments*, ed. M.M. Holland, P.J. Risser & R.J. Naiman, pp. 90–109. New York: Chapman & Hall.

Correll, D. L., T. E. Jordan, & Weller, D. E. (1992). Nutrient flux in a landscape: Effects of coastal land use and terrestrial community mosaic on nutrient transport to coastal waters. *Estuaries*, **15**, 431–42.

Correll, D. L. & Weller, D. E. (1989). Factors limiting processes in fresh-water wetlands: an agricultural primary stream riparian forest. In *Freshwater Wetlands and Wildlife*, ed. R.R. Sharitz & J.W. Gibbons,

pp. 9–23. Aiken: Savannah River Ecology Laboratory.

Faulkner, S. P., Patrick Jr., W. H., Gambrell, R. P., Parker, W. B. & Bood, B. J. (1991). *Characterization of Soil Processes in Bottomland Hardwood Wetland-Nonwetland Transition Zones in the Lower Mississippi River Valley*, Contract Report WRP-91-1, Vicksburg: United States Army Corps of Engineers, Waterways Experiment Station.

Gallegos, C. L., Jordan, T. E., & Correll, D. L. (1992). Event-scale response of phytoplankton to watershed inputs in a subestuary: Timing, magnitude and location of blooms. *Limnology and Oceanography*, **37**, 813–28.

Gilliam, J. W. & Skaggs, R. W. (1988). Nutrient and sediment removal in wetland buffers. In *Proceeding of National Wetland Symposium: Wetland Hydrology*, ed. J.A. Kusler & G. Brooks, pp. 174–7. Berne: Assoc. State Wetland Mgrs. Assoc. State Wetland Mgrs.

Haycock, N. E. & Pinay, G. (1993). Nitrate retention in grass and poplar vegetated riparian buffer strips during the winter. *Journal of Environmental Quality*, **22**, 273–8.

Jordan, T. E., Correll, D. L., Miklas, J. & Weller, D. E. (1991a). Long-term trends in estuarine nutrients and chlorophyll, and short-term effects of variation in watershed discharge. *Marine Ecology Progress Series*, **75**, 121–32.

Jordan, T. E., Correll, D. L., Miklas, J. & Weller, D. E. (1991b). Nutrients and chlorophyll at the interface of a watershed and an estuary. *Limnology and Oceanography*, **36**, 251–67.

Jordan, T. E., Correll, D. L. & Weller, D. E. (1993). Nutrient interception by a riparian forest receiving agricultural runoff. *Journal of Environmental Quality*, **22**, 467–73.

Labroue, L. & Pinay, G. (1986). Epuration naturelle des nitrates des eaux souterraines: possibilites d'application au reamenagement des lacs de gravieres. *Annals Limnologie*, **22**, 83–8.

Lowrance, R. R., Todd, R. L., Fail, J. Jr., Hendrickson, O. Jr., Leonard, R. & Asmussen, L. (1984). Riparian forests as nutrient filters in agricultural watersheds. *Bioscience*, **34**, 374–7.

Malone, T. C., Crocker, L. H., Pike, S. E. & Wendler, B. W. (1988). Influence of river flow on the dynamics of phytoplankton production in a partially stratified estuary. *Marine Ecology Progress Series*, **48**, 235–49.

Officer, C. B., Biggs, R. B., Taft, J. L., Cronin, L. E., Tyler, M. A. & Boynton, W. R. (1984). Chesapeake Bay anoxia: origin, development, significance. *Science*, **223**, 22–7.

Peterjohn, W. T. & Correll, D. L. (1984). Nutrient dynamics in an agricultural watershed: observations of the role of a riparian forest. *Ecology*, **65**, 1466–75.

Peterjohn, W. T. & Correll, D. L. (1986). The effect of riparian forest on the volume and chemical composition of baseflow in an agricultural watershed. In *Watershed Research Perspectives*, ed. D.L. Correll, pp. 244–62. Washington: Smithsonian Press.

Pinay, G. & Décamps, H. (1988). The role of riparian woods in regulating nitrogen fluxes between the alluvial aquifer and surface water: A conceptual model. *Regulated Rivers: Research & Management*, **2**, 507–16.

Pinay, G. & Labroue, L. (1986). Une station d'epuration naturelle des nitrates transportes par les nappes alluviales: l'aulnaie glutineuse. *C. R. Académie Sciences de Paris*, **302**, (III), 629–32.

Schnabel, R. R. (1986). Nitrate concentrations in a small stream as affected by chemical and hydrologic interactions in the riparian zone. In *Watershed Research Perspectives*, ed. D.L. Correll, pp. 263–82. Washington: Smithsonian Press.

Turner, R. E. & Rabalais, N. N. (1991). Changes in Mississippi River water quality this century. *Bioscience*, **41**, 140–7.

21 Stable nitrogen isotope tracing of trophic relations in food webs of river and hyporheic habitats

R.C. WISSMAR*, J.A. STANFORD** & B.K. ELLIS**

* School of Fisheries, Wh-10, and Center for Streamside Studies, AR-10, University of Washington, Seattle, Washington 98195, USA

** Flathead Lake Biological Station, University of Montana, Polson, Montana 59860, USA

ABSTRACT Natural isotopic abundances of N ($d^{15}N$) in organic matter are used to examine the trophic relations in river and hyporheic habitats of the Flathead River, Montana. The $d^{15}N$ of biofilms and Plecopteran-dominated food webs are compared for river and hyporheic waters. We examine the concept that the isotope ratio of an animal undergoes both the food source effect and *in vivo* metabolic effects. Metabolic effects for nitrogen isotopes in animal tissues relates to isotopic fractionation, which exhibits progressive enrichment of the $d^{15}N$ in body tissues at higher trophic levels. These $\Delta d^{15}N$ enrichments are in excess of isotopic abundances of food sources. Comparisons show greater enrichment for the consumer trophic levels of the hyporheic than river channel habitat. The $d^{15}N$ contents of the hyporheic food webs suggest influences of different concentrations and $d^{15}N$ contents of NO_3 and NH_4. These inorganic conditions may be controlled by physical mixing of various water masses, N-transformations associated with biofilm microbiota, and feedbacks of NH_4 through excretion by animals. Our observations indicate that stable isotopes may be powerful tools for determining nitrogen source and process information in hyporheic habitats.

INTRODUCTION

Little is known about the trophic ecology of food webs within the hyporheic ecotones of river ecosystems (Hendricks, 1993). In our study, stable nitrogen isotopes measurements ($d^{15}N$) are used to examine the trophic relations in river and hyporheic habitats of the Flathead River, Montana. The $d^{15}N$ of biofilms and Plecopteran-dominated food webs are compared for river and hyporheic waters. The almost total coverage on most rock surfaces by biofilms suggests that attached microbiota comprise a dominant food resource in both the river channel and hyporheic habitats. The biota of the river provides a food web structure that is primarily based on autotrophic food sources while that of the hyporheic belowground environment is considered non-photosynthetic. Differences in the $d^{15}N$ values of river channel and hyporheic food web components are used to explore two questions: (1) what are the isotopic compositions of nitrogen in the diets of primary consumers, and (2) how do isotopic compositions change between different trophic levels?

The general concept underlying this study is that the isotope ratio of an animal undergoes both the food source effect and *in vivo* metabolic effects. The primary metabolic effects for nitrogen isotopes in animal tissues relate to isotopic fractionation during assimilation, growth, and excretion (Peterson & Fry, 1987). Most aquatic animals commonly exhibit progressive enrichment of the $d^{15}N$ in body tissues at higher trophic levels (Minagawa & Wada, 1984; Fry, 1991).

METHODS

The hyporheic zone's network of habitats, hydrology, and food web structure is defined using information from Stanford & Ward (1988). Samples analyzed for nitrate nitrogen (APHA/AWA/WPCF, 1975) were prefiltered through acid-washed membrane filters. Dissolved organic carbon (DOC) concentrations were determined according to Menzel & Vaccaro (1964) and chlorophll *a* measurements after Strickland & Parsons (1972). Bacterial biomass and composition were estimated using the procedures of Porter & Feig (1980).

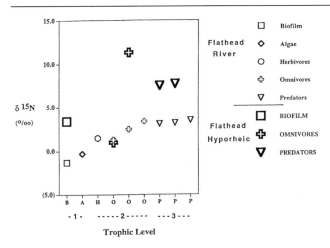

Fig. 1 The d^{15}N values for biofilms and consumers of trophic levels from the river channel and hyporheic habitats of the Flathead River, Montana. Trophic levels denoted as follows: H=herbivores; O=omnivores; and P=predators.

The stable nitrogen isotopic compositions of biofilms, algae, and consumer organisms were sampled in the Flathead River channel and hyporheic waters. Sample sizes for isotopic determinations, 2 to 3 samples at each trophic level, were limited because of logistics and expenses. The consumers were predominantly insect species of primary, secondary, and tertiary trophic levels. The incubation of gravel substrate in wells provided surfaces for sampling biofilm food materials. Samples preparation in the field and laboratory followed procedures of Simenstad & Wissmar (1985). The trophic levels for consumers (herbivores, omnivores and predators) were assigned using reference collections.

The d^{15}N values are reported relative to nitrogen in air as parts per thousand (‰). All samples are combusted to N$_2$ gases prior to being measured for N isotopic values. The d^{15}N values are determined on a Finnigan 251 stable isotope mass spectrometer provided by Coastal Science Laboratory, Inc., Austin, Texas. The isotopic values are expressed as follows:

$$d^{15}N = [R_{sample} - R_{standard}/R_{standard}] \times 10^3 \qquad (1)$$

where $R = {}^{15}N/{}^{14}N$.

The expressions, enriched d 15N values and depleted d 15N values, describe the observed isotopic compositions. Enriched d15N values indicate an increase in 15N relative to 14N. Deplete d15N values indicate a reduction 15N relative to 14N. The isotopic standard, atmospheric nitrogen for d15N, has by definition a d value of 0 ‰ (Fry, 1991).

RESULTS AND DISCUSSION

Hyporheic habitat and food web

An extensive hyporheic network of flowing subsurface paleo-river channels exists within an alluvial aquifer of a sixth-order gravel-bed segment of the Flathead River, Montana. Hyporheic habitats of the Flathead River floodplain are created by Pleistocene-glacial and recent alluvium of the aquifer. Hyporheic waters, 2 km or more from the river channel, are influenced by the river-flow patterns. The average flow rates for the river and hyporheic habitats are 340 m^3 s^{-1} and 0.7 m^3 s^{-1}, respectively. Water and materials flow within a network of high porosity paleo-channels (max. hydraulic conductivity=10 cm s^{-1}) creating a hyporheic habitat with an average depth of 10 m and an average width of 3 km. The size of the hyporheic habitats relates to porosity of the alluvium and the seasonal flow regime of the river. The average volume of the hyporheic habitat approaches 300×10^6 m^3, or 2400 times greater than the river channel volume (125×10^3 m^3). Levels of dissolved oxygen in well waters of the hyporheic are always greater than 50% saturation (Stanford & Ward, 1988).

Average nitrate concentrations (145 µg l^{-1}±3%) in the hyporheic are fourfold greater than in river waters. The potential supply of nitrate in the phreatic waters of the Flathead floodplain is indicated by nitrate concentrations averaging 916 µg l^{-1}, nitrate levels 6 times greater than the hyporheic and 24 times those in the river channel. Average DOC concentrations in Flathead River hyporheic habitats (1.81 mg l^{-1}±8%) are similar to those of the river channel. Minimal inputs of autotrophic organic matter from the river channel to the hyporheic are implied by low chlorophyll a concentrations (ave.=4 ng l^{-1}±15%) in the hyporheic. Chlorophyll concentrations in the hyporheic range from 44 to 112 times lower than the river channel.

Hyporheic habitats of the Flathead River, Montana, contain diverse invertebrate faunal assemblages of ≥70 species. These assemblages consist of residents, migrants from adjoining river channel habitats (benthic), and migrants from non-flowing or phreatic groundwaters. The majority of the animals are amphibionts and benthos (e.g., semivoltine insects such as Plecoptera), with the remainder (~20 species) being true stygobiont animals (e.g., amphipods, cyclopoid copepods, and isopods).

Evidence for hyporheic food webs being primarily supported by 'heterotrophic microbiota' food sources is provided by observations of higher volumes of bacteria (~twofold) and fungal biomass (10% of the total microbe biomass) in the hyporheic than in river channel habitats of the Flathead River. In the hyporheic habitat, greater bacterial volumes appeared attached to the surface of gravels and cobbles than for free-living bacteria in waters.

VALUES OF D^{15}N AND TROPHIC RELATIONS
Comparison of d^{15}N values of organisms representing trophic levels of the hyporheic and river channel food webs reveals differences in isotopic compositions (Fig. 1). The

Table 1. $\Delta d^{15}N$ enrichments in invertebrates of the Flathead River and hyporheic food webs. $\Delta d^{15}N$ estimated by subtracting $d^{15}N$ of potential diets from the $d^{15}N$ of the consumers body tissue. Trophic levels of consumers denoted as follows: herbivores (H); omnivores (O); and predators (P)

Habitat	Trophic level (H, O, P) and invertebrate consumer	$d^{15}N$ of invertebrate body tissue (A)		$d^{15}N$ of 'diets' (B)	$\Delta d^{15}N$ enrichment (A)–(B)
Flathead River (Mont.)	H *Isoperla fulva* (Insecta: Plecoptera)	+1.5	Biofilm	−1.3	+2.8
			Ulothrix	−0.3	+1.8
	O *Pteronarcella badia* (Insecta: Plecoptera)	+1.3	Biofilm	−1.3	+2.6
	O *Pteronarcys californica* (Insecta: Plecoptera)	+2.5	Biofilm	−1.3	+3.8
	O *Arctopsyche grandis* (Insecta: Trichoptera)	+3.4	Biofilm	−1.3	+4.7
	P *Hesperoperla pacificum* (Insecta: Plecoptera)	+3.1 +3.2	Consumers	+1.3 to +3.4	0 to +1.9
	P *Classennia sabulosa* (Insecta: Plecoptera)	+3.5	Consumers	+1.3 to +3.4	+0.1 to +2.2
Flathead Hyporheic (Mont.)	H *Isocapnia crinita* (Insecta: Plecoptera)	+1.0	Biofilm	+3.4	
	O, P *Stygobromus sp.* (Crustacea: Amphipoda)	+11.3	Biofilm	+3.4	+7.9
	O, P *Paraperla frontalis* (Insecta: Plecoptera)	+7.5 +7.7	Consumers	+1.0	+6.6

$d^{15}N$ values for the hyporheic food web show greater enrichments than those of the river channel. The higher isotopic values occur in both the hyporheic biofilm (+3.4 ‰) and the consumer (+7.5 ‰ to +11.3 ‰) organisms. The $d^{15}N$ values for the predators range from +7.5 ‰ and +7.7 ‰ for the Plecopteran *Paraperla frontalis* to +11.3 ‰ for *Stygobromus* sp. *Stygobromus*, an omnivorous amphipod, can feed at several trophic levels. The only apparent $d^{15}N$ anomaly for hyporheic consumers is +1.0 ‰ for *Isocapnia crinita* (Plecoptera), a primary consumer.

The $d^{15}N$ values for the producer trophic level in the Flathead River channel range from 21.3 ‰ to −0.3 ‰ (Fig. 1). These $d^{15}N$ values are low in comparison with $d^{15}N$ nitrogen in air (0 ‰). The most deplete $d^{15}N$ value (−1.3 ‰) is for the biofilm. The biofilm is a microbial assemblage dominated by algae (e.g., diatoms) attached to gravel substrates. The less deplete value (−0.3 ‰) is for nitrogen photosynthetically assimilated by *Ulothrix*, a filamentous green algal.

Consumers in the river channel habitat display enriched $d^{15}N$ values at all three trophic levels (Fig. 1). These enriched $d^{15}N$ values are higher than those of their food sources. The $d^{15}N$ values for organisms representing herbivores and omnivores ranged from +1.3 ‰ to +3.4 ‰. The $d^{15}N$ values are for species from the Order Insecta, Plecoptera (stoneflies) and Trichoptera (caddisflies). The species include Plecopterans *Isoperla fulva* (+1.5 ‰), *Pteronarcella badia* (+1.3 ‰), and *Pteronarcys californica* (+2.5 ‰) and the Trichopteran *Arctopsyche grandis* (+3.4‰). The $d^{15}N$ values of predators indicate minimal enrichments compared with herbivores and omnivores. The predators include the Plecopteran *Classennia sabulosa* (+3.1 ‰ and +3.2 ‰) and the Trichopteran *Hesperoperla pacificum* (+3.5 ‰).

TROPHIC RELATIONS AND $\Delta d^{15}N$ ENRICHMENTS

Most $d^{15}N$ values for consumers of the river channel and hyporheic food webs exhibit enrichment ($\Delta d^{15}N$) in body tissues with increasing trophic level (Table 1).

$\Delta d^{15}N$ enrichments of trophic level represent the $d^{15}N$ of a consumer's body tissues minus the $d^{15}N$ of potential diets (Minagawa & Wada, 1984). $\Delta d^{15}N$ enrichments of trophic levels for the river range from +1.8 ‰ to +2.8 ‰ for herbivores, +2.6 ‰ to +4.7‰ for omnivores, and +0.1 ‰ to +2.2 ‰ for predators. Predators of the hyporheic exhibit the highest enrichment values of all the consumers observed in the surface and hyporheic waters. $\Delta d^{15}N$ enrichments for *Paraperla frontalis* range from +4.1 ‰ to +4.3 ‰. Although the trophic level of *Stygobromus* is in question, its high $d^{15}N$ value (+11.3 ‰) suggests possible consumption of tertiary consumers.

Table 2. *Comparison of* Δ $d^{15}N$ *enrichments in invertebrates from stream amd river ecosystems similar to the Flathead River.* Δ $d^{15}N$ *estimated by subtracting* $d^{15}N$ *of potential diets from the* $d^{15}N$ *of the consumers body tissue. Trophic levels of consumers denoted as follows: herbivores (H); omnivores (O); and predators (P)*

Habitat	Trophic level (H, O, P) and invertebrate consumer	$d^{15}N$ of invertebrate body tissue (A)		$d^{15}N$ of 'diets' (B)	Δ $d^{15}N$ enrichment (A)–(B)
Lookout Creek (Oregon)[a]	H Caddisfly (Trichoptera)	+1.8	Algae	−1.2, −1.5	+3.2
	H Stonefly (Plecoptera)	+2.0			+3.4
	H, O Snails	+1.8	Algae	−1.2, −1.5	+3.2
Kuparuk River (Alaska)[a]	H Aquatic Insects	+1.0, +4.0	Algae	+2.0, +3.0	+1.5
	P Trout	+8.0	Consumers	+1.0, +4.0	+4.0 to +7.0
Sashin Creek (Alaska)[b]	H Caddisfly (Trichoptera)	+4.8	Algae	−2.0, +1.8	+3.0 to +6.8
	P Stonefly (Plecoptera)	+5.1	Consumers	+4.8	+0.3
	P Rainbow trout	+7.2, +10.0	Consumers	+4.8, +5.1	+5.2, +4.9

Notes:
[a]Fry (1991)
[b]Kline *et al.* (1990).

Primary producers and consumers of the Flathead River channel exhibit $d^{15}N$ values similar to those for other stream and river ecosystems in the Pacific Northwest and Alaska. Table 2 summarizes $d^{15}N$ values and Δ $d^{15}N$ enrichments for food webs of Lookout Creek, a small stream in the Cascade Mountains of Oregon, the Kuparak River in the Brooks Range of Alaska (Fry, 1991) and Sashin Creek in southeast Alaska (Kline *et al.*, 1990). The ranges of isotopic compositions for primary producers include −1.2% to −1.5 ‰ for algae in Lookout Creek, +2.0 ‰ to +3.0 ‰ for epilithon and filamentous algae in the Kuparak River and −2.0 ‰ to +1.8 ‰ for algal periphyton in Sashin Creek These $d^{15}N$ values are comparable with those of the producer trophic level (−1.3 ‰ to −0.3 ‰) in surface waters but lower than the hyporheic biofilm (+3.4 ‰) of the Flathead River (Table 2).

Comparisons of the $d^{15}N$ values for possible producer and herbivorous organisms from the same ecosystems also show comparable Δ $d^{15}N$ enrichments. Potential Δ $d^{15}N$ enrichments for herbivores in the Flathead River (+1.8 ‰ to +2.8 ‰), Lookout Creek (+3.2 ‰ to +3.4 ‰), Kuparak River (+1.5‰) are similar while values for Sashin Creek are more variable (+3.0 ‰ to +6.8 ‰) (Tables 1 and 2).

Δ $d^{15}N$ for omnivores and predators for these ecosystems also displays similar enrichments. Δ $d^{15}N$ for omnivores ranges from +0.1 ‰ to +4.7 ‰ for stoneflies and caddisflies in the Flathead River and +3.2 ‰ for snails in Lookout Creek. For vertebrate-predators, a broader range of Δ $d^{15}N$ enrichments is evident: Δ $d^{15}N$ of +2.4 to +7.0 ‰ for trout and salamanders in Lookout Creek, the Kuparak River, and

Sashin Creek (Table 2). These Δ $d^{15}N$ enrichments are comparable with the +6.6 ‰ for predators from the hyporheic zone of the Flathead River. If the hyporheic amphipod *Stygobromus* consumes tertiary consumers, Δ $d^{15}N$ enrichments approach +7.9‰.

CONCLUSION

The above comparisons show that $d^{15}N$ enrichments are common for the different trophic levels in surface waters of stream and river ecosystems (Tables 1 and 2). However, the Δ $d^{15}N$ enrichments in the hyporheic habitat of the Flathead River, regardless of the small sample sizes, suggest greater enrichment for the consumer trophic levels. Differences in inorganic nitrogen concentrations of waters and $d^{15}N$ values in the river and hyporheic food webs suggest a variety of inorganic nitrogen sources and influences of physical and biological processes (Stanford & Ward, 1988, 1993; Duff & Triska, 1990; Triska, Duff & Ananzino, 1990; Wissmar, 1991).

NO_3 and NH_4 levels of the river channel could be regulated by inputs from the atmosphere, surface and subsurface drainage, and mixing of waters and photosynthetic pathways. For the Flathead River channel, atmospheric precipitation could be an important source of deplete $d^{15}N$ inorganic nitrogen. For example, annual weighted mean $d^{15}N$ for precipitation as NO_3 and NH_4 in 'clean' environments like the Flathead River drainage may approach −3‰

(Heaton & Collett, 1985). Heaton (1986) suggests that the bulk of the NO_3 and NH_4 in precipitation is derived from solution of gaseous (NO_x and NH_3) rather than particulate compounds.

Different $d^{15}N$ compositions of biofilms in the Flathead River channel (-1.3) and hyporheic ($+3.4$) habitats also point to biological pathways that may alter inorganic nitrogen concentrations and $d^{15}N$ values. In the river channel where low inorganic nitrogen concentrations may limit primary production, nitrogen species may be photosynthetically assimilated with no apparent isotopic fractionation (Peterson & Fry, 1987). Such conditions are most prevalent for low NH_4 concentrations (Cifuentes, Sharp & Fogel, 1988). In such cases, algae and other microbiota (e.g., bacteria) preferentially take up NH_4 over NO_3 with little isotopic fractionation (Cifuentes *et al.*, 1989). This situation could exist in the Flathead River channel, where low NH_4 concentrations could be used rapidly by the microbiota.

Low concentrations of ammonium and nitrate could also influence isotopic fractionations during assimilatory and oxidation-reduction reactions (e.g., N-transformations) associated with the microbial biofilms in both the river channel and hyporheic habitats. For example, the ^{15}N deplete biofilm (-1.3) in the river channel not only indicates NO_3 and NH_4 limitation and minimal fractionation during photosynthesis but the occurrence of N_2-fixation by microbiota. Nitrogen limitation in the river channel could lead to fixation of atmospheric N_2 by biofilm microbiota and contribute to the observed $d^{15}N$ deplete values (Fig. 1, Table 1). N_2-fixation exhibits minimal isotopic fractionation (Delwiche & Steyn, 1970), resulting in biomass $d^{15}N$ values near that of atmospheric N_2 gas, for which the isotopic composition is essentially constant at 0‰ (Mariotti, 1983).

For hyporheic habitats, other N-transformations (e.g., nitrification and denitrification) more common to subsurface environments probably influence the $d^{15}N$ values. For example, nitrification and denitrification within contiguous soils (Delwiche & Steyn, 1970) could influence the $d^{15}N$ values of inorganic nitrogen species and, in turn, those of the hyporheic biofilm and consumer organisms.

The $d^{15}N$ compositions of the river channel and hyporheic biofilms could also reflect feedbacks in the food webs. ^{15}N-deplete NH_4 of the waters could be maintained through the release of less enriched organic N and NH_4 by algal and microbes (Cifuentes *et al.*, 1989). Both algae-microbe (e.g., bacteria and fungi) regeneration could be contributing NH_4 to the dissolved inorganic nitrogen pool that is isotopically similar or lighter than the $d^{15}N$ of the biofilm. The result would be ^{15}N deplete biofilm diets of consumers that contribute to the ^{15}N-deplete isotopic compositions of the food web. Another significant feedback is NH_4 excretion by animals (Triska *et al.*, 1984) and the recycling of NH_4 to biofilms.

NH_4 excretion by animals could be of considerable importance in the hyporheic food webs of the Flathead River. Here, the consumer standing crop biomass of the hyporheic can easily exceed the benthic biomass in the river habitat (Stanford & Ward, 1988). Feedbacks of NH_4 to the microbiota through excretion by animals could be a major source of available inorganic nitrogen in the hyporheic environment.

Although our current data indicate isotopic differences of trophic levels (e.g., biofilms and consumers) for hyporheic and river habitats, the variation within habitats and their trophic levels need further definition. Given the small number of samples on hand, these observations need to be verified to give a better determination of the signal-to-noise ratio. If these observations hold, isotopes should prove to be very powerful tools for determining source information versus process information (e.g., assimilation) in hyporheic habitats. Such information will set the stage for process studies that define fractionations under natural conditions.

REFERENCES

American Public Health Association, American Waterworks Association, & Water Pollution Control Federation (1975). *Standard Methods for the Analysis of Water and Wastewater.* 14th ed. American Public Health Association, Washington, D.C.

Cifuentes, L.A., Fogel, M.L., Pennock, J.R. & Sharp, J.H. (1989). Biogeochemical factors that influence stable nitrogen isotope ratio of dissolved ammonium in the Delaware Estuary. *Geochimica et Cosmochimica Acta*, **53**, 2713–2721.

Cifuentes, L.A., Sharp, J.H. & Fogel, M.L. (1988). Stable carbon and nitrogen isotope biogeochemistry in the Delaware estuary. *Limnol. Oceanogr.*, **33**, 5, 1102–1115.

Delwiche, C.C. & Steyn, P.L. (1970). Nitrogen isotope fractionation in soils and microbial reactions. *Environmental Technology*, **4**, 929–935.

Duff, J.H. & Triska, F.J. (1990). Denitrification in sediments from the hyporheic zone adjacent to a small forested stream. *Can. J.Fish. Aq. Sci.*, **47**, 1140–1147.

Fry, B. (1991). Stable isotope diagrams of freshwater food webs. *Ecology*, **72**, 2293– 2297.

Heaton, T.H.E. (1986). Isotopic studies of nitrogen pollution in the hydrosphere and atmosphere: A review. *Chemical Geology*, **59**, 87–102.

Heaton, T.H.E. & Collett, G.M. (1985). *The analysis of $^{15}N/^{14}N$ ratios in natural samples, with emphasis on nitrate and ammonium in prescipitation.* CSIR (Counc. Sci. Ind. Res.), Pretoria, Res. Rept. No. 624. 28 pp.

Hendricks, S.P. (1993). Microbial ecology of the hyporheic zone: a perspective integrating hydrology and biology. *J. N. Am. Benth.Soc.*, **12**, 70–78.

Kline, T.C. Jr., Goering, J.J., Mathisen, O.A. & Poe, P.H. (1990). Recycling of elements transported upstream by runs of Pacific salmon: I. $d^{15}N$ and $d^{13}C$ evidence in Sashin Creek, Southeastern Alaska. *Can. J. Fish. Aq. Sci.*, **47**, 136–144.

Mariotti, A. (1983). Atmosperic nitrogen as a reliable standard for natural ^{15}N abundance measurements. *Nature*, **303**, 685–687.

Menzel, D.W. & Vaccaro, R.F. (1964) The measurement of dissolved and particulate carbon in seawater. *Limnol. Oceanogr.*, **9**, 138–142.

Minagawa, M. & Wada, E. (1984). Stepwise enrichment of ^{15}N along food chains: Further evidence and the relation between $d^{15}N$ and animal age. *Geochimica et Cosmochimica Acta*, **48**, 1135–1140.

Peterson, B.J. & Fry, B. (1987). Stable isotopes in ecosystem analysis. *Ann. Rev. Ecol. Syst.*, **18**, 293–320.

Porter, K.G. & Feig, Y.S. (1980). The use of DAPI for identifying and counting aquatic microflora. *Limnol. Oceanogr.*, **25**, 943–948.

Simenstad, C.A. & Wissmar, R.C. (1985). d^{13}C evidence of the origins and fates of organic carbon in estuarine and nearshore food webs. *Mari. Ecol.Progress Series*, **22**, 141–152.

Stanford, J.A. & Ward, J.V. (1988). The hyporheic habitat of river ecosystems. *Nature*, **335**, 64–66.

Stanford, J.A. & Ward, J.V. (1993). An ecosystem perspective of alluvial rivers: connectivity and the hyporheic corridor. *J. N.Am. Benth. Soc.*, **12**, 48–60.

Strickland, J.D.H. & Parsons, T.R. (1972). *A Practical Handbook of Seawater Analysis*, 2nd ed. Bulletin of the Fisheries Research Board of Canada, 167.

Triska, F.J, Duff, J.H. & Ananzino, R.J. (1990). Influence of exchange flow between the channel and hyporheic zone on nitrate production in a small mountain stream. *Can. J. Fish. Aq. Sci.*, **47**, 2099–2111.

Triska, F.J., Sedell. J.R., Cromack, K., Gregory, S.V. & McCorison, F.M. (1984). Nitrogen budget for a small coniferous forest stream. *Ecol. Mon.*, **54**, 119–140.

Wissmar, R.C. (1991). Forest detritus and cycling of nitrogen in a mountain lake. *Can. J. For. Res.*, **21**, 990–998.

22 La zone hypodermique du sol écotone entre eaux météoriques et eaux souterraines dans l'infiltration des pesticides dissous

C. THIRRIOT & B. CAUSSADE

Institut de Mécanique des Fluides de Toulouse – URA 0005 au CNRS Avenue du Professeur Camille Soula, 31400 Toulouse, France

RÉSUMÉ La zone hypodermique du sol peut être considérée comme un écotone pour les produits en solution dans l'eau provenant des précipitations ou de l'irrigation tellement sont importants dans cette couche limite les phénomènes d'adsorption, d'absorption par les plantes et de dégradation ou décomposition des produits chimiques utilisés en agriculture, engrais ou pesticides.

Après avoir rappelé l'importance de l'emploi des pesticides en agriculture et les risques sanitaires pour les nappes phréatiques, on distingue les différentes conventions d'évaluation des concentrations. Puis l'on propose la solution des équations de migration du pesticide en tenant compte du phénomène d'adsorption–désorption et de la fixation définitive.

Enfin, on donne un exemple qui situe les différentes échelles de temps.

ABSTRACT First, the different symbols of the pesticide concentration in water, on dry matter and global porous media are distinguished. Then, a simple transport model solution is presented with help of asymptotic developments for short times and for large times.

A numerical example is considered to show the characteristic time scales.

INTRODUCTION

Pour l'hydraulicien, la couche superficielle d'un sol agricole sera tout naturellement un écotone, zone d'échange et de 'mutation' entre l'eau qui véhicule fertilisants et pesticides, les grains de terre et les racines des plantes, en n'oubliant pas la faune qui catalyse les échanges et les transformations.

Dans ce qui suit, nous allons restreindre notre point de vue à l'examen du devenir des pesticides.

Les produits phytosanitaires sont certes efficaces pour assurer quantité et qualité de la production agricole, mais ils ne sont pas sans danger. Ces risques, difficilement chiffrables, ont cependant donné lieu à des enquêtes. Aux États Unis, un suivi sur 25 années a conduit à estimer la mortalité par pesticide à une personne pour un million d'habitants surtout par suite d'intoxication professionnelle (Tissut *et al.*, 1979). A l'opposé de ce risque décisif mais rare, existe le danger de la pollution des nappes phréatiques. Dans la surveillance de 76 forages destinés à l'alimentation en eau potable du département du haut Rhin, en 1982, il est apparu que la moitié des puits contrôlés étaient contaminés épisodiquement et que pour 16% des mesures, la teneur en pesticides dépassait les normes européennes de potabilité (Belami & Giroud, 1990). La teneur en triazine pouvait atteindre 0.4 µg/l alors que la norme européenne admise est de 0.1 µg/l par élément.

En effet, la question est vitale de savoir si toute la quantité de pesticides demeure ou est décomposée dans la couche superficielle du sol, ou bien si une partie non négligeable percole en profondeur vers la nappe phréatique.

L'objectif ici est de donner, au moins sommairement par le calcul analytique, une valeur approximative de l'épaisseur de cette couche qui peut être assimilée à une couche limite.

LA QUESTION DES CONVENTIONS

L'étude du transport en milieu poreux nécessite, de par sa difficulté, un aller et retour entre le point de vue microscopique à l'échelle du pore et du grain de sol, et le point de

Tableau 1. *Correspondance entre les différents concepts de concentration suivant les phases considérées*

Référence	1 Eau	2 Solide	3 Milieu poreux
Pesticide dans l'eau en mouvement	C_1 g/g	C_2 g/g	C_3 g/cm³
Pesticide associé au solide	S_1 g/g	S_2 g/g	S_3 g/cm³

vue macroscopique où les phases liquide, gaz et solide en présence sont considérées comme formant un milieu fictif continu.

Pour évaluer les proportions de pesticides en solution (C) ou fixé sur les particules (S), on peut prendre l'un des trois points de vue de référence suivants: eau en mouvement (1), partie sèche (2), volume de milieu poreux (3) (tableau 1).

Pour établir la correspondance entre C et S, deux approches peuvent être envisagées.

a. Bilan de matière

Soit m_s la masse adsorbée dans un volume Ω_t de milieu poreux assez petit pour que l'on puisse considérer la répartition comme homogène, P la porosité, ρ_f la masse volumique de l'eau et ρ_s la masse volumique du matériau solide constituant l'architecture du milieu poreux.

Par définition,

$$S_3=\frac{m_s}{\Omega_t} \quad S_2=\frac{m_s}{\rho_s(1-P)\Omega_t} \quad S_1=\frac{m_s}{\rho_f P\Omega_t}$$

D'où les relations entre S_1, S_2 et S_3

$$\rho_f PS_1=\rho_s(1-P)S_2=S_3 \tag{1}$$

Evidemment les mêmes relations existent pour C.

b. Equilibre thermodynamique

Les grandeurs C et S sont des indices de stocks, qui peuvent jouer le rôle de paramètres de qualité. Les ions en solution se déplacent vers les parois par diffusion. Lorsqu'ils se fixent sur les parois solides, il y a adsorption, dans le cas contraire désorption.

L'état des particules adsorbées est caractérisé par un potentiel thermodynamique spécifique ψ_s qui dépend de leur accumulation S; celui des particules en solution avec une concentration C est caractérisé par un potentiel ψ_c. Souvent ψ_c et C sont confondus. Pour obtenir la condition d'équilibre entre les proportions C_e et S_e pour l'égalité des potentiels $\psi_c=\psi_o$, il faut définir la relation $\psi_s(S)$.

Avec $\psi_c=C$, il vient:

$$C_e=\psi_s(S_e) \tag{2}$$

L'indice e signifie équilibre.

Si les phénomènes d'approche et de fixation sont très rapides, les temps d'établissement de l'équilibre peuvent être négligés, dans le cas contraire la vitesse d'accumulation est supposée dépendre directement de la différence des potentiels ψ_c et ψ_s suivant une loi généralement linéaire:

$$\frac{dS}{dt}=h(\psi_c-\psi_s) \tag{3}$$

avec h coefficient d'échange, inverse d'un temps, caractéristique de la cinétique. Les autres quantités (voir tableau 1) sont définies à partir de C_1 et S_2 par simple règle de trois.

Ainsi donc, en considérant la quantité artificielle S_1 comme analogue à une concentration C_1, à l'équilibre, pour assurer l'équilibre des potentiels, il viendrait:

$$S_1=C_1 \quad \text{c'est-à-dire} \quad S_2=\frac{\rho_f P}{\rho_s(1-P)}C_1 \tag{4}$$

Mais, les résultats de mesure montrent qu'à l'équilibre S_2 est souvent supérieur à la valeur $\rho_f PC_1/\rho_s(1-P)$. La relation formelle entre C_1 et S_2 n'est utile que pour les bilans de masse. Par contre, la relation entre concentration et potentiel va dépendre de l'arrangement des molécules adsorbées à la paroi, de la nature des liaisons (physique, électrique ou chimique) et de la surface spécifique du support solide. On est donc amené à faire appel à l'expérience et à l'empirisme pour décrire la cinétique et l'équation d'équilibre.

LES ÉQUATIONS

Dans un souci de clarté et pour éviter des difficultés au plan analytique, nous allons considérer des situations simples.

Les conditions hydrodynamiques

On considère que l'écoulement de l'eau est permanent, saturé ou non saturé, mais uniforme dans un milieu poreux supposé homogène. Le seul paramètre est alors la vitesse moyenne de pore u.

La représentation biochimique des pesticides

Les pesticides sont en concentration $C=C_1$ dans l'eau en écoulement. Ils sont en concentration équivalente $C'=S_2/\alpha$ dans l'eau figée, dans les zones mortes et sur les parois des grains de sol. C'est cette partie C' qui va être active en particulier dans la lutte microbienne.

Il y a échange entre la concentration mobile C et la concentration immobile C' suivant une cinétique linéaire. Le flux de matière de la partie mobile vers la partie immobile est supposé représenté par:

$$\Phi_1 = h(C - C') \tag{5}$$

Le flux de retrait de matière (assimilation par les plantes, les microbes, etc.) est pris en compte par un effet de puits caractérisé par un flux Φ_2 tel que:

$$\Phi_2 = -h'C' \tag{6}$$

La variation de la population C' peut finalement être représentée par l'équation de continuité:

$$\alpha \frac{\delta C'}{\delta t} = \phi_1 + \phi_2 \tag{7}$$

L'équation de transport

L'eau vive transporte les pesticides comme un soluté. Sans échange et sans diffusion longitudinale chimique ou hydrodynamique, le bilan de matière conduit à l'équation d'évolution de la concentration C suivante:

$$\frac{\delta C}{\delta t} + u \frac{\delta C}{\delta x} = 0 \tag{8}$$

En tenant compte de l'échange avec les zones de stockage inertes, il vient:

$$\frac{\delta C}{\delta t} + u \frac{\delta C}{\delta x} = -\phi_1 = -h(C - C') \tag{9}$$

Le modèle retenu

Il est composé des deux équations, déjà présentées, gérant les évolutions conjointes des concentrations C et C'

$$\frac{\delta C}{\delta t} + u \frac{\delta C}{\delta x} + h(C - C') = 0 \tag{10}$$

$$\alpha \frac{\delta C'}{\delta t} + h(C' - C) + h'C' = 0 \tag{11}$$

Le milieu est supposé semi-infini, la coordonnée spatiale x est comptée suivant la verticale descendante.

Condition à la frontière

A la surface du sol, $x=0$, l'épandage et l'infiltration du pesticide sont supposés se passer en un temps très court par rapport aux échelles de temps caractéristiques du phénomène. L'épandage sera représenté par une fonction de Dirac. Nous prendrons comme conditions initiales $C(x,0) = C'(x,0) = 0$.

Soient M_0 la masse épandue par unité de surface du sol et Δ_t la durée supposée petite de l'épandage. Le flux de matière

transportée est $\rho q C_o = \rho u P C_o$, C_0 étant la concentration à l'entrée.

En supposant que la matière en solution ou en suspension dans le liquide est uniformément répartie durant l'intervalle de temps Δ_t, il vient:

$$\rho q C_0 \Delta t = M_0 = \rho u P C_0 \Delta t$$
$$\text{soit} \quad C_0 = M_0 / \rho u P \Delta t \tag{12}$$

Si Δ_t tend vers 0, la concentration est marquée par un pic de Dirac d'intensité $M_0 / \rho u P$.

LA RESOLUTION

Recherche d'une solution par la transformation de Laplace

L'image $\hat{C}(x,p)$ de la fonction originale $C(x,t)$ est donnée par la transformation de Laplace L

$$L(C(x,t) = \hat{C}(x,p) = \int_0^\infty e^{-pt} C(x,t) dt \tag{13}$$

Avec les règles classiques de la transformation intégrale, les équations transformées tenant compte des conditions initiales deviennent:

$$u \frac{d\hat{C}}{dx} + (p+h)\hat{C} - h\hat{C}' - 0 \tag{14}$$

$$(\alpha p + h + h')\hat{C}' - h\hat{C} = 0 \tag{15}$$

Si l'on considère la fonction $\hat{C}(x,p)$ comme inconnue principale, il vient comme équation résultante:

$$u \frac{d\hat{C}}{dx} + \left(p + h - \frac{h^2}{\alpha p + h + h'}\right)\hat{C} = 0 \tag{16}$$

En tenant compte de la condition à la surface du sol $\hat{C}(0,p) = \hat{C}_0(p)$ on obtient la solution image:

$$\hat{C}(x,p) = \hat{C}_0(p) \exp\left[-\left(p + h - \frac{h^2}{\alpha p + h + h'}\right)\frac{x}{u}\right] \tag{17}$$

L'équation (12) permet alors d'obtenir l'image de $C'(x,t)$

$$\hat{C}' = \hat{C}_0(p) \frac{h}{\alpha p + h + h'} \exp\left[-\left(p + h - \frac{h^2 x}{\alpha p + h + h'}\right)\frac{x}{u}\right] \tag{18}$$

Retour à l'original

L'image s'écrit en séparant les exposants

$$\hat{C}(x,p) = \hat{C}_0(p) e^{\frac{-hx}{u}} e^{\frac{-px}{u}} e^{\frac{h^2 x}{u(\alpha p + h + h')}} \tag{19}$$

Le premier terme exponentiel est un simple coefficient d'amortissement en fonction de la profondeur atteinte, le deuxième caractérise un retard, le troisième est le terme spécifique de l'image.

Faire le changement de variable $p'=p+a$, revient à multiplier l'original par $e^{-at'}$. On peut faire apparaître la translation sur p caractérisée par $\alpha=h+h'/\alpha$ et il reste à trouver l'original de

$\dfrac{b}{e^p/p}$ avec $b=h^2x/u\alpha$

Soit L^{-1} l'opérateur de retour à l'original. D'après Angot (1972), il vient

$$L^{-1}\left(\frac{1}{p}e^{\frac{b}{p}}\right)=I_0(2\sqrt{bt'})\tag{20}$$

donc que

$e^{\frac{b}{p}}=pL[I_0(2\sqrt{bt'})]$

La règle sur les dérivées permet d'écrire que:

$$L^{-1}\left(e^{\frac{b}{p}}\right)=I_0(0)\delta(t')+\frac{d}{dt'}(I_0(2\sqrt{bt'}))$$

la variable temporelle étant $t'=t-x/u$

EXPRESSION DE LA CONCENTRATION DANS L'EAU VIVE

Avec une condition d'impulsion à la surface à l'instant initial, l'image $\hat{C}_0(p)$est l'intensité de la fonction de Dirac, $(M_0/\rho uP)$, l'expression de la concentration est donc

$$C(x,t)=0\quad\text{si}\quad t<\frac{x}{u}\tag{21}$$

$$C(x,t)=\frac{M_0}{\rho Pu}e^{-\frac{h}{u}x}e^{-\frac{h+h'}{\alpha}\left(t-\frac{x}{u}\right)}$$

$$\left[I_0(0)\delta\left(t-\frac{x}{u}\right)+\frac{d}{dt}\left(I_0\left(2\sqrt{\frac{h^2x}{\alpha u}\left(t-\frac{x}{u}\right)}\right)\right)\right]\quad\text{si}\quad t\geq\frac{x}{u}\tag{22}$$

L'utilisation de développements asymptotiques va nous permettre de concrétiser l'évolution de C. Mais, d'ores et déjà, on voit apparaître (ou confirmer) certains comportements:

(a) le phénomène de convection mis en évidence par le retard x/u

(b) l'atténuation de la concentration en cours de percolation

(c) la propagation atténuée de la fonction impulsion créée par l'épandage quasi instantané à la surface du sol, l'amortissement de la fonction de Dirac étant d'autant plus rapide que le coefficient d'échange h est élevé

(d) l'effet de puits ou de disparition des pesticides marqué par l'amenuisement exponentiel $e^{-h'/\alpha(t-x/u)}$.

EXPRESSION DE LA CONCENTRATION FIXÉE

Dans le cas d'un épandage quasi instantané on peut encore écrire la solution image:

$$\hat{C}'(x,p)=\frac{M_0}{\rho uP}e^{-hx/u}e^{-px/u}\left(\frac{h}{\alpha p+h+h'}\right)e^{\frac{h^2x}{(\alpha p+h+h')u}}\tag{23}$$

L'expression de $C'(x,t)$ est donc finalement

$$C'(x,t)=0\quad\text{si}\quad t<\frac{x}{u}\tag{24}$$

$$C'(x,t)=\frac{M_0h}{\rho uP\alpha}e^{hx/u}e^{-hx/u(-x/u)}I_0\left(2\sqrt{\frac{h^2x}{\alpha u}\left(t-\frac{x}{u}\right)}\right)\tag{25}$$

Construction de développements asymptotiques

COMPORTEMENT POUR T' PETIT

Utilisons l'expression de développement en série de $I_0(Z)$ rapportée par Angot (1972)

$$I_0(Z)=\sum_{r=0}^{\infty}\frac{((1/2)Z)^{2r}}{(r!)^2}\tag{26}$$

$$I_0(0)=1\tag{27}$$

$$\frac{d}{dt}\left[I_0\left(2\sqrt{\frac{h^2x}{\alpha u}\left(t-\frac{x}{u}\right)}\right)\right]=\frac{d}{dt}\left[\sum_{i=0}^{\infty}\frac{1}{(r!)^2}\left(\sqrt{\frac{h^2x}{\alpha u}\left(t-\frac{x}{u}\right)}\right)^{2r}\right]$$

$$=\sum_{r=0}^{\infty}\frac{1}{(r!)^2}\left(\frac{h^2x}{\alpha u}\right)^r\frac{d}{dt}\left(t-\frac{x}{u}\right)^r=\sum_{r=1}^{\infty}\frac{1}{(r!)^2}\left(\frac{h^2x}{\alpha u}\right)^r\cdot r\left(t-\frac{x}{u}\right)^{r-1}\tag{28}$$

Si $\dfrac{h^2x}{\alpha u}\left(t-\dfrac{x}{u}\right)$est petit, le développement en série est très convergent du fait du dénominateur en$(r!)^2$.

On montre que pour chaque valeur de x, au-delà d'un seuil, il peut apparaître un maximum de la concentration à une date différente de x/u, qui correspond à l'apparition du pic de Dirac résiduel provenant de la propagation de l'impulsion originelle.

COMPORTEMENT POUR T' GRAND

Toujours d'après Angot (1972), pour $Z\gg1$

$$I_0(Z)=\frac{e^z}{\sqrt{2\pi Z}}\left\{1+\frac{1}{8Z}+\frac{9}{128Z^2}+\ldots\right\}\tag{29}$$

La solution est alors pour t grand, avec $t_2=\alpha u/h^2x$

$$C(x,t)=\frac{M_0}{\rho Pu}e^{-h/ux}e^{-h+h'/\alpha(t-x/u)}$$

$$\left[\delta\left(t-\frac{x}{u}\right)+\frac{e^2\sqrt{\frac{t-x/u}{t_2}}}{t_2\sqrt{4\pi}}\left\{\left(\frac{t_2}{t-\frac{x}{u}}\right)^{3/4}-\frac{3}{16}\left(\frac{t_2}{t-\frac{x}{u}}\right)^{5/4}\right.\right.$$

Tableau 2. *Exemple – Valeurs caractéristiques de l'infiltration d'un pesticide suivant le type de sol caractérisé par sa conductivité hydraulique K*

paramètre	Porosité $P=1/3$	Sol léger	Sol intermédiaire	Sol lourd
	$\alpha=1, 5, \beta=2$			
	$t^*=26$ mn, $t^{*\prime}=13$mn	$K=12$ m/j	$K=1,2$ m/j	$K=0,12$ m/j
	Vitesse moyenne d'infiltration	0,5 m/h	5 cm/h	5 mm/h
A la	Temps d'arrivée du premier maximum	24 mn	4 h	40 h
	Intensité relative résiduelle de l'impulsion	40%	1.2×10^{-4}	8×10^{-40}
profondeur	Temps d'apparition du second maximum	24 mn	7 h	103 h
	Concentration relative maximale dans l'eau vive	37%	1.4×10^{-3}	1.4×10^{-27}
20 cm	Temps d'apparition du maximum adsorbé	24 mn	9 h	105 h
	Concentration relative maximale adsorbée	40%	4×10^{-4}	4.5×10^{-28}
A la	Temps d'arrivée du premier maximum	1h 12 mn	12 h	120 h
	Intensité relative résiduelle de l'impulsion	6%	10^{-12}	10^{-120}
profondeur	Temps d'apparition du second maximum	1h 12 mn	29 h	323 h
	Concentration relative maximale dans l'eau vive	17%	2.7×10^{-9}	6.7×10^{-82}
60 cm	Temps d'apparition du maximum adsorbé	1h 12 mn	31 h	325 h
	Concentration relative maximale adsorbée	6%	8.9×10^{-10}	1.9×10^{-82}

$$-\frac{15}{152}\left(\frac{t_2}{t-\frac{x}{u}}\right)^{7/4}\cdots\right\}\right] \tag{30}$$

Cette expression, apparemment compliquée, permet néanmoins de faire apparaître des traits physiques intéressants. Nous n'insistons pas sur l'effet d'amortissement évident dans les exponentielles à exposants négatifs pour la partie liée à l'impulsion ou fonction de Dirac. A la date approximative:

$$t_{Cmax}=\frac{x}{u}\left(1+\alpha\left(1-\frac{3}{2}\frac{u}{hx}\right)\right) \tag{31}$$

l'estimation approchée de la concentration afférente au deuxième maximum s'écrit:

$$C_{max}=\frac{M_0}{\rho P\alpha\sqrt{4\pi}}\sqrt{\frac{h}{xu}}e^{-\frac{h'}{h}\left(\frac{hx}{u}-\frac{3}{2}\right)}x\left(1-\frac{3}{8}\frac{u}{hx}+\frac{3}{10}\frac{u^2}{h^2x^2}\right) \tag{32}$$

On observe que, pour une profondeur x donnée, la concentration résiduelle maximale dépend nettement de la vitesse d'infiltration.

EXEMPLE D'APPLICATION

La bibliographie parait pauvre en données expérimentales. Cependant, Balayannis (1988) présente les résultats d'expériences, fort intéressantes pour le mécanicien des fluides, qui conduisent à l'évaluation du coefficient que nous avons appelé α. Il montre que pour l'atrazine et un certain sol:

$$S_2=K_dC_1 \quad (\text{avec} \quad K_d=26) \tag{33}$$

Nous avons choisi comme exemple trois cas de sol (qui peuvent être considérés comme représentatifs), et deux profondeurs ($x=20$cm et $x=60$cm) qui 'encadrent' le système racinaire des cultures. Les résultats obtenus sont synthétisés dans le tableau 2.

Tout d'abord on observe une grande diversité de comportement suivant la vitesse d'infiltration. Par exemple, en sol léger, il n'y a dans l'eau vive qu'une onde de choc à la vitesse de convection suivie d'une traînée. L'adsorption et l'épuisement amortissent beaucoup les concentrations mais les traces d'atrazine à de profondeur sont encore importantes. Pour le sol lourd, tout est amorti dans les premiers centimètres. Les temps théoriques de passage des maximums deviennent très grands, plus de 13 jours à 60cm dans l'eau vive.

Les chiffres du tableau 2 montrent que, déjà pour les sols intermédiaires mais surtout pour les sols lourds, la couche de sol superficielle joue parfaitement son rôle d'écotone puisque tous les phénomènes de transformation (signifiés ici par l'absorption) vont s'y dérouler et les zones profondes vers la nappe seront à l'abri des pesticides. Mais, dans la réalité, il ne faut pas sous-estimer le risque de 'shunt' que peuvent présenter les fentes de retrait dues à la sécheresse ou les parties sableuses dues à une hétérogénéité accidentelle.

Il est évident que c'est l'orage violent et intense qui sera le plus dangereux pour l'entraînement rapide des pesticides.

Comme nous l'avons déjà dit, si l'intensité de la pluie ou de l'irrigation est insuffisante pour saturer le sol, la vitesse de convection dans les sols légers sera faible et de l'ordre de celle observée lors de l'infiltration en sols lourds.

CONCLUSION

Ecotone pour les écologues, couche limite pour le mécanicien des fluides, la couche hypodermique du sol joue un rôle décisif sur le devenir des pesticides dissous. Zone de passage si la vitesse d'infiltration est élevée, elle est la zone de fixation lorsque l'advection est lente et permet la 'mutation' du produit chimique.

La justification d'une étude analytique forcément simpliste d'un phénomène aussi complexe que celui du transfert des pesticides dans le sol peut se concevoir de deux manières. D'abord à cause de l'incertitude qui règne sur la description mathématique du phénomène réel et à cause de l'imprécision sur la valeur des paramètres lorsqu'on a choisi un modèle. Ensuite par l'intérêt présenté par la vision globale que permet l'étude analytique qui facilite la distinction des comportements asymptotiques et une meilleure compréhension des comportements intermédiaires. Les résultats analytiques permettent de dessiner le 'portrait robot' du phénomène physique. Ils guident l'interprétation de l'effet des différents paramètres essentiels: vitesse du fluide vecteur, force de l'adsorption, cinétique des échanges, épuisement du pesticide. On retrouve bien sûr pour une bonne part des conclusions de bon sens: le transfert vers la nappe dépend pour beaucoup de l'intensité de la vitesse d'infiltration qui accompagne ou suit accidentellement l'épandage, la rapidité de l'absorption ou de la transformation du pesticide dans les couches superficielles très actives par les racines diminue le risque de contamination. Mais, certains résultats peuvent paraître plus nouveaux. Par exemple, l'immobilisation du deuxième maximum du fait de l'arrêt de l'infiltration pourrait concourir à l'explication de l'accumulation de pesticide à une profondeur d'une cinquantaine de centimètres, accumulation signalée par plusieurs observateurs (Moreale & Van Bladel, 1983; Caussade, 1993). L'étude analytique permet aussi la mise en place des échelles de temps et d'espace caractéristiques du phénomène, avec en particulier la distinction des vitesses de déplacement du premier maximum et du second maximum de concentration.

Cette approche analytique est certe handicapée par la condition d'opérateur linéaire mais l'on peut aussi examiner, au prix d'une complication calculatoire, l'effet des non linéarités des isothermes d'adsorption, par exemple à l'aide de méthodes asymptotiques.

RÉFÉRENCES

Angot, A. (1972). *Compléments de Mathématiques à l'usage des Ingénieurs de l'Électrotechnique et des Télécommunications.* Masson et Cie, Editeurs à Paris.

Balayannis, P. (1988). The prediction of the mobility of pesticides in soil, based on thermodynamics characteristics of the system pesticide–soil. *Symposium Methodological aspects of the study of pesticides behaviour in soil.* INRA Versailles, 16–17 juin.

Belami, R. et Giroud, S. (1990). 'Les pollutions liées à l'utilisation des pesticides'. *Perspectives agricoles*, n° 146 – pp. 52–56, avril 1990.

Caussade, B. (1993). *Programme de recherche du Groupe Hydrodynamique de l'Environnement de l'IMFT, sur la migration des pesticides dans un sol sous culture.* Rapport IMFT-HYDRE n° 116.

Moreale, A. et Van Bladel, R. (1983). Transport vertical de solutés vers les eaux souterraines. Une approche sur colonne de sol non perturbé. *Revue de l'Agriculture* n° 6, vol. **36**, nov–déc., pp. 1669–1676.

Tissut, M. Severin, F., Benort-Guyot, G.L., Gachet, H., Mallion, J.M., Degrance, C. et Boucherle, A. (1979). Les pesticides oui ou non? Document Formation Continue – Université Scientifique et Médicale – Presse Universitaire de Grenoble.

LISTE DES SYMBOLES

C	concentration dans l'eau mobile
C'	concentration équivalente afférente à la matière adsorbée
$\hat{C}$	image de C par la transformation de Laplace
h, h'	coefficient d'échange
I_0	fonction de Bessel
L	opérateur de la transformation de Laplace
m_s	masse solide du milieu poreux
p	variable de Laplace
P	porosité du milieu poreux
S	concentration (ou saturation) de la matière adsorbée sur la partie solide
t	temps
u	vitesse moyenne d'infiltration de l'eau
x	profondeur à partir de la surface
α	coefficient caractéristique de l'adsorption à l'équilibre
δ	fonction de Dirac
Δt	intervalle de temps
Ψ_c	potentiel dû à la concentration C dans l'eau mobile
Ψ_s	potentiel dû à la concentration (ou saturation) dans la partie adsorbée à la surface solide
ϕ	flux de matière
ρ_f	masse volumique du fluide
ρ_s	masse volumique de la partie solide du milieu poreux
Ω_t	volume total du milieu poreux

23 Soils of the north-eastern coast of the Caspian Sea as the zone of sea water/groundwater interaction

T.A. GLUSHKO

Water Problems Institute, Russian Academy of Sciences, 10 Novaya Basmannaya Str., P.O. Box 524, Moscow 107 078, Russia

ABSTRACT   The north-eastern coast of the Caspian Sea is a flat plain formed after the sea level drop in 1930–1977. For this shallow part of the Caspian Sea wind-induced surges are usual. They are developed under the strong winds blowing during several days landward. As a result of the wind-induced surges the sea level can rise up to 2 m, flooding vast areas of the dry bottom. When the sea recedes again the halophytic plants cover the grey-brown solontchak-like desert soils. Besides that, saline sea water invades the upper soil horizon due to landward movement of the area of wind-induced surges. The process of the interaction of groundwater and sea water in the soils manifests itself in the salt profiles of the soils.

GENERAL INFORMATION ON THE CASPIAN SEA

The Caspian Sea is the largest (length 1200 km) enclosed sea on Earth with the sea level at -27 m. Located on the boundary of Europe and Asia the Caspian Sea crosses zones of deserts and semideserts of temperate climatic belt and humid and dry subtropics (Fig.1). The relief of the adjacent territory is variable. The shallow Northern part lays in the Pricaspian lowland. The Middle Caspian Sea borders in the West with the Big Caucasus Mountains, in the East with the Kendirli-Koyasan and the Mangyshlak Plateaus. The deepest Southern Caspian Sea meets with the Kura Lowland in the West and the West-Turkmenian Lowland in the East, both located in the zone of Alpine folding.

The water balance of the Caspian Sea depends to a great extent on the Volga runoff giving up to 85% of the total runoff to the Caspian Sea. The climatic changes in the Volga basin, which has an area of 1 380 000 km², determine annual fluctuations of the sea level. During last two centuries the range of the fluctuations reached 7 m (from -22 m in 1800–1830 to -29 m in 1977). From 1929 to 1977 the sea level decreased from -26 m to -29 m, then it began to rise very fast and it is at the level -27 m nowadays. Such fluctuations lead to essential changes of the coastal line.

CHANGES OF THE COASTAL LINE

The greatest changes of the coastal line occurred in the North Caspian Sea due to its shallowness (Fig.2). Open sea area reduced in 1929–1977 by 35 000 km². Between the River Volga and the River Ural the strip of dried bottom was 10–20 km, along the eastern coast from the River Emba to the Komsomolets Bay (Karaton region) the strip was 35–40 km wide. The Karakichu, Kaidak and Komsomolets Bays dried out in 1941 (Badamshin & Chernoscutov, 1947).

NATURAL FEATURES OF THE NORTHERN CASPIAN SEA

A natural boundary separating the Northern Caspian Sea from the other parts, intersects the Island Chechen near the western coast and the Cape Tjub-Karagan (Fig.1) on the eastern coast. The Northern part occupies about 25% of the whole sea surface and only 0.5% of the entire sea water volume; 68% of the area of the Northern part has a depth of 0–5 m (Leontyev *et al.*, 1977). The salinity of the eastern part of the Northern Caspian Sea varies from 8 to 14%.

The dried bottom of the Caspian Sea is a flat uncut plain covered with loam-sandy sediments. The climate of the territory is continental with low winter (average $-7\,^{\circ}$C months^{-1}) and high summer temperatures (average $25\,^{\circ}$C months^{-1}). In winter from

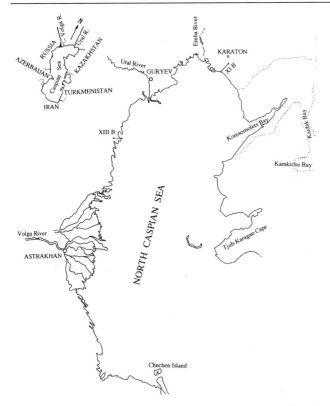

Fig. 1 The Caspian Sea: general situation (scale 1:30 000 000) and Northern Caspian Sea (scale 1:2 500 000) with location of profiles XI B and XIII B.

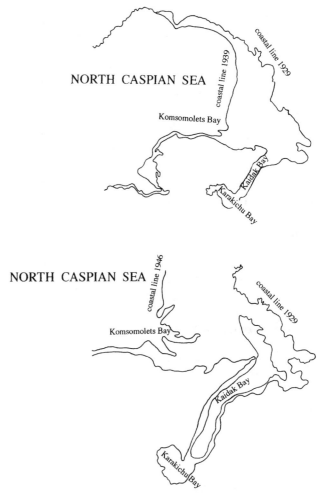

Fig. 2 Changes of the coastal line in the northeastern part of the Caspian Sea (after Badamshin & Chernoscutov, 1947).

December to April ice cover forms in the Northern Caspian Sea. Precipitation is 141 mm yr^{-1} in Atirau and 175 mm yr^{-1} in Astrakhan. The main part of the precipitation falls in the Autumn-Winter season. Evaporation reaches here 1132 mm yr^{-1}.

In this region, eastern winds blow in winter; western, north-western and north-eastern winds blow in summer. The average wind velocity is 5.3–7.0 m s^{-1}. Due to small offshore and coastal declivity, the wind regime promotes the development of wind-induced surges. On the Karaton coast the highest surges of 1.2–2.3 m occurred under strong (24–35 m s^{-1}) western and south-western winds, blowing for 2–4 days. During such surges the width of the inundated coast reached 35–40 km (Gershtanskii, 1977). For example, as a result of high inundation in April 1987, the width of water covered land was as much as 30 km.

Coastal groundwaters are highly mineralized. Mineralization increases from 20 to 180 g l^{-1} in the landward direction.

FORMATION OF THE LANDSCAPES ON THE DRIED BOTTOM OF THE NORTH-EASTERN COAST OF THE CASPIAN SEA

Within the coastal lowlands with a near-surface water table accumulative hydromorphic geochemical landscapes are formed (Mann, 1982). These landscapes receive and accumulate an essential amount of matter with water, solid and chemical flow. But some part of the most mobile matter is carried away with waters of surface and underground flow.

The hydromorphic transitional landscapes are located higher. The groundwaters are close to the surface there. Through the fringe of capillary-film solutions groundwaters influence the soil-forming processes and are vital for the activity of the plant cover (Kovda, 1985).

Let's analyse two sites of the coast: straight open coast – Karaton region (profile XI B, Fig.1) and ingression coast – the Komsomolets Bay. In these sites the ecological profiles with levelling were made from the coastal line in a landward direction. Plant associations were described; soils and groundwaters were sampled and analysed. The preparation of soil for chemical analysis was as follows: 50 g of dry soil were dissolved in 250 ml of water; after 3 minutes of shaking the sample was filtered and obtained water extraction was analysed.

Table 1. *Salt content of water extraction of soils (meqv/100g of soil)*

Profile	depth of sample	pH	HCO₃ meqv/100g of soil	SO₄ meqv/100g of soil	Cl meqv/100g of soil	Ca meqv/100g of soil	Mg meqv/100g of soil	Na meqv/100g of soil	salt content %	geochemical type of soil
XI B 1	0–14	7.7	0.7	10.6	0.6	9.5	1.4	1.0	0.81	S Mg-Ca
	14–35	7.7	0.4	5.1	0.4	4.7	1.0	0.3	0.40	S Mg-Ca
	35–53	7.8	0.5	3.7	1.3	2.4	1.8	1.3	0.36	Cl-S Mg-Ca
	53–71	7.9	0.3	18.5	5.1	14.0	3.0	7.0	1.57	Cl-S Mg-Ca
	71–83	7.6	0.3	8.2	6.0	3.5	2.5	8.4	0.92	Cl-S Ca-Na
	83–103	8.1	0.3	9.2	7.9	4.2	3.7	9.5	1.09	Cl-S Ca-Na
	103–115	8.4	0.4	30.2	44.9	14.3	13.4	47.8	4.62	S-Cl Ca-Na
XI B 2	0–5	8.4	0.5	30.1	52.8	18.6	16.6	48.1	5.04	S-Cl Ca-Na
	5–17	7.6	0.4	4.5	5.9	3.0	2.0	5.9	0.67	S-Cl Ca-Na
	17–39	8.2	0.3	13.5	19.8	7.9	5.1	20.5	2.07	S-Cl Ca-Na
	39–54	8.9	0.4	9.6	19.8	3.3	5.7	20.7	1.80	Cl-Na
	54–82	8.7	0.3	25.9	27.3	13.5	10.5	29.9	3.34	S-Cl Ca-Na
	82–103	8.1	0.4	28.7	31.4	13.2	13.3	34.2	3.74	S-Cl Mg-Na
	103–130	8.2	0.3	15.2	31.3	8.0	10.3	28.4	2.80	Cl Mg-Na
XI B 3	0–3	8.6	0.6	38.1	41.9	16.9	15.6	48.1	4.99	S-Cl Ca-Na
	21–37	8.5	0.3	18.0	36.7	8.6	9.8	36.7	3.32	Cl Mg-Na
XI B 4	0–10	8.4	0.4	33.1	41.0	17.0	16.0	41.4	4.56	S-Cl Ca-Na
XIII B 1	0–3	8.7	0.6	41.3	65.8	16.7	28.3	62.7	6.48	S-Cl Mg-Na
	3–7	8.6	0.4	9.2	11.8	4.1	5.5	11.8	1.31	S-Cl Mg-Na
	30	7.6	0.2	25.6	13.3	14.7	7.8	16.7	2.50	Cl-S Mg-Ca
	60	7.8	0.3	25.1	15.5	14.8	7.1	18.9	2.59	Cl-S Mg-Ca
XIII B 5	0–3	8.8	0.7	45.7	97.1	10.8	35.9	96.9	8.57	Cl Na
	3–19	8.2	0.3	0.9	3.1	0.7	1.4	2.2	0.26	
	19–59	8.1	0.4	9.3	16.7	2.6	5.5	18.4	1.61	S-Cl Na
	59–82	8.1	0.3	6.0	12.8	2.0	4.3	12.8	1.15	Cl Na
XIII B 6	0–2	8.6	0.6	21.2	49.7	9.0	17.0	45.5	4.25	Cl Mg-Na
	2–18	8.4	0.4	4.7	9.6	1.9	4.2	8.7	0.89	Cl Mg-Na
	18–88	7.8	0.3	27.3	21.5	15.3	10.6	23.3	3.07	Cl-S Mg-Ca
	88–118	8.2	0.4	10.8	17.6	4.4	6.7	17.7	1.75	S-Cl Mg-Na
	118–148	7.6	0.3	23.8	15.4	13.6	8.3	17.7	2.45	Cl-S Mg-Ca

In the Karaton region on the dried bottom of the sea the following natural systems were formed from the water boundary landward (profile XI B): associations of *Salicornia europaea* and *Halocnemum strobilaceum* on marsh soils. The salt content of the surface horizon of soils is 4.5% (profile XI B station 4, Table 1), the geochemical type of soil is sulphate-chloridic calcium-sodium. This natural complex is replaced by the Halocnemum strobilaceum mono-association on marsh solontchaks, where the salt content of the upper horizon is 4.73%, decreasing at the depth of 30–40 cm to 0.9% (profile XI B station 3).

Moving inland from the coast, groundwaters become deeper and the process of desalination of soils occurs (profile XI B station 2). Meadow-solontchak soils are forming. The horizon of maximal salt accumulation moves to the depth 80–100 cm just above the water table which is here at the depth of 1.5 m. Some changes occur in plant cover: in *Halocnemum strobilaceum* associations *Limonium gmelinii* and *Atropis distans* appear. Under further deepening of the groundwater table (profile XI B station 1) the horizon of highest salt content (4.6%) reaches the depth 100–115 cm, the geochemical type of soil is sulphate-chloridic calcium-sodium. The upper horizons contain 0.3–0.8% of salts, the geochemical type of soil is sulphate-magnesium calcium. In this place meadow-solontchak soils with *Lepidium perfoliatum*, *Eremopyron orientale* and *Halocnemum strobilaceum* are found. Further inland this natural complex is replaced with *Artemisia* spp. communities on grey-brown desert soils.

Such successions could be developed in conditions of natural change of hydromorphic soils into automorphic ones due to the movement of the shoreline and sinking of the water table. But as we remarked above, this region of the coast is affected by the wind-induced sea level fluctuations. Soil, being a mirror of a landscape, reflects the interaction of surge waters and groundwaters.

The first two natural complexes (associations of *Salicornia europaea* with *Halocnemum strobilaceum* and mono-associations of *Halocnemum strobilaceum*) are under water almost all the time. The next belt of *Halocnemum strobilaceum* with *Limonium gmelinii* and *Atropis distans* is frequently flooded during the surges (its height above sea level being 0.2–0.9 m). Finally the last belt with *Halocnemum strobilaceum*, *Lepidium perfoliatum* and *Atropis distans* is only rarely flooded.

INTERACTION OF SURGE WATERS AND GROUNDWATERS

Interaction of surge waters and groundwaters is well reflected in the soil of profile XI B station 2. As we remarked before, due to lowering of the groundwater table to 1.5 m, salts from the surface horizon were carried away steadily to the depth. But as shown in Table 1 the surface horizon contains 5% of salts at the moment of sampling just after the wind-induced surge (the geochemical type of soil is sulphate-chloridic calcium-sodium). This phenomenon depends on additional amount of salts incoming with the sea waters during wind-induced surges.

Mineralization of the sea water coming over the coast increases gradually due to dissolution and transfer of the soil salts into the water. Chemical analyses showed that the salt content of surge water may increase by 1.5 times within the distance of 1 km.

As the surge spreads further from the average coastal line, the mineralization of the sea water reaches at the water/land boundary. Consequently the salt content of the surface horizon of soils being covered by surge waters later than others will be the highest.

Under further rise of sea level and groundwater and the landward movement of the surge zone, the salt content of the upper horizon in profile XI B station 2 decreases by 3.5%. Its lower boundary will move down. Simultaneously the deep horizon of salt accumulation will lift due to the water table rise. The salt content of this layer will decrease due to decreasing groundwater mineralization caused by landward-moving underground flow from the sea. Finally these two horizons will meet into one as in the soil of profile XI B station 3 located near the water boundary.

Granulometric composition of soils influences the accumulation of salts. In the River Volga and the River Ural inter-stream area, profile XIII B was made at the spit with clay-loam grounds. The salt content in the similar natural complexes is more: 6.4% near the water boundary (profile XIII B station 1 located in 500 m from the coastal line, see Table 1) and 8.5% on the plots frequently flooded by surges (profile XIII B station 5).

On the ingression coast of the Komsomolets Bay a salt depression was formed as a result of fast drop of sea level. It has sparse vegetation and different types of solontchaks and a different degree of salt content. This bay dried totally in 1941. In the mouth part of the extensive bay solontchaks with salt content of upper horizons 4.5–11% were developed. In the central part chloridic and sulphate wet and crust-wet solontchaks with salt contents up to 42% were developed.

In spite of fast drainage of the bay, it hasn't dried out permanently because of its depth and the presence of numerous salt groundwater springs in the central part of the salt depression. Their mineralization is 73–109 g l^{-1} (see Geomorphology of the peninsulas . . ., 1966).

During wind-induced surges, water enters the bay via the deepest branches, later flooding vast amounts of territory. Covering solontchaks, surge water dissolves the salts in the upper horizons of soils in the mouth part of the bay, reducing their salt content by 5% (that is 8–10 times less in comparison with the inner parts of the bay).

CONCLUSION

The above examples indicate that the soils of the sea coasts having undergone wind-induced surges can be used as evidence of sea water/groundwater interaction. Depending on local conditions, the surge waters may either cause the accumulation or dissolution of soil minerals.

In soils with a transitional hydromorphic water regime located along the open straight coasts, the effect of surges will be the salinization of the upper soil horizons. When the sea level rises and the soil is often exposed to frequent surges, the salt content in the soils decreases. This is particularly important on the ingressional coasts with extensive salt plains where progressive desalination occurs in upper horizons.

REFERENCES

Badamshin, B.A. & Chernoscutov, I.A. (1947). Zalivy Mertvii Kultuk i
Kaidak v nastoyashee vremya (The Mertvii Kultuk Bay and the
Kaidak at present). *Izvestia VGO*, **79**, 2, 159–174.

Geomorfologia i chetvertichnaya geologia poluostrovov Mangyshlak i
Buzachi (Geomorphology of the Mangyshlak and the Buzachi peninsulas) (1966). Fund of Institute of Geology, KAZ SSR, Alma-Ata,
175 p.

Gershtanskii, N.D. (1977). Issledovania i raschet sgonno-nagonnych
kolebanii urovnya vody Severnogo Kaspiya (Investigation and
calculation of wind-induced fluctuations of the North Caspian Sea
level) (1977). Candidate dissertation. Astrakhan, 230 p.

Kovda, V.A. (1985). Biogeokhimiya pochvennogo pokrova
(Biogeochemistry of soil cover). Moscow: Nauka.

Leontyev, O.K., Mayev, E.G. & Rychagov, G.I. (1977). Geomorfologia
beregov i dna Kaspiiskogo morya (Geomorphology of the coasts and
bottom of the Caspian Sea). Moscow: Moscow State University,
208 p.

Mann, K.H. (1982). *Ecology of coastal waters: a systems approach.*
Berkeley: University of California Press.

IV

Management and restoration of groundwater/surface water interfaces

24 Ecotones and problems of their management in irrigation regions

V.S. ZALETAEV

Water Problems Institute, Russian Academy of Sciences, 10 Novaya Basmannaya Str. P.O. Box 524, Moscow 107 078, Russia

ABSTRACT Irrigation is a powerful factor of ecological differentiation of the environment. Changes in hydrological and hydrogeological regimes of a territory lead to the intensification of successions of biotic complexes and the formation of numerous interlinked irrigation ecotones, the functional core of which is the surface water/groundwater ecotone. The concept of irrigation zone ecotone is viewed as 'a series of multistepped interlocking subsystems' and considered on the multidimensional point of view.

INTRODUCTION

Irrigation is one of the most ancient activities of people in arid and semiarid regions of the world and it has always been a source of difficult problems (Postal, 1990, 1993; Singh, 1985, 1993; Worthington, 1977). They began six and a half or seven thousand years ago, and by now have become urgent. The area of irrigated land of the world is 222 million hectares and according to the FAO it was 223 million hectares in 1975. There are forecasts that by the end of the twentieth century the area of irrigated lands of the world might reach 400 million hectares. Irrigation is a powerful factor of transformation of the environment (land, water, biotic complexes, ecosystems and landscapes), strengthening its heterogeneity, ecological fragmentation and contrasts (Kassas, 1977). One of the main bases of these phenomena is seepage of water from canals and the creation of surface water/groundwater interactions, influencing the characteristics of land biocomplexes.

About 20 million hectares of land are under irrigation in Central Asia, Kazakhstan (Kostukovskiy, 1988) and the South of Russia. The total length of irrigation and drainage canals is about 180 000 km in Uzbekistan. Very few of the canals (about 12%) have impermeabilized beds, and under such conditions water loss through seepage is about 45–55% of the whole volume of water sources. In this case the groundwater table reaches the rhizosphere horizon and leads to considerable changes in vegetation cover, zoocomplexes and halogeochemical environment. Automorphic ecosystems are transformed into hydromorphic ones.

Management of a natural system's development in rapidly changing environmental conditions needs to be based on an understanding of the ecological essence of environmental transformations in irrigated lands, on the discovery of the mechanisms of the ecosystem transformation, and on forecasts of their changes (Kovda, 1984; Shankar & Kumar, 1993). This concerns the creation and dynamics of ecotones (Naiman & Decamps, 1989).

Ecotonal conception began from the investigation of Clements (1928) where ecotones were understood to be contact 'microzones' between neighbouring plant communities. Then the idea of a contact zone was developed on the border territories between ecosystems and landscapes (Sochava, 1978, 1978a; Zaletaev, 1979, 1989, 1993; Romney & Willace, 1980; Scanlan, 1981, and others), then on the boundaries between land and water (Decamps & Naiman, 1990; Risser, 1990; Pinay *et al.*, 1990; Holland *et al.*, 1991), between surface water/groundwater (Gibert *et al.*, 1990; Gibert, 1992) and between neighbouring geographical zones (Walter & Box, 1976).

The aim of this paper is to define the processes that take place due to irrigation in the contact zones between desert and irrigated land, to study the impact of seepage from canals and the rise in the groundwater table on the vegetation cover and the zoocomplexes in the surface water/groundwater ecotones, and finally to propose a multilevel view of the ecotone in irrigated regions.

IRRIGATION AS A FACTOR OF ENVIRONMENTAL DIFFERENTIATION: PROCESSES

Irrigation leads to different processes, but what exactly are they?

First of all, heat and moisture balances change on vast arid and semiarid territories. They bring about a chain of complicated qualitative environmental modifications. When drainage waters were derived from oases in Central Asia, more than 300 000 hectares of desert lands were flooded and 530 000 hectares became salty and marshy. Thus, ecosystems of the solonchak type developed over a large area (4.06 million hectares). Desert vegetation was replaced by halophytic communities, when the groundwater level rose to 1–1.5 m. Halophytic shrubs and semishrubs: *Salsola*, *Suaeda*, *Climacoptera*, *Calidium*, *Salicornia*, *Halostachys caspica*, *Salsola dendroides*, *Tamarix hispida*, *Nitraria schoberi*, *Halocnemum strobillaceum*, etc . . . were predominant. Change in vegetation type, when desert Artemisia-shrub communities are replaced by meadow-bog and solonchaks, occurs in 5–6 years. The Aral Region, where the area of irrigated lands increased up to 3.23 million hectares and discharge of saline drainage waters entailed the formation of new water bodies is a good example of the scale of the process. There are about 2341 drainage impoundments in the Aral Sea Basin, Amu- Darya and Syr-Darya valleys, with a total area of 7065.9 km². Partially mineralized drainage water flows into these rivers. Hydrogenic successions rapidly develop around drainage water impoundments with water salinity of 1–5 g l⁻¹.

Spatial redistribution of ground and underground waters of river basins leads to the development of different forms of desertification. Desertification due to irrigation is the most important form, and it can be defined as the degradation of natural ecosystems and landscapes with decreasing bioproduction.

The main elements of the irrigation landscape are groundwater/surface water ecotones with a number of different subsystems. Among them are land/inland water ecotones, ecotone systems of infiltration lakes, water bodies/impoundments of drainage waters, ecotone systems of saline depressions, and ecotone systems of former irrigation lands, as well as desert/oasis ecotones.

THE IMPACT OF SEEPAGE FROM CANALS AND A RISING GROUNDWATER TABLE ON THE VEGETATION COVER OF ECOTONES IN IRRIGATED REGIONS

One of the most important environmental factors that determines the development of biotic communities in irrigated regions, is seepage from canals. This causes essential changes

Table 1. *Water losses due to seepage into ground (mainly sandy) and evaporation from Kara-Kum canal in its part from the Amu-Darya river to the river Murgab (320 km) during the first ten years of its exploitation*

Water discharge and losses	Years					
	1959	1961	1963	1968	1969	1970
Water discharge into canal (m³/s)	107.57	130.17	152.57	251.47	265.03	265.03
Water losses (m³/s)						
a – due to seepage	57.76	59.84	55.60	89.45	60.64	70.77
b – due to evaporation from water surface	6.52	7.32	6.65	5.26	3.13	4.08
Total water losses %	59.75	51.6	40.8	37.66	24.06	26.95

in the groundwater state: in the dynamics of its level, in its volume and quality. Change in the groundwater table is the most ecologically active factor; it causes hydrogenic and halogenic successions of vegetation that lead in turn to alterations in the composition and structure of plant communities and to changes in the composition and species ratio in zoocomplexes.

The most effective impact of these forces can be observed in zones where the canals do not have impermeabilized beds. In such conditions the ecological activity of the surface water /groundwater ecotone is especially high. Several aspects of this ecotone can be studied: 1) the surface water and groundwater contact itself, migration of organisms, contaminating substances and bacterial contamination; 2) the direct and indirect impact of the surface water/groundwater ecotone on the terrestrial vegetation and zoocomplexes, the development of halogeochemical processes, soil formation and the structural genesis of biotic complexes and terrestrial ecosystems.

The example of the Kara-Kum canal is very significant. It cuts across the sandy desert of Kara-Kum in the southern regions of Turkmenistan. Water losses due to seepage at the beginning of its exploitation from 1959 till 1970 on the first part of its transect of 320 km were about 1400 million m³ per year (Table 1). During the first year (1959) seepage was 1800 million m³. The groundwater table rose by 10–12 m and even by 16–23 m on some patches of sandy desert near the canal as the result of seepage. But while 50 m from the canal the groundwater table rose by 21.6–23.9 m, 5 km away it rose by only 1 m. However the impact zone appeared to be rather broad: up to 5–6 km on the left bank of the canal, and up to 10–20 km on the right bank. And in the region on Kelif Uzboy this zone is 30–40 km wide (Kuznetzov, 1978).

So the 'surface water/groundwater' ecotone quickly occupied the vast area in the zone of the Kara-Kum canal and can be identified by the colonization of hygrophytes and phreatophytes (Fig. 1). On the main part of the Kara-Kum canal

Fig. 1 Area occupied by hygrophytes and phreatophytes that indicate the zone of land ecological activity of surface water/groundwater ecotone. Kara-Kum canal and flood plain of Kelte-Beden (South-Eastern Kara-Kum desert, Turkmenistan). Infiltration lakes could be seen (1965) with their colonization by hygrophytes and increasing area occupied by phreatophytes (1972).

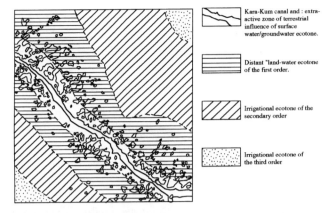

Fig. 2 Ecotone viewed as a set of interlocking sub-systems in the zone of the Kara-Kum canal influence. Kara-Kum canal and extra-active zone of terrestrial influence of surface water/groundwater ecotone are shown.

from the river Amu-Darya to the river Murgab (about 320 km) the whole area of newly formed water surface is up to 130.6 km², including the surface of infiltration lakes; the area of those lakes to which the waste water in discharged – is 61.1 km², the zone occupied by reed (*Phragmites australis*) and cat's-tail *(Typha latifolia, T. angustifolia, T. minima)* – is 89 km², the zone occupied by phreatophytes – *Alhagi persarum, Karelinia caspica* – and by reed that grows on land is about 92km², the zone under *Tamarix ramosissima, Tamarix hispida* – is 39.3 km².

Therefore, the area of the surface/groundwater ecotone and its adjacent plant communities spreads over more than 220 km² (Kuznetzov, 1978). The area of vegetation successions caused by the construction of the canal is more than 300 km².

We determined direct correlation between the groundwater table and the composition and structure of plant communities (Zaletaev & Kostukovskiy, 1980; Zaletaev, 1989) (Table 2). The shallow waters of the canal, its floodplain and infiltration lakes form patches of reed thicket formation (*Phragmites australis*) that are as high as 3–4 m, and of cat's-tails *(Typha latifolia, T. angustifolia, T. minima)*.

The spatial distribution of the above-mentioned plant communities in the zone of the Kara-Kum canal and irrigation canals in Central Asia reflects the area of active ecological impact of the surface water/groundwater ecotone on the land biotic communities (Fig. 2).

When the groundwater table is at a depth of 5–6 to 10–15 m, *Tamarix* is typical for the plant communities, but other phreatophytes disappear. In this marginal belt the impact of the canal on the biota has practically vanished and in the vegetation cover the typical desert groups of psammophytes dominate: *Astragalus villosissimus, Agriophyllum latifolium, Tournefortia sogdiana, Corispermum lehmannianum, Carex physodes* and shrubs: *Calligonum setosum, C. microcarpum, C. caput-medusae, Salsola richteri* with attendant *poacea* plants – *Anisantha tectorum, Bromus dantonii, Schismus arabicus* and *Eremopyrum orientale*.

In those places where the groundwater is more than 12–15 m deep the impact of the canal on the composition and structure of desert vegetation is negligible.

Research on the ecological impact of the Irtish-Karaganda canal is also carried out by the Laboratory of Dynamics of Terrestrial Ecosystems with the Water Problems Institute (Russian Academy of Sciences). The impact zone of the Irtish-Karaganda canal can serve as an example. It is constructed in the subzone of dry steppes and is 451 km long. In the dry steppes and semideserts of Kazakhstan the influences of the impact of the surface water/groundwater ecotone on the vegetation cover revealed as seepage from canals and

Table 2. *Terrestrial impact of surface water/groundwater ecotone: dominant species of plants on the patches with different depths of groundwater*

Depth of Groundwater	Dominant Species of Plants	
	Hydrophytes	Halophytes
From 0 to 50–80 cm	*Phragmites australis, Paramicro-rhynchusprocumbers, Polygonum amphibium, Atriplex tatarica, Chenopodium album, Solanum nigrum, Cynodon dactylon, Aster tripolium*	*Salicornia europea, Suaeda transoxana, Bassia hyssopilfolia, Climacoptera lanata, Aeluropus hyssopifolia, Climacoptera lanata, Aeluropus littoralis, A. repens, Cressa cretica*
From 50–60 cm to 200 cm	*Karelinia caspica, Phragmites australis (0.7–1.5 m high), Tamarix ramosissima, Imperata cilindrica, Erianthus ravennae, Alhagi persarum (rare), Zygophyllum fabago*	*Limonium gmelinii, Tamarix hispida, Halostachys caspica*
From 200 cm to 500–600 cm	*Alhagi persarum, Karelinia caspica, Salsola pelucida, Tamarix ramosissima, Artemisia scoparia, Erianthus ravennae, Erigeron canadensis.* All these species form thickets.	
From 5–6 m to 10–15 m	Phreatophytes disappear, but *Tamarix ramossissima* can be met Psammophytes dominate.	

reservoirs and as changes in the groundwater level, appear to be the same, but the composition of plant communities on the patches with different groundwater table levels and successional rows appears to be different (Table 3).

In conditions of dense clay-loam ground the width of the belt of changed vegetation along the canal that indicates the area of land influence of the surface water/groundwater ecotone is considerably smaller in comparison with sandy patches and is 10–30 to 100–300 m wide,depending on the relief on each side of the canal. In some places only, where the canal crosses the old river-bed of the Shiderta-river, the width of the belt of changed vegetation in 5 km.

So, the above-mentioned lists of plant species could be used as indicators of the zone of terrestrial ecological activity of the surface water/groundwater ecotone: in the first case for irrigated regions of Central Asian deserts, in the second case – in dry steppes and semi-deserts.

MOISTURE CONTENT OF SOIL AND ITS IMPACT ON THE COMPOSITION AND DISTRIBUTION OF ZOOCOMPLEXES IN THE ZONE OF SURFACE/GROUNDWATER ECOTONE IMPACT

Seepage from canals and reservoirs in irrigated regions and water accumulation in soils as the result of watering (includ-ing the use of water-sprinklers) extend the zone of ecological influence of the surface water/groundwater ecotone. The main active factor of this influence is the increase in the moisture content of soils and subsoils. In the sphere of this influence are the rodents: *Rhombomis opimus, Meriones meridianus, Meriones lybicus, Cricetelus migratorius, Nesokia indica, Mus musculus,* that are widely spread in the biotopes of irrigated regions; squirrels: *Spermotophilopsis leptodacty-lus,* and insect-eating mammals: *Crocidura hyrcana.*

In these conditions various species of small mammals react in various ways to the changes in the moisture content of soil and subsoil. For some species *(Rhombomis opimus, Meriones meridianus, M. lybicus, Spermotophilopsis lepto-dactylus)* moisture increases in some soil layers and corre-sponding moisture increases in their burrows appear to be unfavourable factors and cause changes in biotopes or changes in the distribution of species.

For others *(Nesokia indica* and *Mus musculus)* this factor appears to be attractive and promotes their settling in the regions of irrigation. For a third group of animals *(Cricetulus migratorius* and *Crocidura hyrcana)* short-term changes in soil moisture do not influence their distribution.

As an example of soil moisture dynamics during one year on the sandy territory in the zone of the surface water/ groundwater ecotone influence we can use the measurements made close to the Kara-Kum canal to the north of the station Giaur in southern Turkmenistan (Table 4). The indices of soil

Table 3. *Connection between the depth of filtration waters and vegetation in the zone of Irtish-Karaganda canal (Kostukovskiy, 1988)*

Depth of seepage waters	Characteristics of vegetation and soil	Plant communities
Water on the surface	Infiltration lakes. Water and water-riparian vegetation	*Potamogeton lucens, P. perfoliatus, P. crispus, Myriophyllum spicatum, Polygonum amphibium, Phragmites australis, Typha angustifolia, Juncus gerardii*
0–0.2	Riparian thickets; marshy meadow-marshy soils	*Salix caspica, S. triandra, Phragmites australis, Typha angustifolia, Carex acuta, Sonchus asper, Ranunculus repens, Juncus gerardii, Artemisia abrotanum*
0.2–0.5 (sweet)	Damp poacea-herbaceous meadows, meadow soils	*Elytrigia repens, Agropyron pectinatum, Juncus gerardii, Artemisia abrotanum, Poa pratensis, Cirsium arvense, Carex acuta, Potentilla anserina, Filipendula ulmaria*
0.2–0.5 (salt)	Damp saline meadows, meadow salinized soils	*Tamarix ramosissima, Elytrigia ramosum, Pulcinella distans, Tripolium vulgare, Limonium gmelinii, Artemisia sieversiana*
0.5–1.0 (sweet)	Meadow marshy-herbaceous vegetation, meadow-chestnut soils	*Elytrigia repens, Agropyron pectiniforme, Poa pratensis, Medicago sativa, Artemisia dracunculus, Galium verum, Lathyrus pratensis*
0.5–1.0 (salt)	Solonchaks with halophytes, meadow-solonchak and solonchak soils	*Salicornia europea, Suaeda salsa, Aeluropus littoralis, Limonium gmelinii, Artemisia maritima, Atriplex verrucifera*
1.0–2.5	Steppe meadows, transition to the primary steppe associations	*Leymus sp., Festuca valesiaca, Stipa lessingiana, S. joannus, S. capillata, Artemisia austriaca, Calamagrostis epigeios, Achnatherum aplendens, Galium verum, Astragalus onobrychus*
2.5	Zonal steppe vegetation, zonal soils	*Festuca valesiaca, Stipa lessingiana, S. joannus, S. capillata, Artemisia austriaca, Artemisia frigida, Artemisia dracunculus*

moisture dynamics during one year on the top of a sandy dune in the desert far from the zone of the canal's influence are given for comparison (Table 5).

In the sands adjacent to the mass of irrigated fields the increased soil moisture can be observed in the 60–140 cm layer in different seasons for different reasons (Table 4): in April as the result of rain water accumulation and seepage from the canal to the neighboring field (layer 60–100 cm), in May and June as the result of watering of the neighboring field (60–140 cm layer). In that mass of sands the burrows of *Meriones meridianus* Pallas were inhabited throughout the year, but they were no more than 80 cm deep. Meanwhile, the old colonies of *Rhombomis opimus* Pallas near the irrigated fields were abandoned and there were no burrows of *Spermotophilopsis leptodactylus*.

On the second patch in the barkhanes (density of sward up to 15%), where the influence of the canal and irrigated fields on the moisture of sandy soil was not detected (maximum moisture content of sand was no more than 5.0–6.86%) (Table 5), we found colonies of *Rhombomis opimus* Pallas. On the slopes and tops of barkhanes there were colonies of *Meriones meridianus* and burrows of *Spermotophilopsis leptodactylus*.

Similar observations were made in the zone of the Kara-Kum canal influence at the scientific station Kulbukan (in the region of Annau Settlement) of the Desert Institute of Turkmenistan Academy of Sciences. We measured the sand moisture content in the walls of the burrows of *Rhombomis opimus* in the hillocks and sand ridges not far from the canal. It was found that the *Rhombomis opimus* burrows were not inhabited completely on the ridge slopes. The burrows in the

Table 4. *Moisture dynamics of soil and water resources in rolling lowland sands adjacent to gated fields near Kara-Kum canal (Giaur region of Turkmenistan)*

N	Depth of soil cm	April %	April mm	May %	May mm	June %	June mm	October %	October mm
1	0–20	6.63	18.4	2.47	6.84	2.14	5.97	1.84	5.09
2	20–40	9.4	27.28	6.21	18.05	8.61	25.49	4.28	12.54
3	40–60	8.44	25.00	6.51	20.18	8.65	26.81	7.69	23.85
4	60–80	15.38	47.94	7.06	22.04	11.93	37.21	9.55	22.79
5	80–100	10.08	31.44	5.82	18.15	15.09	47.09	14.22	44.39
6	100–120	2.16	6.74	12.68	39.55	11.71	36.54	3.04	9.48
7	120–140	2.58	8.04	10.86	33.89	2.96	9.25	2.43	7.59
8	140–160	1.85	5.75	2.23	7.64	2.23	6.96	2.05	6.38
9	160–180	2.18	6.78	2.31	7.19	2.97	9.27	2.01	6.28
10	180–200	1.89	5.98	2.51	7.84	3.54	11.04	2.46	7.87

Notes:
1–3, 8–10: dry sand
4, 5: wet sand, as the result of field watering in June
6, 7: wet sand, as the result of field watering in May

Table 5. *Moisture dynamics of soil and water resources in scarcely-vegetated barkham sands beside the Kara-Kum canal and irrigated fields impact (on top of barkham in the Giaur region of Turkmenistan)*

N	Depth of soil cm	April %	April mm	May %	May mm	June %	June mm	October %	October mm
1	0–20	6.31	19.56	4.79	13.68	0.66	2.03	0.62	1.93
2	20–40	5.11	15.84	2.48	7.68	3.35	10.38	0.95	2.92
3	40–60	2.65	8.00	1.49	4.48	5.02	14.75	1.35	4.06
4	60–80	1.98	6.02	1.94	5.70	5.33	15.66	1.79	5.26
5	80–100	2.06	5.84	2.36	6.70	4.19	11.88	1.92	5.44
6	100–120	2.28	6.46	2.33	6.60	5.40	15.33	2.25	6.38
7	120–140	2.31	6.56	2.30	6.52	5.03	14.28	2.17	6.16
8	140–160	2.57	7.28	2.47	7.00	4.52	12.83	2.39	6.79
9	160–180	2.76	7.84	2.68	7.60	3.30	9.36	2.71	7.68
10	180–200	3.14	8.93	2.83	8.00	2.77	7.87	2.59	7.34

lower part were abandoned. The reason was the high moisture content of sands in the holes at a depth of 30–40 cm (25.4–27.4%). In the middle part of the colony where the moisture content of sand is lower (128% in the 40–60 cm layer and 4.22–5.67% in the upper 0–40 cm layer) the holes of *Meriones meridianus* were found; they had settled in the colony of *Rhombomis opimus* and used its burrows. *Rhombomis opimus* itself inhabited only the upper part of the colony, where the sand was absolutely dry (in the 0–20 cm layer – 3.5%, in the 20–40 cm layer – 5.3%, in the 40–60 cm layer – 6.31%, in the 60–80 cm layer – 6.39%, in the 80–100 cm layer – 6.30%, in the 100–120 cm layer – 2.69%).

So we can observe the cohabitation in one colony of two different species of Gebrillidae, which became possible as the result of the differentiation of microbiotopes because of differences in the moisture content in the burrow walls.

We must mention that on the neighboring patches near the banks of infiltration lakes where the moisture content of sand is high (26.7–32.5%) large colonies of *Nesokia indica* were observed. It settled actively along the Kara-Kum canal in the biotopes characterized by active terrestrial influence of the surface water/groundwater ecotone. In these conditions the colonies of *Nesokia indica* are formed on the banks of filtration lakes and occupy an area of 6–10×9–17 meters. The colonization of this species is accompanied by mass reproduction located in the narrow belt along lake banks.

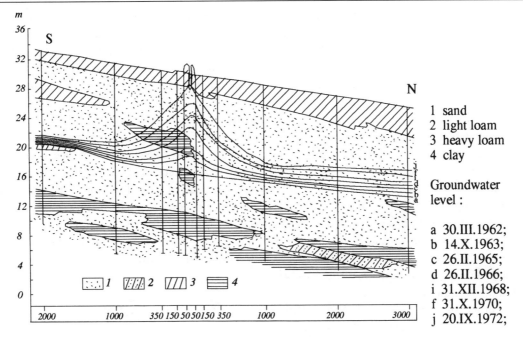

1 sand
2 light loam
3 heavy loam
4 clay

Groundwater
level :

a 30.III.1962;
b 14.X.1963;
c 26.II.1965;
d 26.II.1966;
i 31.XII.1968;
f 31.X.1970;
j 20.IX.1972;

Fig. 3 Groundwater table rise under the Kara-Kun canal. 1: sand; 2: light loam; 3: heavy loam; 4: clay; groundwater levels: 30 March 1962; 14 October 1963; 26 February 1965; 26 February 1966; 31 December 1968; 31 October 1970; 20 September 1972.

In similar biotopes with high density of sward of hygrophytes and hygromesophytes the number of *Mus musculus* is very high. On the territories occupied mostly by phreatophytes (irrigation ecotone of the second order) the common species are *Cricetulus migratorius, Meriones lybicus, M. meridianus, Ellobius talpinus* and *Crocidura pigmea*.

Thus the spatial distribution of small mammals in irrigated regions is determined by the ecological influence of the surface water/groundwater ecotone. Knowledge of this distribution and of reproduction dynamics is important for medico-biological purposes.

ECOTONES FROM IRRIGATED REGIONS VIEWED AS INTERLOCKING SUBSYSTEMS

The groundwater/surface water ecotone is obviously the main active functional body of the whole ecotone system in irrigated regions. The interaction of groundwater and surface waters, that determines the water supply of the territory and thus the living conditions of organisms and the possibility of the formation and development of biotic communities and natural systems, initiates the **multistep links** between abiotic factors of the environment and biotic agents of environment-forming processes. Gibert *et al.* (1990) created a functional classification of these ecotones, based on the dynamics of water masses, and studied the

fluctuations of ecotone boundaries and some biological aspects of the problem.

While considering a groundwater/surface water ecotone as a functional 'nucleus' of a whole complicated set of various biotic communities we determined four main pathways: (1) direct hydrodynamic influences (by means of seepage into the soil and subsoil); (2) changes in moisture content and hence the formation of particular microclimatic conditions in irrigated regions; (3) chemogenic impact on soils and biotic complexes by means of the intensification of halochemical processes (mainly owing to activation of matter buried in soil and subsoil and to contamination from the surface waters); (4) biogenic impact owing to the colonization of organisms, regrouping of species in the communities, formation of new biogeocoenotic chains and their spatial expansion. The last pathway will lead to the secondary expansion of the zone of the central surface water/groundwater ecotone influence.

Our experience in investigating terrestrial ecotones in various irrigated regions of Central Asia enabled us to determine several gradations of the influence zone of the complex ecotone 'nucleus'. They are pronounced in the zone of the Kara-Kum Canal in Turkmenistan and are related to the degree of groundwater level rise due to seepage from the canal (Figs. 1, 2 and 3).

1. Surface water/groundwater ecotone which is a natural 'catalyst', activating environment-forming processes in the set of ecotone communities comprising interlocking subsystems.

2. Extra-active zone of terrestrial influence of surface water/groundwater ecotone which is a narrow strip including the shore of a water body (canal, reservoir,

infiltration lake) and a narrow strip of terrestrial ecosystems that are under the impact of water table changes and periodically populated or visited by shallow-water inhabitants; it is populated by a specific amphibian community and hydrophytes (depth of groundwater 0–80 cm).

3. Distant 'land-water ecotone', which is a comparatively wide belt, including the patch of water-logged land adjoining it Exfiltration of groundwater and formation of infiltration lakes are frequent. This is the zone of active hydraulic impact on groundwater under the surface waters seeping from the canal. Microclimatic parameters sharply differ on this patch from the neighboring ones, so the hydrophytes and mesohydrophytes can live here. Active hydrogenic and halogenic successions in the biotic complexes take place. Plant density is high (depth of groundwater 60–200 cm).

4. Irrigation ecotone (sub-system) of the first order. This is formed in the zone of weak hydraulic impact. Groundwaters lie at a depth of 1.0–2.5 m. The ecotone community includes mesophytes, mesohygrophytes and phreatophytes. Plant density varies within wide ranges.

5. Irrigation ecotone of the second order. This is in the range of weak hydraulic influence of seeping water. It forms sand ridges and hillocks. Ecotone communities have a more xerophylious appearance and are represented by a combination of phreatophytes in association with mesohygrophytes, halophytes and the original psammophyte community. Halophytes penetrate into psammophyte communities if there is salinization of underlying sands (depth of groundwater 200–600 cm).

6. The irrigation ecotone of the third order is situated in the zone of a considerably decreased hydraulic influence of seeping waters. This type of ecotone is formed in depressions and is supplied with water from seepage and precipitation. The combination of these two sources of water supply ensures increased soil wetting, the combination in the ecotone of halophytes and psammophytes and the regrouping of species in the zoocomplexes (depth of groundwater from 5–10 to 15 m).

7. The marginal (remote) ecotone is situated between the desert and irrigated region. It is formed on marginal areas of the irrigation region due to the prolongation of biocoenotic links and limited expansion of animals and plants from the irrigation region. This ecotone preserves the appearance of a desert landscape, but species, new for arid dwelling (mesophyles and hydromesophyles) penetrate into biotic complexes. As a rule, these new species are not numerous. But their presence gives the ecotone character to desert communities. This pertains, especially, to certain species of Lepidopteres and some birds.

The concept of irrigation zone ecotone viewed as 'a series of multistepped interlocking subsystems' might be useful for the understanding of the most important functional role of the groundwater/surface water ecotone, that is for the determination of the degree of its influence on natural systems, natural-anthropogenic systems and agrosystems and for ecotone regionalization in irrigated areas. This regionalization is important for the determination of the area of land plots, occupied by different types of ecotones, and for working out differentiated irrigation and drainage systems of disturbed lands, and for conservation. Conception of ecotone may prove to be very useful and perspective in ecology. It will be an application for ecological investigations of irrigation regions where a wide complex of differents types of ecotones are formed.

ACKNOWLEDGEMENTS

I thank the Organizing Committee of the International Conference, Conference General Chairperson Professor Janine Gibert and Division of Ecological Sciences of UNESCO, Dr Frédéric Fournier, for the attention to my Report, and the financial support that gave me the opportunity to take part in the Conference.

REFERENCES

Clements, F.E. (1928). *Plant succession and indicators*. N Y.
Decamps, H. & Naiman, R.J. (1990). Towards an ecotone perspective. In *The Ecology and Management of Aquatic-Terrestrial Ecotones*, ed. R.J.Naiman & H. Decamps, pp. 1–5. Paris-London, Man and the Biosphere Series. Vol. 4. UNESCO, Parthenon Publ.
Gibert, J. (1992). Ground water ecology from the perspective of environmental sustainability. *Am. Water. Res. Assoc.*, 3–13.
Gibert, J., Dole-Olivier M.J., Marmonier, P. & Vervier, Ph. (1990). Surface water-groundwater ecotones. In *The Ecology and Management of Aquatic-Terrestrial Ecotones*, ed. R.J. Naiman & H.Decamps, pp. 199–225, Paris-London, Man and the Biosphere series. Vol.4. UNESCO, Parthenon Publ.
Holland, M.M., Naiman, R.J., Risser, P.G. (1991). (ed.) *Ecotones* (Monograph). Washington.
Kassas, M. (1977). *Arid Land Irrigation in Developing Countries: Environmental problems and effects*, ed. E.B.Wortingtoned, pp. 335–340, Oxford, Pergamon Press.
Kostukovskiy, V.I. (1988). *Central Kazakhstan Landscape Dynamic under Water Management*, ed. V.S.Zaletaev, 160 pp, Moscow, Publ.House 'Nauka'.
Kovda, V.A. (1984). *Problems of the combat against desertification and salinity of irrigated soils*. 304 pp., Moscow, Publish House 'Kolos'.
Kuznetzov, N.T. (1978). *Kara-Kum Canal and the Changing of Environment in Zone of its Influence*, pp. 134–167, Moscow., Publish House 'Nauka'.
Naiman, R.J. & Decamps, H. (1989) *Role of Aquatic-Terrestrial Ecotones in Landscape Management*. London.
Pinay, G., Decamps, H., Chauvet, E. & Fustec, E. (1990). Function of Ecotones in fluvial systems. In *The Ecology and Management of Aquatic-Terrestrial Ecotones*, ed. R.J.Naiman & H.Decamps, pp. 141–170, Paris-London, Man and the Biosphere series. Vol.4. UNESCO, Parthenon Publ.

Postal, S. (1990). *Saving Water for Agriculture. State of the World*, pp. 39–58, London, W.W.Norton and Company, N.-Y.

Postal, S. (1993). Water for Agriculture: Facing the Limits. In *Arid Land Irrigation and Ecological Management*, ed.S.D.Singh, pp. 1–41, Jodhpur (India), Scientific Publishers.

Risser, P. (1990). The Ecological Importance of Land-Water ecotones. In *The Ecology and Management of Aquatic-Terrestrial Ecotones*, ed. R.J. Naiman & H.Decamps, pp. 7–23, Paris-London, Man and the Biosphere series. Vol.4. UNESCO, Parthenon Publ.

Romney, E.M. & Willace, A. (1980). Ecotonal distribution of salt-tolerant shrubs in the Northern Mojave Desert. *Great Basin Natur. Mem.*, **4**.

Scanlan, M.J. (1981). Biogeography of forest plants in the prairieforest ecotone in Western Minnesota. *Ecol.Stud.*, **41**, 97–124.

Shankar, V. & Kumar, S. (1993). Ecological III-Effects of Arid Land Irrigation and some Combating Measures. In *Arid Land Irrigation and Ecological Management*, ed.S.D.Singh, pp. 351–400, Jodhpur (India), Scientific Publishers.

Singh, S.D. (1985). *Irrigated agricultura in arid areas*. WAPCOS, pp.442–454, New Delhy.

Singh, S.D. ed. (1993). *Arid Land Irrigation and Ecological Management*. Jodhpur (India), Scientific Publishers, 401 p.

Sochava, V.B. (1978). *Introduction into the Teaching of Geosystems*. Publ.House 'Nauka', Novosibirsk. 319 p.

Sochava, V.B. (1978 a). Borders on the geobotanical maps and buffer wegetation communities. In *Geobotanical mapping*. Leningrad, Publ.House 'Nauka'.

Walter, H. & Box, E. (1976). Global classification of natural terrestrial ecosystem. *J. Vegetation*, **32**, 2.

Worthington, E.B. (1977). *Arid Land irrigation in developing countries: Environmental Problems and effects*. Oxford, Pargamon Press.

Zaletaev, V.S. (1979). About some mechanism of biogeocoenosis stability on the ecologically transitional territories in South deserts. *Problems of Desert Development*, **6**, 38–44.

Zaletaev, V.S. (1989). *Ecologically destabilized environment. Ecosystems of Arid Zoned under Changing Hydrological Regime*. Moscow, Publ. House 'Nauka'. 150 p.

Zaletaev, V.S. (1993). Arid Land Irrigation and Desertification. In *Arid Land Irrigation and Ecological Management*, ed.S.D.Singh, pp. 401–441, Jodhpur (India), Scientific Publishers.

Zaletaev, V.S. & Kostukovskiy, V.I. (1980). Successional processes on arid territories under changing hydrological regime. In *Biogeographical aspects of nature management*, pp. 97–108, Moscow, Publ. House 'Misl'.

25 Hydrochemistry and ecohydrology of the transition area of the Netherlands Delta and the Brabantse Wal

T. W. HOBMA

Vrije Universiteit Amsterdam, Faculty of Earth Sciences, De Boelelaan 1085, 1081 HV Amsterdam, The Netherlands

ABSTRACT Groundwater/surface water interactions are often characterised by hydrological and hydrochemical processes which result in complex abiotic and biotic gradients. Therefore a study was made of the hydrochemistry and ecohydrology of a groundwater discharge area in a low coastal area, lying in front of a higher sandy recharge area in the southwest of the Netherlands.

A hydrochemical facies analysis was applied to identify and map the major factors accounting for variations in hydrochemical processes. Research was concentrated in a relatively small fresh water upwelling zone in the discharge area, to obtain insight into the actual interrelationship between groundwater, surface water, aquatic ecology and human influence. During 1992 surface water samples and groundwater samples were taken along four regional transects. In the upwelling zone, sampling of shallow groundwater and surface water was combined with detailed mapping of aquatic vegetation. In total, 43 species were found and over 275 water samples were analysed in the laboratory. The two main water systems, the Brabantse Wal hydrosome and the polder hydrosome, are described by seven chemical water-types, redox potential index, eutrophication potential index and calcite saturation index. The regionally derived maps of ground and surface water quality correspond fairly accurately. In the fresh water upwelling zone four types of aquatic vegetation (noda) have been discerned. Each nodum has distinct ecological amplitudes for alkalinity, pH and salinity of the surface water. These do not match general indicative values for the Netherlands, as described in literature. The hydrochemical characteristics of the ground and surface water in the upwelling zone deviate to a large extent. This points to an increasing contribution of polluted groundwater from a shallow groundwater flow system. The interrelationship between excessive groundwater exploitation of the recharge area and changing ecohydrological conditions in the fresh water upwelling zone is difficult to identify, since agricultural land-use and maintainance of the ditches have been intensified during the years.

INTRODUCTION

In the transition area of the Holocene polders of the Netherlands Delta and the Pleistocene higher grounds, a steep cliff-face is situated: the Brabantse Wal. The cliff-face is north-south directed and reaches a maximum altitude of approximately 20 m. At the southern end, the cliff-face changes abruptly from a north-south to a northwest-southeast direction and becomes less steep (Fig. 1). The Brabantse Wal area constitutes a recharge area that is connected with several discharge areas in the lower coastal zone to the west

by groundwater flow patterns. These upwelling zones form groundwater dependent ecosystems which used to be considered very valuable for their richness in rare plant species, caused by the complex abiotic gradient. Previous studies showed that the wet ecosystems are endangered, due to changes in the regional hydrological system. Reduction of the discharge, caused by large scale deep groundwater extractions at the Brabantse Wal for public water supply, is considered to be responsible for this. Notably, the disturbance of the groundwater flow patterns in the deep aquifers and thus the intensity and the hydrochemical

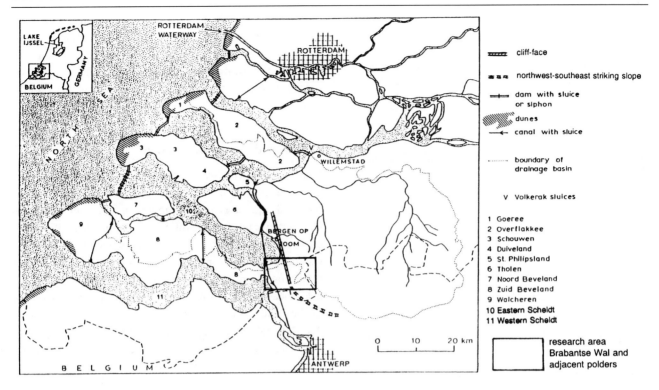

Fig. 1 Situation of the research area.

characteristics of the upward seeping water are changing. In particular, the contribution of the deep regional groundwater flow system to the waterbalance of the upwelling zone has decreased and is replaced by water from a local polluted system. From several hydrological studies in the Netherlands it can be concluded that changing groundwater flow patterns can be quantified and that the hydrochemical evolution of groundwater can be assessed but the translation of abiotic effects into biotic consequences is still a very uncertain step. Can a relation be determined between changing seepage water intensity and quality and changing vegetation patterns, or are these influences obscured by other factors, like changes in land-use? The paper focusses on this aspect.

STUDY AREA

The study area is situated in the transition zone between the higher sandy area and the marine clayey soils of the coastal plain. The Pleistocene sands dip to the west below the Holocene deposits, and at the boundary between the two areas an erosional cliff with a maximum height of 20 m +MSL (mean sea level) has developed. The edge of the cliff, south of the city of Bergen op Zoom, is called the Brabantse Wal (Fig. 1).

The medium grained fluvial pleistocene deposits are covered by eolian sands that give the area a slightly undulating appearance with height differences of up to 3 m. The area forms a recharge area where the precipitation excess (200–300 mm year^{-1}) partly infiltrates, is discharged by subsurface flow, and partly runs off by surface water flow, towards the polder area. The groundwater component discharges in the polder as diffuse upward seepage. The polder area in turn discharges its precipitation surplus and the recharge from the Brabantse Wal by means of a dense drainage system by ditches. Because of its flat and low position, the polders are dewatered by power pumping.

East of the Brabantse Wal the combination of pasture and cornfield indicates stock-farming. On top of the Brabantse Wal there are forests, moorlands, some agricultural land and built-up areas. The fresh water upwelling zone consists of wet agricultural grasslands. Further west in the polder area there are agricultural lands and orchards.

Regional planning authorities are aware of the deterioration of wet ecotopes close to the Brabantse Wal since the vegetation surveys by Cools (1986). In the fresh water upwelling zone small water courses and ditches, imbedded in agricultural grasslands, function as refuges for various endangered mesotrophic plant species. This development is connected to the large scale groundwater extractions on the Brabantse Wal (Stuurman et al., 1990). The relations between changes in the water regime and the effects on ecosystems through a chain of abiotic and biotic links (with 'short cuts') are shown in the diagrams of Fig. 2, for a general case and for the Brabantse Wal case.

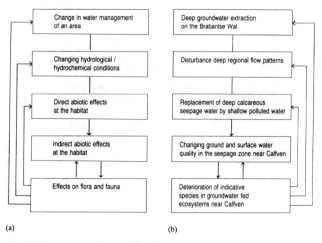

(a)   (b)

Fig. 2 Diagram of the relations between changes in the water regime and the effects on ecosystems, in general (a) and applied to the Brabantse Wal and adjacent polders (b).

METHODS

Regional Hydrochemical Facies Analysis

The Hydrochemical Facies Analysis (HYFA) is a procedure to map and diagnose the major factors accounting for regional variations in hydrochemistry, the results from environmental pollution and hydrological disturbances (Stuyfzand, 1993). HYFA has been applied to the Brabantse Wal and adjacent polders near Calfven in five successive steps: (1) gathering and selection of data on regional hydrology and hydrochemistry along four east-west directed transects; (2) objective determination of the hydrochemical characteristics (facies) of each ground- and surface water sample; (3) identification of its origin; (4) construction and description of maps and cross-sections presenting the spatial distribution of the two discerned hydrosomes (water bodies, each with a distinct origin); (5) interpretations of maps and cross-sections, leading to the recognition and understanding of evolution lines (facies chains) in the direction of groundwater flow within each hydrosome.

Groundwater was sampled with manual equipment in over 100 existing observation wells, with filters up to 119 m below the surface (April-May 1992), and additionally in 35 piezometric tubes, that were installed along 5 water courses in the fresh water upwelling zone (June-July 1992). Surface water was sampled in the water courses spread over the area in two periods (April, August 1992). In total over 275 water samples were analysed in the laboratory.

The facies has been determined in this study by integration of four independent facies-parameters according to Stuyfzand (1989): (a) the chemical watertype, which includes in one code, chlorinity, alkalinity, the dominant cation and anion, and a base exchange index; (b) the semi-empirical

redox index, as deduced from the concentrations of NO_3, Mn, Fe-t, and SO_4 (altered after Stuyfzand, 1989); (c) the eutrophication potential index (EPI), based on the concentration of orthophosphate; (d) the calcite saturation index (SI), the most relevant mineral saturation index for the system.

Local aquatic ecological research

Based on the preliminary results of HYFA, seven small water courses in the fresh water upwelling zone were selected for the local aquatic ecological approach. With an additional field survey a good impression was obtained from the distribution of plant species in the upwelling zone. The selected water courses were considered to represent the largest variety of vegetation in the area. In July 1992, vegetation surveys, field measurements on the abiotic characteristics of the ditches, and ground and surface water sampling were done. Any aquatic or terrestrial plant species that was present within 0.2 m (hor.) from the water line and 0.1 m (vert.) above the water line was registered. A total of 27 vegetation surveys were made, distributed over the seven ditches depending on abiotic and floristic heterogeneity. Representative surface water samples from each surveyed plot were obtained by taking mixed samples. The phreatic groundwater was sampled from piezometric tubes that were installed along the ditches.

The vegetation data were classified into four types of vegetation (noda), based on floristic characteristics. A nodum is an abstract vegetation unit without any rank or status (Wheeler & Giller, 1982). A mean hydrochemical characteristic and the deviation have been calculated per nodum. Distinct ecological amplitudes for the most relevant surface water quality parameters can be established for the noda. These hydro-ecological relations were evaluated with Ellenberg's indicative values (Ellenberg et al., 1991) and response tables for waterplants related to water quality and condition of the soil in the Netherlands (De Lyon and Roelofs, 1986). Finally, some correlations have been calculated and conclusions have been drawn on the interrelationship between groundwater, surface water, aquatic ecology and human influence in the upwelling zone.

RESULTS

Hydrogeology

The groundwater considered, is contained in predominantly marine sands, clays and peat of Quarternary and Tertiairy age. Deposited in the period Middle-Oligocene – Holocene, the marine (sandy clay) Rupel, (sandy) Breda, (calcareous sandy) Oosterhout, (sandy clay) Tegelen and (silty clay and peat) Westland Formations and the eolian (fine coversand) Twente Formation are found in the area. From several geological cross-sections (to a depth of 15 m below the surface) it can be concluded that the cliff-face of the Brabantse Wal is the result of successive marine (Emien) erosion and fluvial erosion by the Scheldt river (Kasse, 1988; Caris et al., 1989; Hobma et al., 1993).

The geohydrological schematisation of the Brabantse Wal area consists of three sandy Pleistocene and Tertiairy aquifers, separated from each other by two less pervious clay layers, and with the impervious Tertiairy marine deposits at the base (110 m −MSL; Fig. 4).

South of Bergen op Zoom (Fig. 1) perched watertables occur frequently. Piezometric heads of the phreatic groundwater range from 16.60 m +MSL, on the Brabantse Wal, to 0.20 m −MSL, in the adjacent polders. The direction of groundwater flow in the middle deep and the deep aquifer is still east-west: piezometric heads of the middle deep and deep groundwater range from 14.00 m +MSL to 0.20 m −MSL and from 12.00 m +MSL to 0.23 m −MSL respectively. This flow is discharged in the polder area by diffuse upward seepage. The piezometric levels of the middle-deep and deep groundwater have declined, compared to the situation in 1967. Consequently, the declined upward gradient of heads in the phreatic and middle deep groundwater (d(dh dz^{-1})=0.50 m) resulted in a reduction of the subsurface discharge into the polder area. This change in groundwater regime can be subscribed to the increased groundwater extractions on the Brabantse Wal since the 1960s (Stuurman et al., 1990; Hobma et al., 1993).

In the upwelling zone, only fresh groundwater from the Brabantse Wal is discharged by the ditches. Further west (1 km), a transition zone exists where both fresh and brackish and brackish-salt groundwater types occur. West of this transition zone, groundwater is exclusively brackish and brackish-salt, due to upwelling water originating from the Western- and eventually the Eastern-Scheldt. The variation in hydrochemistry in the area is very wide: pH ranges from 3.5 to 7.9, chlorinity ranges from 11 to 11 000 mg.l^{-1} and alkalinity is highly variable. As a result of this variation a clear characterisation of watertypes is possible.

Types of groundwater and processes

The areal distribution of the seven discerned types of groundwater is presented in a horizontal section at about 20 m −MSL (Fig. 3) and in two cross-sections (III and IV; Figs. 4a and 4b). Typical chemical analyses of the watertypes are listed in Table 1.

Calcareous Brabantse Wal water, with low chlorinity (30<Cl<150 mg. l^{-1}), high alkalinity (122<HCO_3<488 mg. l^{-1}) and NaKMg-equilibrated is present deep in the second

aquifer of the Brabantse Wal and a major extension is situated in the polders close to the Brabantse Wal. Its richness in Ca and HCO_3 (−1.10<SI<−0.01) is a result of the interaction with the Oosterhout and Breda Formations, which are rich in shell fragments.

Acid Brabantse Wal water, with low alkalinity (HCO_3<122 mg. l^{-1}) and low chlorinity (18<Cl<40 mg. l^{-1}) covers the upper 50 m of the Brabantse Wal. The low alkalinity is the result of infiltration of polluted water into the decalcified Twente and Tegelen Formations (SI<−1.0) and oxidation of iron sulfides (pyrite) and/or organic matter, due to the lowering of groundwater tables.

Salinising polder water, with high alkalinity (122<HCO_3<977 mg. l^{-1}), high chlorinity (1000<Cl<10000 mg. l^{-1}) and NaKMg-deficit is encountered at various locations in the polder. Typical is the NaKMg-deficit, as a result of cation exchange with Ca, that originates from intruded salt Western Scheldt water (Cl=15 590 mg. l^{-1}) or Eastern Scheldt water (Cl=44 120 mg. l^{-1}), flowing through the Oosterhout and Breda Formations (−0.40<SI<0.34). In the salinisation process the deposition of $CaCO_3$ is prevented by simultaneously occurring sulphate reduction (Appelo and Geirnaert, 1983).

Freshening polder water, with high alkalinity (244<HCO_3<1953 mg. l^{-1}), relative lower chlorinity (Cl<1100 mg. l^{-1}) and NaKMg-excess is encountered in the top layer in the polder, as a result of mixing with rain water. The typical NaKMg-excess is a result of cation exchange with Ca, present in fresh water. In the (re-)freshening process $CaCO_3$ dissolution is taking place.

Brackish equilibrated water, with high alkalinity (488<HCO_3<1952 mg. l^{-1}), high chlorinity (980<Cl<4570 mg. l^{-1}) and NaKMg-equilibrium occurs mainly in the deep aquifer in the northern half of the polder. The equilibrium exists probably due to the decreasing salt upwelling flow from the Eastern Scheldt as result of the damming of Lake Markiezaat (Fig. 3). The soil pores have been flushed by the same types of mixed water for a long time.

Freshening water at the fresh/brackish interface, with a wide variety of compositions; high alkalinity (488<HCO_3<1952), highly varying chlorinity (12<Cl<394 mg. l^{-1}) and NaKMg-excess. This watertype occurs at the interface of fresh intruding groundwater from the Brabantse Wal with the salt-brackish polder water. Cation exchange of Na, K and Mg against Ca, present in the fresh Brabantse Wal water, is a dominant process. The following freshened watertypes are subsequently encountered along a flow line at the fresh/brackish interface (cross-section IV): B3–NaCl+, f4–NaHCO$_3$+, F4–NaHCO$_3$+, g3–NaHCO$_3$+, g3–CaHCO$_3$+. High alkalinity and very low nitrate content in the polder water indicate denitrification under reducing conditions. Low sulphate content compared to high

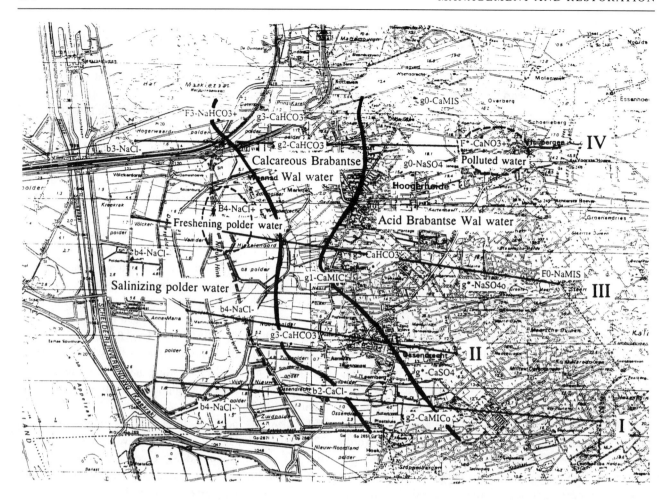

Fig. 3 Horizontal section over the research area, showing the location of the regional geohydrochemical transects (I–IV) and the areal distribution of groundwater types in the first aquifer, at a depth of 20 m $-$MSL (April-May 1992).

chlorinity, high alkalinity and frequently occurring peat layers indicate sulphate reduction by organic matter, in a reducing environment.

Polluted groundwater, acid water, with NO_3 as most important anion is encountered only in the upper 5 m of the Brabantse Wal. The occurrence of this watertype indicates the infiltration of manure down to the groundwater.

Groundwater/surface water relations on a regional scale

In the area both fresh and salt(-brackish) surface water is encountered. The upwelling of fresh groundwater into the surface water system of the polder area can be measured as far as 1 km west of the Brabantse Wal. Corresponding to the above groundwater classification, five types of surface water have been discerned. The derived maps of hydrochemical facies in ground and surface water correspond fairly accurately (Hobma *et al.*, 1993). Salinising surface water does

not occur in the area. Most of the surface water samples appear to have a SO_4-excess, while most of the polder groundwater samples have a SO_4-deficit, relative to a conservative mixture. These differences can be ascribed to the larger influence of polluted acid rainwater on the surface water composition.

Hydrochemical Facies Analysis

Two hydrosomes can be recognised: the Brabantse Wal hydrosome and the Polder hydrosome. Their areal distribution can be derived from the Figs. 3 and 4. The *Brabantse Wal hydrosome*, consists of mesotrophic to eutrophic groundwater (2<EPI<4). The water can be undersaturated (SI<0) or in equilibrium with calcite (SI=0). The hydrosome is present in the sub-surface of the Brabantse Wal and as far as 1 to 1.5 km west into the polder. The *Polder hydrosome*, consists of eutrophic to hypertrophic groundwater (3<EPI<6). The water can be saturated or over-saturated with calcite SI>0). The hydrosome is encountered in the main part of the polder area. In half of the groundwater samples the redox level is reduced, in the other half it can not be determined, due to mixing. Only a few samples appear to be oxic.

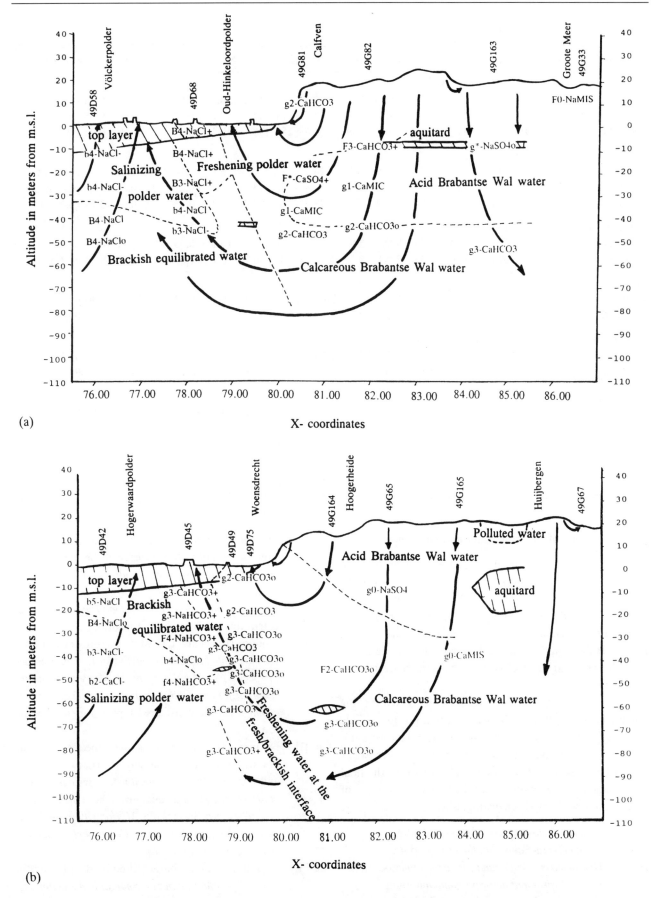

Fig. 4 East-west cross-sections III (a) and IV (b) of the research area, showing the areal distribution of groundwater types (April-May 1992).

Table 1. *Typical chemical analyses of the seven discerned types of groundwater*

FIELDCODE	EC	pH	Na	K	Mg	Ca	NH_4	Cl	HCO	SO_4	NO_3	Mn	Al	PO_4	Si	watertype
CALCAREOUS BRABANTSE WAL WATER																
49G82VI	303	7.01	8.14	1.4	0.05	54.5	0.3	13.4	154	11.8	0.08	0.05		0.47	−0.72	g2-HCO_3
49G164IV	459	6.94	8.58	1.27	0.05	91.8	0.86	13.3	278	14.2	0.8	0.83		0.51	−0.35	g3-HCO_3
49G82IV	812	6.41	39.9	48	17.4	104	0.66	31.7	256	148	10.1	0.9		0.2	−0.95	F3-$CaHCO_3^+$
ACID BRABANTSE WAL WATER																
49G88II	431	4.13	14.7	5.95	12.2	5.9	1.2	24.7	1	139	6.9	0.24	1.85	0.89		g*-$MgSO_4^+$
49G85I	1186	3.42	15.1	51.5	20.3	114	1.1	52.8	1	414	5.6	0.21	1.03	0.5		F*-$CaSO_4^+$
49G65	414	5.1	24.2	4.98	0.64	16.4	1.3	27.2	38	95.7	0.3	0.58		0.63	3.78	g0-$NaSO_4$
SALINISING POLDER WATER																
49D60IV	2340	6.77	5720	25.4	322.5	909.8	8.86	9657	195	958	10.1	0.05	0.16	0.81	−0.24	b2-$NaCl^-$
49D70II	6395	7.13	975.9	21.6	99.43	295.8	11.13	2090	568	60.7	0.4	1.07	0.05	1.3	0.34	b4-$NaCl^-$
FRESHENING POLDER WATER																
49D60I	4998	7.2	757.1	148	148.2	137.2	11.42	1100	1410	49.3	0.2	0.88	1.75	11.5	0.49	b5-$NaCl^+$
49D68III	2514	7.54	442	24.8	41.6	33.6	2	598	472	49	1.1	0.13		1.3	−0.12	B3-$NaCl^+$
BRACKISH EQUILIBRATED WATER																
49D42II	8814	6.65	1602	44.7	162	212	3.7	2739	1263	73.4	0.2	0.87		0.61	0	b5-NaCl
49DS8IV	3662	6.93	582	10.1	67.5	150	6.1	977	790	1.1	0.4	0.12		0.63	0.08	B4 NaCl
FRESHENING WATER AT THE FRESH/BRACKISH INTERFACE																
40D49IV	1681	7.9	326	26.8	17.5	15.1	1.2	394	382	6.4	0.2	158		1.1	−0.14	B3-$NaCl^+$
49D45I	6629	7.05	95.8	8.17	4.6	50.1	3.2	63.5	363	4.7	0.05	0.23		0.26	−0.41	F3-$NaHCO_3^+$
49G84V	1080	7.19	42	54	4.57	143	0.75	59.9	545	70.6	13.4	0.15	0.36	1.2	0.29	F4-$CaHCO_3^+$
POLLUTED WATER																
49G50	980	4.43	12.8	38.3	13.1	129	0.4	33.3	29	72	381	1.14		0.12	−3.73	F*-$CaNO_3^+$

Aquatic ecology of the fresh water upwelling zone

In the upwelling zone a considerable variation in ground and surface water quality between and within the surveyed ditches has been observed. The hydrochemical variation appears to be reflected in the aquatic vegetation which could hardly be estimated from the regionally derived HYFA results. HYFA gained only a first zonation of acid and calcareous ground and surface water types in the fresh water upwelling zone.

Fortythree plant species have been registered in the ditches, including *Ranunculus aquatilis*, that was found only once during the reconnaissance survey. The data were initially classified using the alkalinity of the surface water (Wiegleb, 1978a and 1978b; Zonneveld, 1988). In the final classification, based on floristic characteristics, four types of vegetation (noda) have been distinguished.

I.   *Juncus effusus-Agrostis stolonifera* nodum
II.   *Berula erecta-Equisetum fluviatile* nodum
III. *Hydrocharis morsus-ranae-Potamogeton* nodum
IV.   *Rumex hydrolapathum-Sparganium* nodum

ad I.)   Nodum I is characterised by an often high coverage by *Juncus effusus* and *Agrostis stolonifera*. In addition, a number of species can be regarded as characteristic for this nodum, for example: *Alopecurus geniculatus*, *Polygonum hydropiper*, *Juncus bufonius* and *Lotus uliginosus*. To a lesser extent this holds for *Holcus lanatus*.

ad II.) Nodum II is characterised by the presence of both *Berula erecta* and *Equisetum fluviatile*. The presence of species like *Mentha aquatica*, *Myosotis palustris*, *Sium latifolium*, *Galium palustre* and *Veronica beccabunga* can be considered typical for this nodum too.

ad III.) Nodum III contains some species of the Potamogeton family (namely *P. crispus*, *P. natans* and eventually *P. pusillus*). Besides these, *Hydrocharis morsus-ranae* can be considered a typical species. Species like *Myriophyllum verticillatum*, *Elodea canadensis* and *Ceratophyllum demersum* also belong to this unit.

ad IV.) Nodum IV can be characterised by the presence of *Rumex hydrolapathum* and *Sparganium erectum ssp. erectum* and/or *Sparganium emersum*. Furthermore

Table 2. *Estimated and measured mean values of discriminating surface water quality parameters for the four noda. The difference is expressed as a factor. Standard deviations are between brackets*

Nodum		pH	HCO$_3$	Ca	Mg	Na	salinity
				(in mmol/l)			
I	estimat.	7.27	2.88	0.76	0.54	1.53	8.33
	measur.	6.29 (0.55)	1.17 (0.75)	0.97 (0.43)	0.15 (0.10)	0.59 (0.14)	4.74 (1.06)
	factor	1.2	2.5	1.3	3.6	2.6	1.8
II	estimat.	7.44	2.95	0.86	0.57	1.81	9.32
	measur.	7.07 (0.17)	2.76 (0.53)	1.71 (0.44)	0.48 (0.36)	0.61 (0.12)	7.54 (1.36)
	factor	1.1	1.1	2.0	1.2	3.0	1.2
III	estimat.	7.55	2.79	0.84	0.61	1.97	9.03
	measur.	7.19 (0.35)	5.02 (0.54)	3.52 (0.49)	1.06 (0.06)	1.17 (0.11)	16.4 (1.86)
	factor	1.1	1.8	4.2	1.7	1.7	1.8
IV	estimat.	7.52	3.19	0.93	0.67	2.06	10.3
	measur.	7.49 (0.08)	6.38 (0.97)	4.25 (1.01)	1.43 (0.17)	1.79 (0.65)	19.9 (3.73)
	factor	1.0	2.0	4.6	2.1	1.2	1.9

species such as *Ranunculus sceleratus, Lycopus europaeus* and *Polygonum amphibium* occur in this unit.

Some of the species, such as *Glyceria maxima, Glyceria fluitans, Phragmites australis, Lemna minor* and *Callitriche platycarpa* occur frequently throughout the upwelling zone and do not reflect accurately enough the subtle variations in hydrochemistry. Therefore they are considered to be associated species.

The noda appear to have an increasing affinity for pH, alkalinity, calcium, magnesium and salinity of the surface water (Table 2), expressed in the ecological amplitude (max.-min.). The measured ecological amplitudes of the noda for alkalinity, pH, salinity and for a combination of the three parameters are shown in Fig. 5. In most cases only a small overlap in the amplitudes exists. The quasi 3d-chart shows that vegetation zonation in the ditches is still determined to a large extent by a combination of (interdependent) surface water quality parameters.

Ecohydrological analysis

Given the above ecological amplitudes of the noda for surface water quality parameters, two methods have been applied to estimate water quality parameters based on presence/absence of the individual plant species in the 27 vegetation surveys. The mean estimated parameter values per nodum, using Ellenberg's indicative values for acidity and inorganic nitrogen (Ellenberg *et al.*, 1991) and response tables for waterplants related to water quality (De Lyon & Roelofs, 1986), are not in line with the mean hydrochemical characteristic based on field measurements (Table 2). In

general, the response tables gain lower calcium and higher sodium contents (Table 2). The measured hydrochemical variation in alkalinity, magnesium and salinity does not correspond with the estimated values at all. From the analyses it followed that the species with the lowest estimated indication for pH, HCO$_3$, Ca, Mg, Na and salinity (*Juncus effusus*) is characteristic of nodum I, while characteristic species of nodum IV (such as *Ranunculus sceleratus* and *Rumex hydrolapathum*) gain the highest values (Hobma *et al.*, 1993). Because the surveys contained many identical associated plant species, these differences are not expressed in the final estimation.

Groundwater/surface water relations in the upwelling zone

From the combined samples of ground and surface water correlation coefficients between the hydrochemical parameters have been calculated (Table 3). Moderate to high correlations are expected to reflect the natural conditions in groundwater/surface water interactions. The many low correlations may indicate recent disturbance of the natural groundwater flow pattern or the impact of land and water management on the drainage system.

Large differences in correlation of ground and surface water parameters per ditch are shown in Table 3. Parameters such as NO$_3$, PO$_4$, NH$_4$, SO$_4$, pH and HCO$_3$ are not likely to be strongly correlated, since they take part in many geohydrochemical processes. The large differences in correlation of the chloride content between the ditches are remarkable, since chloride is not expected to play any reactive role in the fresh water zone. Groundwater/surface water interactions at the ditches A and H seem to be dominated by the presence of a thick peat layer in the sub-soil. The highest correlations are

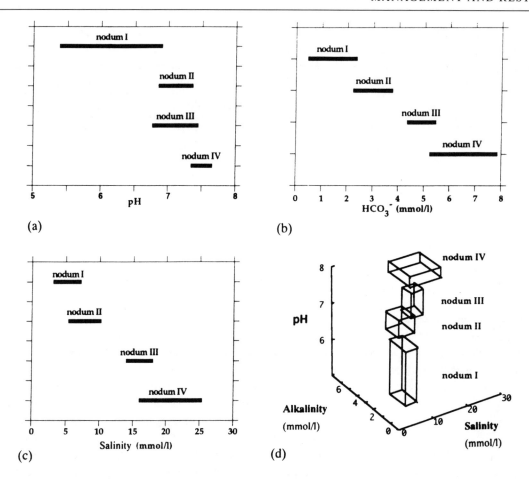

Fig. 5 Ecological amplitudes of the four noda for: (a) alkalinity, (b) pH, (c) salinity and (d) for a combination of the parameters in the fresh water upwelling zone between Woensdrecht and Ossendrecht.

found in ditch H, where the samples were taken only five days after a rainy period.

DISCUSSION AND CONCLUSION

With a regional hydrochemical facies analysis, the major factors accounting for the complex hydrology and hydrochemistry of the transition area of the Netherlands Delta and the Brabantse Wal have been determined and mapped to a satisfactory degree for a local ecohydrological study. Changes in regional hydrology strongly influence ground- and surface water quality and the wet ecosystems in the fresh water upwelling zone close to the Brabantse Wal. Since the 1960s, a considerable inflow of deep calcareous groundwater from the Brabantse Wal seems to be replaced by an expanding (sub-) regional acid, freshening and polluted system, fed by rainwater.

The hydrochemical variation in the upwelling zone is still reflected by the zonation of aquatic vegetation in a limited number of ditches. Ground and surface water composition at the ditches deviate to a large extent, indicating disturbance of the natural groundwater flow pattern. Intensified agricultural land-use upstream and maintainance of the small water courses in the area is also a major factor in the deterioration of wet ecosystems. Effective measures for conservation and regeneration of the complex abiotic and biotic gradients along the Brabantse Wal should include: (i) restoration of deep calcareous groundwater flow towards the upwelling zone by reducing groundwater extraction (which would have significant financial consequences), (ii) extensification of land-use near the cliff-face on the sandy Brabantse Wal and (iii) as a vital basic condition, improvement of integral water management to eliminate pollution.

Due to regional variation, response tables for aquatic vegetation in the Netherlands by De Lyon and Roelofs, and Ellenberg's indicative values for vegetation in Central Europe, can not be used in subregional and local ecohydrological studies without complementary field measurements. Research on hydrology, hydrochemistry and ecology that integrates the results of regional and local studies is essential if understanding and restoration of groundwater dependent ecosystems is concerned. Methodology development in ecohydrology that is based on hydro-ecological relations, which are derived for vegetation

Table 3. *Correlation coefficients (r) between the major hydrochemical parameters of ground and surface water in the fresh water upwelling zone, arranged per ditch (A–J). Only correlations >0.35 are depicted; correlations >0.5 are bold-faced; correlations >0.75 are in italics and bold type*

parameter	A	C, D, E	H	I	J
			r per ditch		
pH	0.37	−0.66	*0.98*	−0.54	*0.87*
EGV	*−0.83*		*0.88*	0.54	*0.78*
NH₄		−0.70	0.50		
Cl	*0.97*	0.64	*0.98*	*−0.83*	−0.53
HCO₃	*0.75*		*0.95*		0.65
NO₃	−0.35		0.73	*0.80*	
Al	−0.38	0.44		0.64	
Ca	*0.76*		*0.75*		*0.83*
Fe	−0.50	0.48		−0.73	
K		−0.63	*0.99*	−0.43	
Mg	0.50		0.66	−0.63	0.44
Na	0.49	−0.36	*0.91*	*−0.93*	
SO₄		0.52	*0.88*	0.74	*0.94*
PO₄	0.72	−0.36	−0.46	0.66	*0.78*
Mn			*0.87*	*0.83*	
Si	0.66			*0.85*	*−0.85*

types in regional or local surveys, may be preferred to the use of generalised indicative values of individual plant species.

ACKNOWLEDGEMENTS

I am indebted to J.J. de Vries, D. Newcombe and an anonymous reviewer for their valuable comments on the manuscript. I also thank M. v.d. Graaf, B. Hoogeboom and M. v.d. Leemkule, who did an excellent job during their final practical work under my supervision. Financial support from the Cornelis Lely Stichting of Rijkswaterstaat and the Netherlands Organisation for Scientific Research (NWO) is gratefully acknowledged.

REFERENCES

Appelo, C.A.J. & Geirnaert, W. (1983). Processes accompanying the intrusion of salt water. *Geol. Appl. Idrogeol.,* **18**, II, 29–40.

Caris, J.P.T., Thewessen, T.J.M. & Felix, R. (1989). Genesis of the cliff-face near Bergen op Zoom in the southwest of the Netherlands. In *Geologie en Mijnbouw,* **68**, 277–284.

Cools, J.M.A. (1986). *A floristic survey of the Brabantse Wal and Lake Markiezaat. Natuur, Milieu en Faunabeheer,* Report NMF, Tilburg (in Dutch).

Ellenberg, H., Weber, H.E., Düll, R., Wirth, V., Werner, W. & Paulissen, D. (1991). Zeigerwerte von Pflanzen in Mitteleuropa. *Scripta Geobotanica* 18. Verlag Erich Goltze KG, Göttingen.

Hobma, T.W., Graaf, M. v.d, Hoogeboom, B. & Leemkule, M. v.d. (1993). *Ecohydrology and hydrochemistry of the Brabantse Wal and adjacent polders near Calfven.* Institute of Earth Sciences, Free University Amsterdam, Report VU, Amsterdam (in Dutch).

Kasse, C. (1988). *Early Pleistocene tidal and fluviatile environments in the Southern Netherlands and Northern Belgium,* Ph D Thesis, Free University Amsterdam.

Lyon, M.J.H. de & Roelofs, J.G.M. (1986). *Waterplants related to water quality and condition of the soil.* Parts 1 & 2. Laboratory for Aquatic Ecology, Katholieke Universiteit Nijmegen, Report KU, Nijmegen (in Dutch).

Stuurman, R.J., van der Meij, J.J., Engelen, G.B., Biesheuvel, A. & van Zadelhoff, F.J. (1990). *The hydrological systems analysis of western Noord-Brabant and surroundings.* DGV-TNO/Free University Amsterdam, Report DGV-TNO, OS 90–25–A, Delft (in Dutch).

Stuyfzand, P.J. (1989). *A new hydrochemical classification of watertypes.* Proc. IAHS Third Sci. Ass.Baltimore USA, 10–19 May 1989, IAHS-publ. 182, 89–98.

Stuyfzand, P.J. (1993). *Hydrology and hydrochemistry of the coastal dune area of the Western Netherlands.* Report KIWA I11 Nieuwegein/PhD-Thesis, Free University Amsterdam.

Wheeler, B.D. & Giller, K.D. (1982). Status of aquatic macrophytes in an undrained area of fen in the Norfolk Broads, England. *Aquatic Botany,* **12**, 277–296.

Wiegleb, G. (1978a). Untersuchungen über den Zusammenhang zwischen hydrochemischen Umweltfaktoren und Makrophytenvegetation in stehenden Gewässern. *Arch. Hydrobiol.,* **83**, 443–484.

Wiegleb, G. (1978b). Der soziologische Konnex der 47 häufigsten Makrophyten der Gewässer Mitteleuropas. *Vegetatio,* **38**, 165–174.

Zonneveld, I.S. (1988). Establishing a floristic classification. In *Vegetation mapping,* A.W. Küchler en I.S. Zonneveld, Kluwer Academic Publishers, Dordrecht.

26 Cautious reforestation of a wetland after clearfelling

G. JACKS*, A. JOELSSON**, A.-C. NORRSTRÖM* & U. JOHANSSON***

* Land and Water Resources, Royal Institute of Technology, S-100 44 Stockholm, Sweden

** Halland County Board, S-310 86 Halmstad, Sweden

*** Tönnersjöheden Experimental Forest, Swedish University of Agricultural Sciences, S-310 38 Simlångsdalen, Sweden

ABSTRACT A small forest catchment in Sweden is drained via a peatland. The mature spruce stand (108 years) was clearcut after which a younger stand (60 years) on the peatland was accidentally stormfelled the next year. The clearcutting causes nitrification in the upland. To maintain the nitrogen reduction function of the peatland, planting on mounds was practiced instead of the conventional fishbone drainage pattern. During the fourth and fifth years after the clearcutting the nitrogen flux from the upland to the peatland was 17 kg ha^{-1}.a while only 7 kg left the peatland. While about 85% of the nitrogen entering the peatland was in the form of nitrate, the fraction of nitrate in the runoff was only 40%. The establishment of new forest stands on peatland soils by planting on mounds gives considerable environmental advantages as compared to conventional drainage. Both fluxes of nutrients and suspended matter are decreased. Spontaneous regeneration of the wetland spruce stand results in loss of about ten years of growth and a younger stand on the wetland than on the upland with the risk of the stormfelling being repeated after the upland forest has been harvested.

INTRODUCTION

Most of the Swedish forest land is drained by a system of ditches. Draining has been done for the purpose of increasing forest production. A peak in the draining operations was seen in the 1930s and it has again increased during the last decades (Löfroth, 1991). A maximum of 12 000 km of drains have been dug yearly, equal to the distance from Stockholm to Vladivostok and back again, much through the same terrain. However, due to concern over species conservation and the need for wetlands as nutrient sinks the official attitude towards drainage of forest land has become more strict. Also, landowners have become aware of the importance of wetlands, especially fens, as critical habitats for forest birds.

A new incentive for the protection of forest wetlands has been initiated in southern Sweden where atmospheric nitrogen deposition is high. It is believed that in the order of 5–10 kg ha^{-1}.a can be retained in the forests with conventional management (Johnson, 1992). The current deposition is about 20 kg ha^{-1}.a. Nitrogen saturation has occured in the forest land in the Netherlands, resulting in nitrification, nitrate leaching and acidification (Stams et al., 1991). This scenario transplanted to southern Sweden would be a very serious threat to the adjacent marine areas where algal blooms are already occurring due to the nitrate brought in by the rivers from agricultural lands (Fleischer & Stibe, 1989).

Where the groundwater level is high, denitrification may be a major sink for the excess deposition (Tietema & Verstraten, 1991). The nitrate leaching from forest land in southern Sweden is so far moderate, in the order of 5 kg ha^{-1}.a in the county of Halland (Fleischer & Stibe, 1989) or approximately twice the background value. Higher nitrate losses from forest land have been observed only as a result of forest disturbances such as stormfelling and clearcutting. This is a normal feature in all forest management. It seems, however, that the nitrate pulses after forest disturbances in southern Sweden have grown in magnitude and duration. Wiklander et al. (1991) have observed nitrogen losses in the order of 15 kg ha^{-1}.a after the stormfelling of a spruce forest. The nitrogen losses decreased after a few years as regrowth progressed. Signs of increased nitrogen accumulation in south Swedish forest soils are also reflected by changes in plant species composition (Falkengren-Grerup, 1992; Rosén, 1992).

A partly clearcut catchment has been observed as part of a larger study of the function of wetlands as nitrogen sinks. A new management was used which represents a compro-

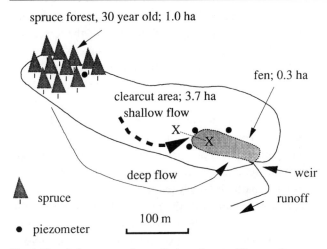

spruce forest, 30 year old; 1.0 ha

fen; 0.3 ha

clearcut area; 3.7 ha
shallow flow

X

X

deep flow

weir

runoff

▲ spruce

● piezometer

100 m

Fig. 1 Sketch figure over the studied catchment. Water pathways to the wetland indicated as two categories; shallow emerging from clearcut area and deep coming from a young spruce stand.

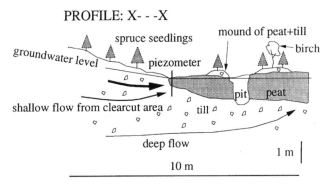

PROFILE: X- - -X

spruce seedlings

mound of peat+till

birch

groundwater level

piezometer

shallow flow from clearcut area

till

pit

peat

deep flow

1 m

10 m

Fig. 2 Section through the upland/wetland interface showing a pit and mounds heaped up around it.

mise between the conservational interests and the interests of optimizing forest production.

MATERIALS AND METHODS

The catchment is situated in the county of Halland about 40 km east of the town of Halmstad (N 56° 42′, E 13° 7′) in Southern Sweden. It is 5.0 ha with a fen of 0.3 ha in the lower part (Fig. 1). It represents a common building block in the northern coniferous forests. The upland soil is a Haplic podsol (USDA classification system) formed in unsorted till. The fen has a peat thickness of about 1 m in its central part. The fen is drained by a short ditch in the southwestern end, overgrown with *Sphagnum* mosses. The peat is a *Sphagnum* peat. A 108-year-old planted spruce stand was clearcut on 3.7 ha of the land adjacent to the bog in 1987. A 60-year-old spruce stand on the bog was left but was stormfelled the year following clearcutting. Peatland downstream from clearcut areas are difficult to reforest as the groundwater level rises

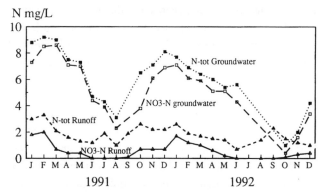

N mg/L

N-tot Groundwater

NO3-N groundwater

N-tot Runoff

NO3-N Runoff

J F M A M J J A S O N D J F M A M J J A S O N D
1991   1992

Fig. 3 Total nitrogen and nitrate-nitrogen in groundwater piezometers at the perimeter of the wetland and in the runoff after the passage of the wetland.

due to decreased evapotranspiration. An earlier practice was to dig a fishbone pattern of drainage ditches. As peat is very impermeable this drainage network has to be very dense (Hillman, 1988). Drainage in combination with clearcutting tend to give increased outflow of nutrients and a drastic increase in suspended solids in the runoff (Ahtiainen, 1988). Thus the reforestation of the bog was done by using heavy equipment to form mounds of peat and underlying mineral soil on which the next spruce generation was planted (Fig. 2).

The runoff from the wetland was measured by means of a V-notch weir supplied with a water level recorder. Groundwater piezometers were installed along the perimeter of the bog to sample the groundwater flowing from the upland. Sampling was done monthly for two years which was found sufficient to follow the seasonal changes in the water quality (Fig. 3). Fluctuations in discharge were gentle due to the damping effect of the vadose zone in the well drained upland of the catchment.

To study the redox conditions in the bog two series of redox indicators were installed from the periphery of the bog towards the centre. One type of indicator consisted of lead-oxide painted steelrods which turn black on reacting with hydrogen sulphide. The other type was uncoated steelrods that rusted on encountering oxygenated conditions but remained metal-shiny in an anoxic environment. When in contact with hydrogen sulphide they will also turn black; this reaction is, however, slower than the formation of lead sul-phide.

RESULTS

The influx of nitrate to the bog turned out to be astonishingly high (6–9 mg. NO_3-N l^{-1} during the dormant season), in view of the time that had elapsed since the clearcutting. The setting of the piezometers means that they will catch mainly the shallow flowlines to the bog (Fig. 1). Research concerning groundwater flow in till concludes that the structure of till soils tends to give preference to near surface pathways (Lundin, 1982; Newton, 1987). This is due to the genesis of the till, to frost-heaving and root penetration increasing the hydraulic conductivity in the near surface sections of till soil profiles.

There is no surface drainage to the peatland. The inflow of groundwater is considered to be proportional to the upland part of the whole catchment or 94% of the discharge measured at the V-notch weir. The groundwater inflow can be considered to consist of shallow inflow caught by the piezometers at the perimeter of the peatland and a deep inflow. The shallow inflow was considered to come from the 3.7 ha clearfelled area while the deep inflow was from the 30-year-old spruce stand on 1.0 ha in the upper reaches of the catchment (Fig. 1). It is known that the runoff from a clear-felled area is about twice that from a closed coniferous stand (Rosén, 1984). Thus the shallow pathways recorded by the piezometers were considered to represent about 84% of the total groundwater inflow, the deeper flow making up the remaining 16%.

The nitrogen contents in the runoff and the mean contents in the shallow piezometers are shown in Fig. 3. A clear seasonal pattern is seen with high contents in the dormant season. The concentration in the deep flow, considered to come from the spruce stand in the upper reaches of the catchment, was taken as the concentration recorded in a piezometer in the spruce stand itself. This was similar to that recorded in the runoff of an adjacent catchment covered by another part of the same spruce stand. The content in this stream is 0.4 mg. l^{-1} in winter time and Tot-N comes up to a maximum of 1.8 mg. l^{-1} This may be slightly higher than it is after the long passage through the groundwater zone. On the other hand, the inflow from the spruce stand is considered to be only 16% of the total groundwater inflow as advocated above and moderate errors in the estimation of the concentrations in the deep inflow will not appreciably affect the budget. Hill (1991) notes that 90% of the NO_3–N influx to a swamp comes via the shallow pathways. The situation in this catchment is similar.

While the runoff water was dominated by organic nitrogen and ammonium, nitrate constitutes about 85% of the nitrogen in the groundwater. As the runoff has a similar seasonal pattern, at least during the last few years when winters were mild, the seasonal transport of nitrogen to the peatland and from the whole catchment is even more pronounced. The transport by groundwater into the peatland in 1991 and 1992 was 17 N kg ha^{-1}a of which 14 kg ha^{-1}.a was NO_3–N. The loss from the catchment was 6.8 kg ha^{-1}a of which 2.9 kg ha^{-1}.a was NO_3–N. Thus the wetland is an efficient sink for nitrate especially in the dormant season. The studies by Wiklander *et al.* (1991) of a catchment in the same region can

serve as a reference. The catchment was extensively drained even before a stormfelling of about 30% of the forest. The export of nitrogen from the whole catchment was about 10–12 kg ha^{-1}·a for 3–4 years after the stormfelling. The discharge from the stormfelled area was calculated at 18 kg ha^{-1}.a. The export of 6.8 kg ha^{-1}.a is only slightly above the regional transport in water courses considered to be about 5 kg ha^{-1}.a (Fleischer & Stibe, 1989). Thus the practice of planting on mounds preserves the nitrogen reduction function of the wetland. It is also most likely that the transport of suspended matter is largely reduced.

The investigation with the redox indicators showed that the uppermost 5 cm of the moss-peat sections were oxidizing, causing the steel rods to rust. Below this zone there was a gradual shift to sulphate-reducing conditions which prevailed in the next 5–10 cm. The blackening of the lead oxide was not uniform but occurred in streaks, interrupted by obviously less reducing environments. The unpainted steel rods were also blackstained but to a lesser degree indicating a slower rate of formation of ferrous sulphide than of lead sulphide. The unpainted steel rods were unaffected below the sulphate-reducing zone indicating absence of oxygen. The reason that the sulphate reduction took place in the upper portion of the peat sections is likely to be due to the presence of more reactive organic material leaking from the surface vegetation. The sphagnum peat is otherwise quite refractory under anoxic conditions (Zehnder & Svensson, 1986).

DISCUSSION

The nature of the nitrate reduction was not investigated. Hemond (1983) examined the nitrogen budget of a bog and postulated two major sinks for nitrate. The foremost one in the bog he studied was reduction to ammonium and incorporation into the exchange complex. The other one was denitrification. Hemond (1983) showed that denitrification was greatly stimulated by the addition of nitrate. The peatland in this investigation is abundantly supplied with nitrate from the upland, especially in the dormant season, and denitrification should be a major pathway for the nitrate reduction. A number of analyses have been done on the gas emitted from this and other peatlands in the area showing that the major constituent was N_2 with traces of N_2O_2 which abundantly supplied with nitrate from the upland and denitrification should be a major pathway for the nitrate reduction. The N_2 content showed peaks in winter time reaching 90% while it was in the order of 60% during the summer season. The groundwater level in the wetland is elevated the year round due to the larger influx of groundwater after the clearfelling. In winter time when the evapotranspiration is low, it is observed that the groundwater reaches close to the ground surface in the wetland. This favours the denitrification as the organic matter is fresher and more reactive near the surface (Koerselman *et al.*, 1993).

Conventionally drained wetlands allow considerable losses of NO_3–N from upland clearcut areas as is shown by Wiklander *et al.* (1991). A spontaneous, natural reforestation of the studied wetland by birch followed by spruce has also considerable drawbacks. It would again result in a younger stand on the wetland than on the upland with the risk of repeating the same story again of stormfelling after the upland stand has been harvested. The loss in economic terms of about 10 years' growth before the spruce spontaneously colonizes the wetland is considerable. An early establishment of a forest stand may also catch some of the nutrients mobilized through mineralization in the clearcut upland and in the wetland itself. It is also possible that the digging of the pits is favourable for the nitrogen retention in the sense that it obstructs the natural channelling through the peatland. Peatlands are found to be drained by pipe flow even when no surface drainages are visible (Kullberg *et al.*, 1992). The velocity of the pipe flow may be as high as 1–2 m min^{-1} (Jacks & Norrström, 1993) which may limit the nitrogen retention. Birch has spontaneously colonized the mounds (Fig. 2) and has grown fast enough to act as a protection against frost for the planted spruce seedlings. Frost, common in May and June, has damaged the upland spruce seedlings while the ones planted on mounds in the wetland are unaffected.

It can be concluded that it is possible to have a fast regeneration of forest stands on peatland and still preserve the nitrogen retention properties of the wetland by planting on mounds dug up from pits through the peat down into the mineral soil. This management is economically feasible and environmentally advantageous as it protects downstream recipients from sediments and eutrophication. Still another way of harvesting with the aim of minimizing nitrogen losses would be a shelterwood system of regeneration. The biology and economy of such a practice in spruce stands is, however, so far not well evaluated.

ACKNOWLEDGEMENTS

This investigation was supported by the World Wildlife Foundation and the Swedish National Board of Environmental Protection.

REFERENCES

Ahtiainen, M. (1988). Effects of clear-cutting and forestry drainage on water quality in the Nurmes study. In *Symposium on the Hydrology of Wetlands in Temperate and Cold Regions*. Joensuu, Finland, June 1988. Academy of Finland, Helsinki, 4/1988, 206–219.

Falkengren-Grerup, U. (1992). *Soil- and florachanges in south Swedish decidous forests* (in Swedish). Swedish National Board of Environmental Protection, Report 4061, 97 p.

Fleischer, S. & Stibe, L. (1989). Agriculture kills marine fish. *Ambio*, **18**, 346–349.

Hemond, H. F. (1983). The nitrogen budget of Thoreau's bog. *Ecology*, **64**, 99–109.

Hill, A. R. (1991). A ground water nitrogen budget for a headwater swamp in an area of permanent ground water discharge. *Biogeochemistry*, **14**, 209–224.

Hillman, G. R. (1988). Preliminar effects of forest drainage in Alberta, Canada on groundwater table levels and stream water quality. In *Symposium on the Hydrology of Wetlands in Temperate and Cold Regions*. Joensuu, Finland, June 1988. Academy of Finland, Helsinki, 4/1988, 190–196.

Jacks, G. & Norrström, A-C. (1993). Water pathways in the ecotones of the HUMEX lake. *HUMOR-HUMEX Newsletter*. Norwegian Institute of Water Research 2/1993, 6–7.

Johnson D. W. (1992). Nitrogen retention in forest soils. *Journal of Environmental Quality*, **21**, 1–12.

Koerselman, W., van Kerkhoven, M. B. & Verhoeven, J. T. A. (1993). Release of inorganic N, P and K in peat soils; effect of temperature, water chemistry and water level. *Biogeochemistry*, **20**, 63–81.

Kullberg, A., Petersen, Jr R. C., Hargeby, A & Svensson, M. (1992). Transfer of octanol soluble organic carbon through the soil/water interface of the HUMEXl lake. *Environmental International*, **18**, 631–636.

Lundin, L. (1982). *Soil moisture and groundwater in till soil and the significance of soil type for runoff* (in Swedish with English abstract). University of Uppsala, Department of Physical Geography, Report 56, 216 p.

Löfroth, M. (1991). *Wetlands and their importance* (in Swedish) Swedish National Board of Environmental Protection, Report 3824, 93 p.

Newton, R. M. (1987). The role of flow paths in controlling stream water chemistry at Pancake Creek in the Adirondack Region of New Yorl State. In *Extended Abstracts from Geomon Workshop*, ed. B. Moldan & T. Paces. Geological Survey, Prague.

Rosén, K. (1984). Effects of clear-felling on runoff in two small watersheds in Central Sweden. *Forest Ecology and Management*, **9**, 267–281.

Rosén, K. (1992). Critical loads for nitrogen in forest lands. In *Nitrogen saturation*. ed. T. Eriksson. Royal Swed. Academy of Agriculture and Forestry, Report 62, 26–37.

Stams, A. J. M., Booltink, H. W. G., Lutke-Schipholt, I. J., Beemsterboer, B., Woittiez, J. R. W. & van Breemen, N. (1991). A field study on the fate of 15N-ammonium to demonstrate nitrification of atmospheric ammonium in an acid forest soil. *Plant and Soil*, **129**, 241–255.

Tietema, A. & Verstraten, J. M. (1991). Nitrogen cycling in an acid forest ecosystem in the Netherlands under increased atmospheric nitrogen input. *Biogeochemistry*, **15**, 21–46.

Wiklander, G., Nordlander, G. & Andersson, R. (1991). Leaching of nitrogen from a forest catchment at Söderåsen in southern Sweden. *Water, Air and Soil Pollution*, **55**, 263–282.

Zehnder, A. J. B. & Svensson, B. H. (1986). Life without oxygen: what can and what cannot? *Experientia*, **42**, 1197–1205.

27 Responses of riparian ecosystems to dewatering of the Aral Sea in the vicinity of the Tedgen and Murgab rivers

T. V. DIKARIOVA

Water Problems Institute, Russian Academy of Sciences, 10 Novaya Basmannaya Str., P.O. Box 524, Moscow 107078, Russia

ABSTRACT Nowadays the ecological sustainability of water bodies is uncertain because of anthropogenic acceleration of environmental change. The Tedgen and Murgab oases in Central Asia are zones of threshold ecological tension because they are near to the zone of Aral ecological crisis. Riparian ecosystems of the rivers Tedgen and Murgab serve as indicators of this tension. The waters of Tedgen and Murgab have high mineralization, especially sulfate, and are polluted by phenol and pesticides. The riparian ecotone reflects impacts of pollution: 51% of riparian ecosystems of these rivers have semisalted soils, 10% – heavy and very heavy salted soils and 38% – nonsalted or little salted soils. Also 30% of all salted soils have the sulphate type of salinization. In the future riparian ecotone may be expected to sequester various pollutants. Parameters for monitoring are: 1) biodiversity of riparian phytocoenoses; 2) presense and amount of pollutants; 3) direction and character of matter flows through ecotones; 4) character of processes in ecotones.

INTRODUCTION

Catastrophic consequences of irrigational activity have occurred in the Aral region during the last 30 years. This sea lost 900 cubic kilometres of water or 16 years of runoff by river inflow. Its level dropped by 15 meters, water surface shrank by 45% and water volume decreased by 65%. Its salinity increased by three times. The sea is now composed of three basins, rather than one. About 70 million tons of salted dust are carried away from the dried bottom of the sea annually. The dust reaches distances of 300–500 kilometres from the sea. The Aral sea lost its significance for economy and life of people. The ecosystems of the Amu Darya and Syr Daria deltas were ruined. About 200 species of flora and fauna were lost for this region (Sadykov & Veselov, 1993). A variety of new land/water ecotones emerged in this quickly changing environment (Zaletaev, 1989) (Fig. 1).

We examined the basin of the Tedgen and Murgab interfluent region, which suffers the impact of intensive irrigative agriculture. The main objectives of the study were to find out the processes in riparian ecosystems effected by irrigation. We detected signs of ecosystem collapse here as well, but think that the situation might be improved and that ground-

water/surface water ecotones may play an important role in this improvement (Fig. 2).

ECOLOGICAL SITUATION

The Tedgen and Murgab basins have been irrigated for 6000 years and have passed through several crises. Today the situation is problematic, along with dewatering. The deficit of drinking water in the region is 550 thousand cubic meters per day. The main consumer of water is irrigated agriculture – 80% of irrigation water is consumed and return flows pollute surface and ground waters.

The rivers have the sulphate type of salinity. The content in Tedgen is 4–6 times higher than maximum advisable limits; mineralization of water is 1.5 times higher than normal. Pollution by phenols is 2 times higher than the maximum advisable concentration, by pesticides – 1.5 times higher. The salinity of waters of the Murgab is 1.3 times higher than normal, and sulphate content is 4.5 times higher than the maximum advisable concentration. Content of phenols is 1–4 times higher than normal and content of pesticides is 2 times higher. The high level of water pollution is

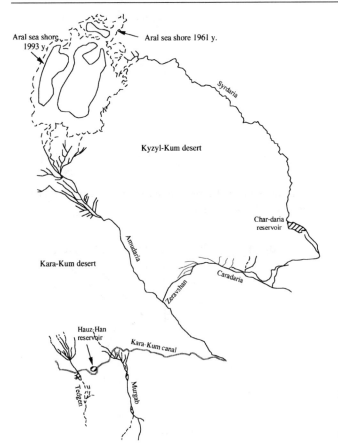

Fig. 1 The Aral sea drainage basin.

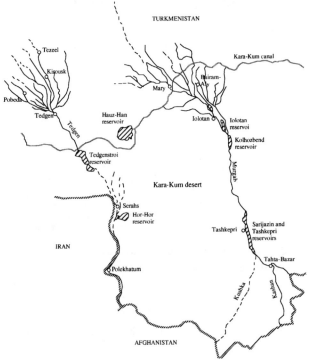

Fig. 2 The Tedgen and Murgab basins – the study area.

reflected by reduced soil fertility. About 57% of agricultural lands are damaged and forest phytocoenoses have declined by half during the last 50 years. The area of secondary, simplified ecosystems has enlarged along the irrigation and drainage canals.

We consider the riparian forests to be the best protector of a water body from pollutants. As it was shown in the research of Orazmuhammedov (Orazmuhammedov, 1990), the flow of pollutants and salt from the floodplain to the rivers is less when there is *a Tamarix* community on the bank, and is at a minimum when there is a *Poplar* community. It happens because these plants can absorb salt and pollutants from the soil. Besides, forests protect banks of the rivers from soil erosion and reduce evaporation.

MATERIAL AND METHODS

We studied the riparian ecosystems of the Tedgen and Murgab rivers according to the concept of aquatic-terrestrial ecotones (Naiman & Decamps, 1989). The study sites were located along a topographical gradient. We took soil samples in the different vegetation communities from the different soil layers and analysed salinity. Measures included: density,

colour, wetting, mechanical composition and inclusions. The depth of groundwaters was measured and their salinity was analysed. We revealed the dominants of vegetation communities and classified them according to these dominants and subdominants. About 300 descriptions of vegetation communities and 165 samples of soils and ground waters were analysed.

RESULTS AND DISCUSSION

We found four types of processes in the riparian ecosystems: salinization, desertification, pasturable degradation and secondary hydromorphysation. These processes are revealed according to the state of vegetation, soils and ground waters. The functioning of groundwater/surface water ecotones is different in the different cases of these processes. We revealed three stages of salinization in the ecosystems of the Tedgen and Murgab interfluent area.

Stage of deep salinization

This stage is typical for the ecosystems of high floodplains of the Tedgen and Murgab, around the lagoons on the first terraces as well as on the high banks of the Kara-Kum canal. Vegetation communities are Poplar-Reed (*Populus euphratica* and *Phragmites australis* as dominants);*Tamarix hispida, Tamarix pentandra* tugais. Soil salinization is deeper than

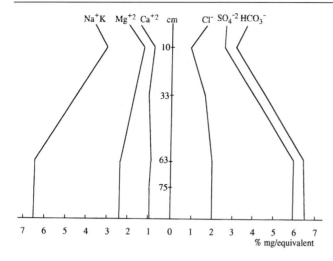

Fig. 3 Distribution of easily dissolved salts on the soil's profile – stage of deep salinization.

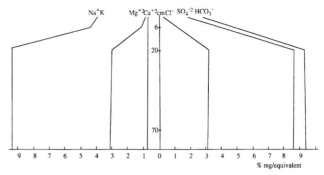

Fig. 4 Distribution of easily dissolved salts on the soil's profile – stage of middle salinization.

70 cm and not very high (Fig. 3). Groundwaters are salted and deeper than 2.5 meters. The ecotones act in this case as a buffer between salted groundwaters and fresh surface waters.

Stage of middle salinization

This stage is typical for ecosystems along the delta's branches and on the reservoir's coasts as well as in some filtration depressions of the reservoirs. This process is reflected in the diagnostic species of vegetation communities. They are communities of *Tamarix ramosissima* and of of Reed-Camel's thorn (*Alhagi-pseudalhagi*). In these associations salinization is not reflected in the dominant species. Soil salinization is high from the depth of 20–30 cm (Fig. 4). Groundwaters are at the depth of 1.5–2 meters and salted. The ecotones act in this case partly as buffer but partly as transmitter of salinization.

Stage of surface salinization

This stage is typical for 'islands' with *Tamarix* on the floodplains, for beach-ridges around reservoirs and in some filtration depressions of reservoirs. The salinization process is reflected in the composition of subdominant species as well as in composition of dominant species. Vegetation communities are Camel's thorn-Soliankas (*Salsola dendroides*), and Cereal-Soliankas communities (*Climacoptera lanata, Salsola dendroides, Suaeda altissima*). Soil salinization is high from the surface, or only on the surface to the depth of 70–80 cm (Fig. 5). Groundwaters are at the depth of 1–2 meters, semi-salted. The ecotones act in this case as transmitters of salinization.

We studied also the ecosystems bare of vegetation in the valleys of Tedgen and Murgab on the analogous relief sites

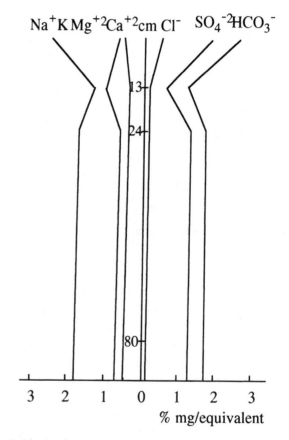

Fig. 5 Distribution of easily dissolved salts on the soil's profile – stage of surface salinization.

with the same groundwater level and found out that the salinization process on those sites is more intensive than on the sites with riparian vegetation. According to our research 51% of riparian ecosystems of the rivers Tedgen and Murgab have semisalted soils (0.25–1% salt content in absolutely dry soil's sample), 10% – heavy and very heavy salted soils (1–2% or > 2%) and 38% – nonsalted or little salted soils (<0.25%). Some 30% of all salted soils have sulphate type of salinity. Content of phenols and pesticides is 1.5–2 times higher than maximum advisable limits in the groundwaters of the valleys.

The process of desertification is described on the contact

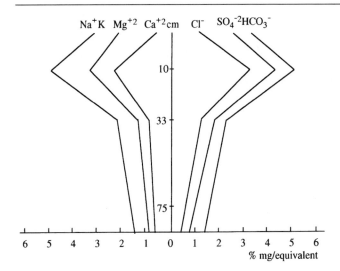

Fig. 6 Distribution of easily dissolved salts on the soil's profile – process of secondary hydromorphysation.

between the upper floodplain and surrounding desert of the rivers and on the upper floodplains of the river Murgab. Vegetation communities are Poplar-Reed – on the upper floodplain. Desertification is reflected here in degradation of the tree layer and abundance of ephemeral cereals (*Bromus dantonii, Anizantha tectorum, Hordeum leporinum, Poa bulbosa*). On the contact of floodplain and desert a community of *Tamarix ramosissima* with ephemeral cereals is described. The level of ground waters is 2.8–3.5 meters, floodings are very rare (once every 3–5 years), soils are non-salted, but dry. The ecotones here are indicators of the desertification process.

The process of pasturable degradation is widely spread in the Tedgen and Murgab region. This oasis is the center of pasturable cattle-breeding. This process is described in the Poplar-Reed communities with *Xanthium strumarium, Cynodon dactylon* and *Artemisetum scoparia*. Floristic composition of these communities is poor, ruderal species are abundant, trampling down is 30–60%. The trees grow dry and the young trees are weak or there are none. Among Camel's thorn-Reed communities we must mention associations with *Heliotropium supinum* and *Cuscutus campestris*, among Camel's thorn communities – association with *Alyssum desertorum*, that also suffer from overpasturing. The upper layers of the soil in these communities are condensed, sometimes salted and dry. But the lower layers are nonsalted, so are the groundwaters. In this case the ecotones act as a buffer for the pasture impact as they prevent soil erosion on the banks of the rivers.

The process of secondary hydromorphysation is the creation of lagoons and ponds on the bottom of dried reservoirs and filtration depressions along canals and reservoirs. These naturally occuring water bodies contain vegetation under succession in response to soils drying and the groundwaters lowering. These are cereal abundant invasions of *Tamarix ramosissima* or *Tamarix hispida* and Camel's thorn-Reed communities with *Heliotropium supinum* and *Cuscutus campestris* are the communities under secondary hydromorphysation. These communities are also under pressure of pasturage, so the processes are superimposed. The soils are nonsalted. In some cases the process of desalinization occurs here (Fig. 6). The level of groundwaters is 0.4–1 meters. They are nonsalted. In this case the ecotones act as a transmitter of the process. We can see that in some cases the ground-water/surface water ecotones act as buffers and in some cases as transmitters of the processes.

MANAGEMENT AND FUTURE WORK

Preliminary advice for management of the riparian ecosystems of the Tedgen and Murgab rivers is recommended. First, impose a ban on the cutting of the riparian forest and improve environment protection of the riparian ecosystems. Afforestation is necessary in places of intensive agriculture along the rivers. Second, reorient agriculture from the monocrop of cotton to cereal and vegetable production on the floodplains. Third, improve water management by collection of rainfall waters in small reservoirs and reconstruction of irrigation systems such as closed and faced canals, and more effective drainage. Fourth, reduce pasture in the riparian forests, organize the cycle of pasturing and soil protection measures. Organise nature reserves in the places of old, well preserved forests. Finally, monitoring of riparian ecosystems is necessary: 1) biodiversity of riparian phytocoenoses; 2) presense and amount of pollutants; 3) direction and character of matter flows through ecotones; 4) character of processes in ecotones and their role in the functioning of ecosystems. By means of well organized monitoring we can obtain the necessary data for working out a management strategy of safe and productive natural usage in a desert zone.

REFERENCES

Naiman, R.J. & Décamps, H. (1989). *Role of Aquatic-Terrestrial Ecotones in Landscape Management.* London.

Orazmuhammedov, A. (1990). Connection of vegetation cover with floodings and ground waters and with soil salinisation in the lower part of the Tedgen delta. *Isvestia of Turkmenia Academy of Sciences. Biological series*, **4**, 19–24.

Sadikov, G.S. & Veselov, V.V. (1993). *Water-ecological situation hanges in the basin of the Aral Sea under the influence of intensive agriculture.* Alma-Ata, in Russian.

Zaletaev, V.S. (1989). *Ecologically destabilized environment.* Nauka, Moscow, in Russian.

28 Water regime management of desertificated ecotone systems in the Amudarya delta (Aral Sea basin)

N.M. NOVIKOVA & I.N. ZABOLOTSKY

Water Problems Institute, Russian Academy of Sciences, 10 Novaya Basmannaya, P.O. Box 524, Moscow 107 078, Russia

ABSTRACT Regularities of the changes of ecosystems of the ecotones of stream and lake banks under the natural evolution of landscapes in the Amudarya delta are studied.

Regimes of flooding and groundwater, typical to each stage of development, are revealed. Investigation of the modern conditions for two types of ecotone ecosystems (wetland and terrestrial) enable us to determine the degree of their desertification. The main concept of water regime management is to protect the biological diversity by means of conservation of all the variants of land – water ecotone systems.

Based on the present-day state-of-the-art of lake and tugai ecosystems in the Amudarya river delta, recommendations on their watering are developed to guarantee optimum functioning or rehabilitation of the ecosystems.

INTRODUCTION

The Aral Sea is situated in the centre of Turan desert area in Central Asia. Its water balance depends on the Amudarya and Syrdarya rivers inflow. Historically, the Aral Sea region has been desertified on several occasions. Recent desertification is a human-induced phenomenon brought about mainly by overuse of river flow in the middle and upper reaches of the Amudarya and Syrdarya rivers for the needs of developing irrigation. That leads to a lack of available water resources in deltas. Decreasing the input of the river water in the Aral sea and in the Amudarya and Syrdarya river deltas reached its critical means in 1960, where the main reservoirs was built and the irrigated area in the Aral Sea basin became 1×10^6 ha. Since this time the Aral Sea level has dropped from the absolute elevation of 53 metres to 30 metres in 1991. The area of the sea has decreased from 68 000 to 37 000 km^3 and the volume is down from 1090 to 340 km^2 (Kust, 1992).

In the Amudarya delta in 1932–1960, the average water inflow was 47.3 km^3 $year^{-1}$ (Fig. 1). The open water surface area was 980 km^3. The area of permanently flooded reeds reached 3456 km^2 and seasonally flooded ones – 4186 km^2. Abrupt ecological changes began to manifest since the

spring of 1974, when the first facilities of the Takhiatash hydraulic structure came into operation. Water inflow to the delta declined in 1975–1980 to 15.0 km^3 $year^{-1}$. Mineralization of the Amudarya river water increased from 0.3–0.7 g l^{-1} to 1–1.4 g l^{-1} and in the channels, up to 2–4.4 g l^{-1}. More importantly, summer floods ceased completely. In some years (e.g. in 1980) the river flow was zero during the growing season. As a result, the groundwater table has dropped from 1–2 metres to 3–5 (10) metres; the area of delta lakes decreased to 214 km^2. The natural river net was changed by a system of irrigation channels (Fig. 2) and irrigation return flow began to play an increasingly important role in the formation of aquatic ecosystems and wetlands. Environmental changes in the Amudarya river delta lead directly to soil drying and salinization, degradation of the riparian (tugai) vegetation and land-water ecotones.

The purpose of our investigations was to study and describe the different ecological patches that have resulted from the changed hydrographs and to work on recommendations for manipulating flood regimes to maintain remnant features of the pristine flood plain and to rehabilitate very damaged habitats on the base the land-water ecotone systems.

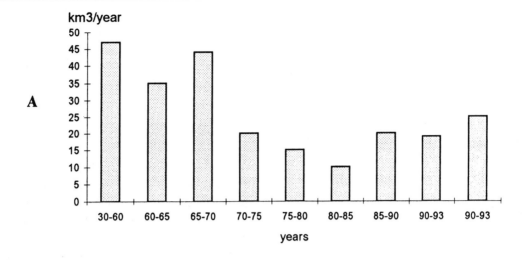

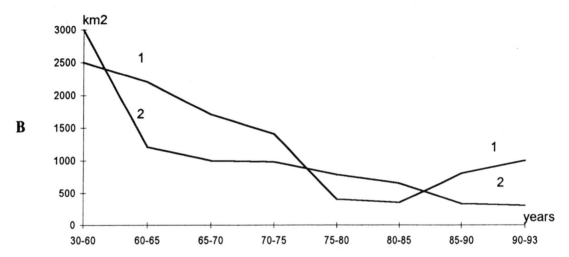

Fig. 1 Changes of the river runoff (A) and the area of hydromorphic ecosystems (B) in the Amudarya river delta (1930–1993). 1 – reeds area; 2 – riparian (tugai) forests area.

MAIN VARIANTS OF LAND-WATER ECOTONES

Ecosystems of delta plains are the parts of the spatial-temporal ecotone systems of the evaluated deltaic landscapes on the direction from water via hydromorphic to desert ecosystems (Table 1). In this paper we touch only the land-water ecotones of the landscapes scale and real time. Within the delta plain these ecotones are on the stream banks and lake shores due to rapid changing of the main ecological factors: duration of flooding, depth and mineralization of ground water, and degree of soils salinity.

Wetland Ecotone Complexes (WECs) compose the landscape of delta plains. By WECs we mean lakes covered with macrophytes, and coastal areas, their ecotones and food webs (Zabolotsky, 1992).

Field observations and the analysis of scientific literature allowed as to single out six main types of WECs (Table 2). Each is determined mainly by the combination of different lake types and serial stage related to frequency and duration of flooding (inundation). Plant species composition is the best characteristic of the WEC conditions at a certain moment of time and changes occurring at different stages of their development. The birds also reflects it. The biological resources are differently used by people at every stage of development.

The components of WECs are interdependent in relation to the straight of hydraulic connectivity in different parts of the delta. Thus, the preservation of valuable biotic components (muskrats, fish, reed communities), as well as the bioresource potential of the delta as a whole is possible only on the basis of the preservation of all six types of the WEC. Here, the controlling factor is their hydrological regimes. In the conditions of the natural evolution of the deltaic landscape all of these six types of ecotone systems existed in the same time in different parts of the delta plain.

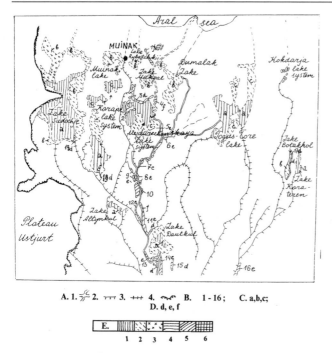

Fig. 2 Contemporary hydrographic scheme of the Amudarya river
delta and recommended measures for water regime management
of ecosystems.
A. 1 – Main river body of Amudarya; 2 – river branches used as
channels; 3 – drainage channels; 4 – borders of plateau Ustjurt.
B. Tugai forests: 1 – Muinak, 2 – Shege, 3 – Zakirkol, 4 – Zair, 5 –
Voroshilov, 6 – Kazakhdarya, 7 – Aspantaj, 8 – Bozatau, 9 –
Kyzyldjar, 10 – Dzerjinskii, 11 – Erkindarja, 12 – Akbashly, 13 –
Dautkul, 14 – Sovkhoz 'Russia', 15 – Chortambaj, 16 –
Nurumtubek, 17 – Botakkol, 18 – Raushan, 19 – Kungrad.
C. Lake ecosystems watering: a – permanently all the year, b –
periodically for a long time, c – periodically for a short time.
D. Situation in the tugais: d – good conserving, e – desertified, f –
degraded.
E. Recommended water regime:
for a lakes: 1 – conserve modern water regime, 2 – stable watering,
3 – stop watering; for a tugai forests: 4 – conserve modern regime,
5 – flooding in autumn or in spring time, 6 – stop flooding.

According to the character of function and pattern peculiarities, we determined several, relatively stable, discrete ecosystems, which compose land-water ecotones on the stream banks (levees). We regard them as the main variants of terrestrial delta ecosystems: riparian forests, meadows, bushes, solonchaks and edaphic zonal ecosystems (i.e. relatively stable ecosystems of intrazonal dwelling places, formed on sand and clayey alluvial deposits and corresponding to the modern zonal climate conditions; Table 1).

At the first stage of water supply of terrestrial ecosystems formation is mainly due to floods. During the low-flow period, near-surface groundwaters serve as additional sources of water. They are constantly replenished due to water seepage from the river channel. The ecotone systems in this time include reeds and young forests; peatbog process dominates in soils. There is little species diversity. At the second stage the low-flow period is more durable and the groundwater table is at the depth 1.5–3 m. Biocomplexes of meadows, tugai forests, bushes form ecotone systems. This stage is characterized by maximum species diversity of biota and plant communities, maximum biological activity of soils and diversified total bio-productivity.

When floods cease on some delta areas, the water supply falls sharply, and the relations between ecosystems and river flow is realised mainly via ground waters, whose mineralization gradually increases. As shallow groundwater evaporates, the salt accumulation processes develop and meadow solonchaks and galophyllows bushes appear, which could be later substituted by typical solonchaks, with a lack of vascular plants.

After cessation of floods, the groundwater level may drop below 10 m. There is the transition from the hydromorphic stage of ecosystem development to the automorphic one. In this case, the ecosystems are supplied by water only by precipitation. Soil desertification begins and phreatophytes are replaced by ombrophytes. Humus mineralization and soil depletion decrease the yield capacity of plant communities by up to 700 times. Zonal desert ecosystems appear instead of azonal riparian ones.

In the conditions of the natural evolution of the deltaic landscape all of these types of ecotone systems on the levees and its slopes existed in the same time in different parts of the delta plain.

MAN-INDUCED CHANGES

Desertification of the environment in the delta of Amudarya not only decreased the area of wetland and tugai ecosystems. It turned the ecotone systems to desertification. Drainage waters began to play an increasingly important role in the formation of aquatic ecosystems. Increasing high salinity has a negative effect on the coastal vegetation and ichthiofauna, sharply reducing their habitats and productivity. High reeds were replaced short ones and lying forms. In desiccated lakes, takyrs and solonchaks appeared. A sharp decrease in the phytomass growth is observed in meadows, and this has an economic significance as natural pastures and hay meadows declined. In the Amudarya river delta saline types of WEC (4 and 5, Table 2) prevail now. Tugai – bush communities were transformed irreversibly into solonchak deserts during 10–15 years (Novikova, 1985). Vast areas of tugai massifs are preserved due to the fact that the main tree species are tolerant of soil salinization and desiccation. They can tolerate 16% salinization in the soil profile (according to some authors – up to 36%) and groundwater mineralization of 10–16 g l⁻¹ (Novikova, 1992).

Table 1. *Ecosystems changes under natural evolution of the deltaic landscape of the Amudarya river*

Stage	Stage of development	Conditions of humidification	Ecosystems on		Types of utilization
			levees	interfluve depressions	
B O G	I	Long surface flooding. Bog regim.	Reed flood plain swamps – 'Plavni', productivity – 40 tons ha^{-1} domination of hydrophytes		Fishery Musk-rat Breeding
		Annual flooding, unconfined ground water – 0.5–3 m	Tugai landscapes with tugai forests, bushes, meadows on alluvial-meadow soils. Productivity 1.0–2.0 tons ha^{-1}	Reed flood plainswamps and reed – forb meadows on meadow-bog, bog or peat bog soils. Productivity 2.0–2.5 tons ha^{-1}	Fishery Haymowing Grazing Musk-rat Breeding
H Y D R O M O R P H I C	II			Forb – cereal meadows on bog, meadow bog drying – up soils. Productivity 0.5–0.8 tons ha^{-1}	
	III	Short and non-regular flooding, drying up of upper soil horizon in the second half of summer	Drying up tugai forest and meadow ecosystems and replacing by bushs with ephemers on alluvial – meadow drying up soils	Galophytic meadows on meadow-solonchak soils. Productivity 1.0–1.8 tons ha^{-1} Solonchaks without plants	'Liman irrigation' for haymowing, Grazing Irrigated agriculture
		Cessation of floods unconfined ground water > 5–6 m		Soil desalinization, residual solonchaks, takyrization, pause in the vascular plants overgrow	
A U T O M O R P H I C	IV	Cessation of river flow. Humidification by precipitations. The depth of unconfined groundwaters are more than 10 m	Domination of *Haloxylon aphyllum, Salsola orientalis* on takyr-like soils Productivity 0.6 tons ha^{-1}	Takyrs, laking of vascular plants Aeolian input of sands from the near sandy deserts, overgrowing by psammophytes Productivity 0.4 tons ha^{-1}	Winter pastures

The field investigation of the state of terrestrial ecosystems on the levees and its slopes allowed us to single out three categories of ecotone ecosystems in the Amudarya delta: degraded, desertificated and rather well-preserved ones. Degraded tugais conserve trees and bush-edificators, herbs are represented by mesohalophytes, such as *Aeluropus littoralis, Sphaerophys salsola*. They are located along the main stream of the Amudarya – Akdarya. Desertification affected the youngest tugais, which formed in the protruding delta of Amudarya in the 1950s along its three main arms. There are

Table 2. *Main types of wetland ecotone ecosystems in natural environment*

N type	Lake characteristics	Aquatic vegetation	Series of coastal ecotone ecosystems	Birds	Economic use
1	Permanently flowing transparent water, Depth-up 6–8 m Transparency-up to 4–5 m	High degree of overgrowing with *Potamogeton* sp. *Chara* sp.	Hydrosere *Typha* sp. → *Phragmites australis*	*Netta rufina* *Aythya nyroca* *Phalacrocorax carbo* *Pelecanus crispus*	Fishery Hunting Grazing Haymowing
2	Permanently flowing turbid water Depth-up 2–4 m Transparency-up to 0.2–0.4 m	Almost total abscence of submerged aquatic plants, area of the *Phragmites australis*	Hydrosere *Typha* sp. → *Phragmites australis* Vast massifs of half-submerged reedplantations	*Larus argentatus* *Egretta garzetta* *Ardea cinerea* Anatinae	Fishery Grazing Haymowing
3	Floodplain lakes: a) permanently connected with main delta branches by fals channels; b) lakes fed by groundwaters during the low-flow period	Rich and diverse submerged plants: *Potamogeton* sp., *Chara* sp.	Xerosere *Phragmites australis* → grass meadows → bushs → takyr	*Anas clypeata* *Anas platyrhynchos* *Netta rufina* *Sterna hirundo* Charadiiformes	Haymowing Grazing
4	a) Lake systems becoming flowing during flood. During dry periods water goes away. b) Certain floodplain lakes, completely or appreciably drying in the low-flow period. Depth 1–2 m, in dry period – 1 m	Diversified aquatic plants dying during dry season *Potamogeton cristatum*, *Chara* sp.	Halosere weakly developed reed → *Halostachys caspica* → solonchaks	*Hymanthopus hymanthopus* *Charadrius alexandrinus* *Charadrius dubius*	Grazing
5	Salinized lakes with groundwater feeding; Depth 1–3 m	Vegetation depends on salinity, *Zannichelia* sp.	Halosere *Salicornia herbacea*, → *Tripolium aster* *Salsolas* sp., *Suaedas* sp. → solonchaks	*Tardona tardona* *Hymanthopus hymanthopus* *Charadrius alexandrinus*	
6	Floodland lakes with stable water feeding all the year	Diversified, but not reach submerged vegetation	Xerosere Reed → grass meadows → tugai forests and bushes	*Phasanius colchicus* Anatinae *Ardea*	Haymowing Grazing Hunting

now only dead trees of *Populus ariana, Elaeagnus turcomanica*. Tugai herbs replaced by ephemeres (*Descurainia sophia, Senecio subdentatus*) and annual salsolas (*Cimacoptera lanata, C. aralensis, Salsola paulsenii*).

Rather well-preserved ecotone systems are only in some places along the main stream in the head of delta.

RECOMMENDED MEASURES

The measures, we recommend, may be ordered in 5 groups:

(a) measures to protect and to improve the conditions of ecosystems, without changes in the hydrological regime;

(b) measures aimed at preserving existing hydrological regimes;

(c) measures to modify water supply or water quality;

(d) measures to restore lacustrine, wetlands and terrestrial ecosystems;

(e) measures to water rate and monitor results of all (a-d) measures.

Results of a remote sensing survey have shown that considerable areas of the lakes Sudochie, Touguz-Tore, Karadjar and Mezhdourechenskie lake systems and a water body in the Muinak Bay, etc. (Fig.2) are now permanently flooded. These areas are a valuable genetic resource for flora and fauna, typical of wetland ecosystems. Rare, endemic and valuable (from the economic point of view) types and communities are preserved here. They can spread from these sites to restored or newly-formed water bodies. Clearly, these areas should be preserved in the regime of permanent flooding.

Seasonal water level rises in lakes and inundation of land plots, occupied by coastal vegetation, are ecologically important. The depth of flooding should not exceed 100 cm depth, because flooding between 10 to 100 cm depth caused riparian grasses to grow intensively, which ensures favourable conditions for spawning and shelter against predators. The recommended period of flood inundation is April–May.

However it is impossible to preserve the entire diversity of WEC ecosystems, including relict and endemic types, without implementing traditional measures such as organization of reserves. That is why we suggest organizing a network of reserves of WEC. Up to 1991, there was no reserve on these lands. Using the species composition of a certain WEC and taking into account the presence of rare and disappearing species (including those throughout the World and locally in the Central Asian republics) as a criterion of its value, and bearing in mind the ability of this natural complex to resist successfully negative environmental changes, we can single out two ecosystems for organising reserves:

– the Sudochie and Kungrad lake systems with WEC – on the left bank of the Amudaria;
– the Tougus-Tore lake and Akpetki sand-solonchak massif – on its right bank.

Watering of tugai complexes is necessary for their preservation. It will be expedient to restore the hydrological regime, close to the natural one, on certain areas of the desertified tugai, located along the Amudarya main channel (the reach between the Kazakhdaria inflow and the Takhiatash

dam). This can be done using a cascade of special dams to increase the water level in the river. This will promote the rise of groundwater, intensification of the alluvial process and the improvement of water supply of coastal tugais. But, in this case it will be necessary to investigate closely variants of possible salinization of intra-channel depression slopes, because groundwaters seeping from the river channel will evaporate. It will be also necessary to determine the optimum height of water level rise in the river.

Outside the delta, in the Amudarya floodplain and upstream at the Takhiatash dams, tugai communities are in better condition. Seed deposition and germination is promoted by active formation of islands, and tugai communities proliferate. The optimum regime of watering mature tugais is a period of up to 20 days in spring and late in summer (July–August). The depth of flooding should be 50–100 cm. The groundwaters should lift up 1.5–3 m.

Moreover the creation of a water conservation zone along the whole Amudarya channel downstream to the Tyuyamuyn dam is needed to restore and protect the river bank by development of riparian forests. The forest belts stretching along the river channel should be 1.5–3 km wide. Land ploughing, cattle grazing, forest cutting, any construction works and waste water discharge into the river should be prohibited within this belt.

In order to preserve the genetic and cenotic fund of tugai ecosystems on the territory of well-watered tugais, it is necessary to decrease the direct anthropogenic load on them by restoring reserve regimes on the Nurumtubek and Nazarkhan tugais. In addition new reserves should be created to protect young tugais in the mouth of the Raushan Canal and on the Kokdarya floodplains (Fig. 2).

ACKNOWLEDGEMENTS

This research was supported by UNESCO and BMF of Germany under project 509/RAS/40 'Aral Sea'.

REFERENCES

Kust, G. (1992). Desertification assessment and mapping in the pre-Aral region. *Desertification Control Bulletin*, **21**, 38–44 (In English).
Novikova, N. (1985). The plant cover dynamics at the delta landscapes caused by the river's flow transformation. In *Biogeographical aspects of desertification*, pp. 31–40, Moscow, MFGO (In Russian).
Novikova, N. (1992). La dégradation de la végétation dans l'actuel delta de l'Amoudarya. *Sécheresse*, **3**, 155–167 (In French).
Zabolotsky, I. (1992). Types of Wetland Complexes in the Ily river delta and their dynamics. *Geography and natural resources*, **4**, 78–86 (In Russian).

V

Conclusion

29 Problems and challenges in groundwater/surface water ecotone analysis

J. GIBERT* & F. FOURNIER**

* Université Lyon 1, URA CNRS 1974, Ecologie des Eaux Douces et des Grands Fleuves, Hydrobiologie et Ecologie Souterraines, 43 Bd du 11 novembre 1918, 69622 Villeurbanne cedex, France

** UNESCO Division des Sciences Ecologiques, 7 Place de Fontenoy, 75700 Paris, France

ABSTRACT   Groundwater/surface water ecology and hydrology are relatively new areas of study that are growing rapidly. Significant points and issues identified in this book relate to a diverse array of groundwater/surface water ecotone characteristics. They include bidirectional coupled fluxes from surface and groundwater environments, high dynamicity, high heterogeneity, low predictability, biodiversity 'hot spots', etc. These points highlight the principal challenges facing biologists and hydrologists and managers both now and in the future. In our rapidly changing world information needs are multiple and complex, while ecotones are becoming increasingly important in the regulation of ecosystem and landscape processes.

INTRODUCTION

Research on groundwater/surface water ecotones has increased our understanding of the structure and functioning of stream, lake and groundwater ecosystems through a broader spatial perspective that takes into account the entire drainage network, and recognises that processes occurring in the surface and groundwater environments are influenced by the riparian and hyporeic zones (Amoros & Petts, 1993; Gibert *et al.*, 1994). These zones consist of environmental and metabolic gradients of different micro- to macroscales and thus can be seen in terms of Landscape Ecology (Holland *et al.*, 1991; Hansen & di Castri, 1992).

This book evaluates the functioning and the role of groundwater/surface water ecotones in functional landscapes with particular reference to processes and to the implications for managing biological diversity and ecological flows. It attempts to identify a set of fundamental principles that could provide a sufficient basis for the understanding of complex transition zones and for the development of a comprehensive body of scientific knowledge for the management of the different sources of water in an overall strategic plan.

Our objective is to identify and evaluate central themes emerging from this book. We first evaluate research on the groundwater/surface water ecotone: synthesis and problems. We consider implications for management and human investment and then present the progress to be expected, as well as future needs and challenges. Finally, we provide some recommendations.

EVALUATION OF RESEARCH ON THE GROUNDWATER/SURFACE WATER ECOTONE: SYNTHESIS AND PROBLEMS

This book presents studies that lead to a better understanding of the observed phenomena and the scientific bases of the study of groundwater/surface water ecotones. Some papers concern models of prediction of water resources. A few papers seek solutions to the problems of water use and finally, a very few are based on new methodologies.

We now know that these ecotones are water flow regulators, permanent or temporary sinks for organic (especially for nitrogen and carbon cycles) and mineral matter and contaminants from watersheds, filter and buffer systems that protect groundwater quality and improve surface water quality. However the relationships between nutrient and sediment retention efficiency, assimilation capacity and buffering capacity remain largely unknown. What are the thresholds of these capacities? What are the resistance and the resilience of these systems?

The importance of biological processes has also been underestimated. In the ecotone there is a diverse interface community (mainly microbial and faunal) that alters the timing and magnitude of nutrient and contaminant flows. It would be useful not only to study the community structures but also the capacity of organisms to integrate environmental factors with time. How is the structure of the ecotone community regulated? What are the links between hydraulics and biodiversity? What are the roles of ecotonal fauna and biofilm at the scale of the drainage basin? Some hypotheses should be put forward concerning flow circulations, kinds of exchanges and water table fluctuations.

Ecotones are viewed as heterogeneous and patchy features in the development of Landscape Ecology. Introducing patchiness to homogeneous hydrological systems could be a helpful approach to an understanding of the function of the ecotone. At the same time the behavior of functional landscapes emerges from the interactions of patches and ecotones.

Explicit consideration of scale is fundamental in the study and understanding of ecotones. However, ecologists and hydrologists tend to work at different scales. Most ecological studies have been carried out at microscale (microhabitat) and mesoscale (local conditions or site) whereas much hydrological information is available at regional (landscape) and global scales. Moreover it is important to take into account the time scale in the hydrogeologic and biogeochemical processes. Groundwater flow can take a thousand years whilst rainfall processes can change in a few minutes, and so it is necessary to be able to transpose information between different spatio-temporal scales. Multiple-scale studies are certainly essential to determine which ecotone properties are scale-dependent and which ones are scale-invariant (Hansen & di Castri, 1992; Malanson, 1993). Matching scales in hydrology and ecology are recommended.

The problem of interfacing the domains of hydrology and ecology is crucial. Hydrologists and ecologists do not work together often enough, although considerable mutual benefit would arise from such cooperation. Each could influence not only the methods and materials but also the conceptual framework used by the other. Each discipline has much to learn from the other and ought to cooperate much more (Osborne *et al.*, 1993).

IMPLICATIONS FOR MANAGEMENT AND HUMAN INVESTMENT

Expansion of human development, industrialisation and exploitation of water resources all pose a serious threat to the protection and conservation of aquatic surface and subsurface ecosystems and groundwater resources as a sustainable source of drinking water. Groundwater/surface water ecotone management is developing in the context of land/water ecotone management, but it must be viewed in a wider perspective that crosses boundaries into surface water and terrestrial ecosystems. At the SEATTLE conference (February 1994) the proposed approach to be adopted by a new generation of Ecotone projects under the title '*Management of Ecotones for sustainability of social and Environmental needs*' was the transfer of information so it could be put into managerial practice, in view of sustainable development by integrating human and environmental needs within the framework of catchment areas and at the landscape level. Development of a strategic plan of action across groundwater/surface water ecotones has been delayed because of the lack of scientific information available to managers and in particular of predictive models linking hydrological and biological information in both surface and groundwater environments. Another problem is the lack of political will and lack of funding, so that a comprehensive strategy that could provide guidelines and ideas on how scientists can work better to influence management (see Round Table 5: Simons & Notenboom, this volume) is not yet in existence. The transfer of information to the decision makers should be of much higher priority.

Different methods and tools exist. However an integrated approach to water protection considering the entire hydrological cycle, surface water, land use, groundwater management, etc . . . has to be promoted. The development of better methods of ecological investigation into matters of environmental management is long overdue. Indubitably better analyses of the whole procedure will emerge (Nachtnebel *et al.*, 1993). Finally it is important to consider ecological research as a primary component of environmental decision-making rather than an increasingly peripheral and marginal procedure.

NEEDS FOR THE THE FUTURE AND CHALLENGES

An important question arises when the environment of the future is considered: are we ready to effectively accommodate the enormous changes in population, consumption, environment, social organisation and technology that will take place early in the next century?

It has to be realised that the population will increase from 5.5 billion to 10–11 billion by 2050, that 90% of the world population is concentrated in the developing countries, that more than 85% of the world population resides in urban settings, that there is an increase in migration rates. The environment has been subjected to the greatest rate of change for milleniums. More than 75% of the water is under control.

There is real biotic impoverishment. Social organisation shows institutional change and a new distribution of wealth and power.

Better information may lead to better knowledge, wisdom and management but technology alone cannot resolve all the issues and there is a need for interdisciplinary approaches and partnerships.

How do the above issues relate to the subject of the present publication? There is certainly a strong relation as water is a strategic resource because more than 60% of the population lives less than 1 km from a source of water. Water consumption rates are increasing rapidly and there is a need for water system restoration and water purification.

Successful approaches to the study of groundwater/surface water ecotone structure, functioning and management should require the following.

1   New data should be obtained because the diversity of ecotones is very high and we have not identified a function restricted number of ecotones. There is a lack of 'hard data' in a field towards which ecologists are only belatedly turning.

2   There be more cooperation and collaboration between hydrologists and ecologists and also human sciences.

3   The study of groundwater/surface water ecotones should integrate theory and predictive capacity. The challenge is to understand the structure and dynamics of communities and ecosystems in order to predict their futures in a changing world. It is to promote an understanding of evolution, both in and of the ecotone.

4   There should be better approches to modelling not only concerning hydraulics, heat flow and biogeochemistry but also community dynamics and physical and biological couplings.

5   New techniques, tools and methodologies should be developed. For example, we should investigate how appropriate methodologies for up/downscaling procedures are. At present there are indications that principles derived from information theory, fractals, network theory, complexity, catastrophe and chaos theory or new thermodynamic approaches to far from equilibrium systems could all contribute to a better understanding of the properties of the ecotone.

6   The ecotone should not be reduced to component mechanisms without losing sight of the essence of the system, that is its interconnectedness. An analysis of the system into its component parts cannot explain the multiscale reality of the whole. We have to move towards a science of complex ecology in a holistic framework for the next century. We have to acknowledge complexity as fundamental and essential in biological existence and make it an absolute focus of investigation.

7   There should be increased effort to build ecological research programs that should have the following fundamental objectives: to dictate possibilities for management, to prevent confusion and inappropriate interpretation of information, to investigate alternatives when the options used for management have failed and to evaluate the processes of management.

Finally, we could recommend the following.

– The promotion and undertaking of scientific intiatives to ensure the protection of the global aquatic environments from toxic contamination and nutrient enrichment by focusing on the interface zones.
– Favouring holistic and experimental methods, develpments and approaches.
– Coalescing and disseminating environmental information through public awareness programs and education.
– The encouragement of international, interdisciplinary and cross-sectional communication concerning ecosystem health and recovery.

CONCLUSION

Groundwater/surface water ecotones are important landscape features because of their physical and biological characteristics, their unique value and their spatial and temporal characteristics. As R.G. Wetzel (1992) wrote 'Interface limnology will certainly dominate many future freshwater studies and will be mandatory for accurate biomanipulation in many aquatic ecosystems'. The past years have been a period of conceptual development and data acquisition in the study of groundwater/surface water ecotones. Our future task is to test these ideas in a wider range of transition zones, to fill the gaps by developing a comprehensive view of how the ecotone system operates, to compare data obtained by ecologists and hydrologists, to develop models for prediction and also reinforce ecotone research in the landscape context.

ACKNOWLEDGEMENTS

Many of these ideas and conclusions emerged from papers and round-table discussions at the International Conference on Groundwater/Surface Water Ecotones, in Lyon, in July 1993. We thank all participants and contributors at this conference. Financial support for the preparation of this book was provided by the Centre National de la Recherche Scientifique (CNRS, France), the Environment Programme of the CNRS, the UNESCO Man and The Biosphere Programme.

REFERENCES

Amoros, C.& Petts, G.E. (ed.) (1993). *Hydrosystèmes fluviaux*. Paris: Masson.

Gibert, J., Danielopol, D. & Stanford, J.A. (ed.) (1994). *Groundwater Ecology*. San Diego, USA: Academic Press.

Hansen, A.J.& di Castri, F. (ed.) (1992). *Landscape boundaries: consequences for biotic diversity and ecological flows*. New York: Springer-Verlag.

Holland, M.M., Risser, P.G. & Naiman, R.J. (ed.) (1991). *Ecotones: the role of landsape boundaries in the Management and Restoration of Changing Environments*. New York-London: Chapman and Hall.

Malanson, G.P. (ed.) (1993). *Riparian landscapes*. Cambridge University Press.

Nachtnebel, H.P., Kovar, K. & Zuidema, Z. (1993). *Hydrological basis of ecologically sound management of soil and groundwater*. Report of the UNESCO-ICGW Working Group to IHP Project M-3-1, UNESCO 71p.

Osborne, L.L., Bayley, P.B. & Higler, L.W. (1993). Lowland stream restoration: theory and practice. *Freshwater Biology, special issue*, 187–341.

Wetzel, R.G. (1992). Concluding remarks I. Limnology now in the future. *Hydrobiologia*, **243/244**, 481–485.

Annex

Round Table 1 Riparian vegetation and water quality improvement

M. TRÉMOLIÈRES*, D. CORRELL** & J. OLAH***

* *Laboratoire de Botanique et Ecologie Végétale, CEREG URA 95 CNRS, Institut de Botanique, 28 rue Goethe, F-67083 Strasbourg cedex, France*

** *Smithsonian Environmental Reasearch Center, P.O. Box 28, Edgewater, Maryland 21037, USA*

*** *Fisheries Research Institute, H-5541 Szarvas, Hungary*

The overall goal of this research on riparian zones is to understand how these habitats function and what factors control their functions. The water quality factors of greatest interest are nutrients and toxic materials. The original publications which reported on these buffering functions of riparian forest were Gilliam *et al.* (1974), and Gambrell *et al.* (1975). These were followed by a series of more focused reports (Lowrance *et al.*, 1984a, 1984b; Peterjohn & Correll, 1984; 1986; Labroue & Pinay, 1986; Pinay & Labroue 1986; Schnabel, 1986; Pinay & Décamps, 1988; Correll & Weller, 1989; Sanchez-Perez *et al.*, 1991, 1993).

Comparative data are needed to understand the role of plant communities in different riparian systems and interaction processes between compartments (Fig. 1). Such questions need to be addressed to ascertain the relative effectiveness of grass, herbs and trees; the width of vegetation needed for effectiveness; the importance of primary production, plant diversity, age structure, depth of root/rhizosphere zones; and how the surface plant community interacts with below-ground microbial communities. These systems should be characterized in respect to their hydrology, geomorphology, biogeochemistry, below-ground conditions, and hyporheic zones. More similarity needs to be developed in approaches and methods in order to produce more comparable data in the future.

GENERAL QUESTIONS FOR FUTURE RESEARCH

a) What are the capacities of these systems for processing nutrients and toxins?
b) Are these systems self-sustaining? For how long?
c) What are the principal mechanisms of water quality affects and what controls their rates?
d) What is the importance of biodiversity in controlling the efficiency of riparian zone processing of man-made chemicals?

More should be known about the impacts of exotic species and new genetic varieties of plants that are appearing in the riparian zones. For example, some produce organic compounds resistant to biological breakdown or toxic to biota. It is very important to examine plant diversity and the differing roles of the individual species in affecting water quality (Fig. 2). Communities with similar diversity but different species sometimes function differently.

The riparian zone needs a better definition. It is the zone of land adjacent to water where there is interaction with groundwater and the soil is waterlogged at least during some months of the year. There is also a transition zone between the uplands and the riparian zone. There might be 20–40 different kinds of riparian systems. There may be no universal classification system, but only specific to a certain region. The descriptions should be basic such as hydrological or geomorphological. The classification system should rely heavily on below-ground conditions and processes.

A long-term perspective is needed when investigating below-ground conditions and responses of riparian systems to manipulations. It may take 80–100 years for below-ground organic matter and biomass to develop or disappear when the conditions of a riparian system have been changed.

The US Forest Service recently published a land manager-oriented booklet with recommendations for riparian forest management in agricultural drainage basins (Welsh, 1991). Among these recommendations was a stream bank buffer of 10 m of undisturbed forest, a second zone of managed forest 10–30 m from the stream, and a third zone 10 m wide of grass buffer for sediment trapping. This is an example of how man-

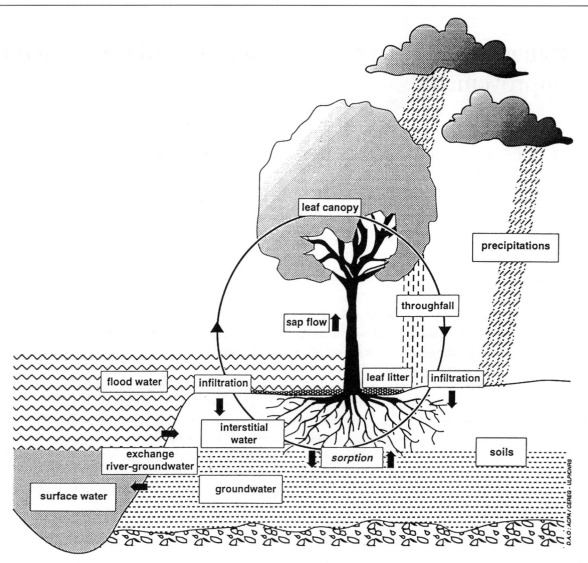

Fig. 1 Compartments and interaction processes in an alluvial forest ecosystem.

agement guidelines are becoming more sophisticated, without being too complex.

We do not adequately understand basic hydrological effects of riparian vegetation or how things like a fluctuating water table affect the vegetation's capacity to remove nutrients and toxins. In order to accomplish this research, coordinated interdisciplinary teams need to be created, in which, for example, ecologists can efficiently interact with hydrologists.

The effect of disturbances should be predicted on basic interactions and metazoan populations in these riparian zones. Techniques such as GIS should be used to more quantitatively describe them and to allow a better use of statistical and mathematical approaches to research.

The most impacted riparian zones are in densely populated countries. More concern should be given to improving their management. There is a crisis in the ability to provide

good drinking water in many parts of the world. Can riparian zones help to improve water quality? UNESCO can act as a catalyst to focus riparian research on key issues and move this field forward more rapidly. Before any effective recommendations can be made for the restoration and improved management of these riparian areas, a better mechanistic understanding of their functioning is needed. Some 60–70% of the world's population live near water in or adjacent to riparian zones.

RECOMMENDATIONS FOR UNESCO

UNESCO/IHP should sponsor a small workshop with the goal of constructing a better framework for comparative riparian vegetation zone studies. This workshop could begin the process of defining parameters that should be measured in riparian studies to increase their comparability. These would include both parameters that describe the type of riparian system and parameters that describe

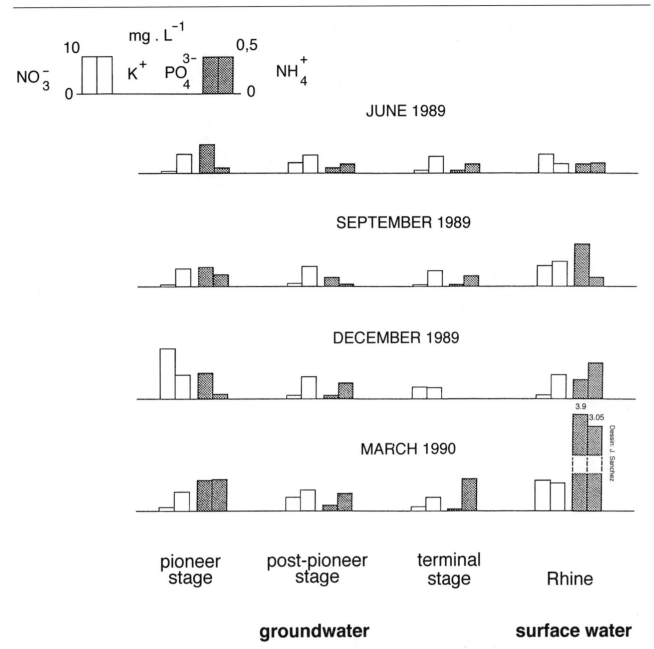

Fig. 2 Variations of nutrient concentrations in groundwater under different stages of the alluvial forest succession (case of the Rhine floodplain in Alsace, France) (from Sanchez-Perez *et al.*, 1991).

the water quality functions of the riparian systems under study.

A synthesis of existing studies would be valuable for management. Managers need to make decisions now and are often unaware of existing studies and their results. A small group of experts should be requested to compile a comprehensive review of the published literature in non-technical language. UNESCO should publish this literature synthesis.

REFERENCES

Correll, D.L. & Weller, D.E. (1989). Factors limiting processes in fresh water wetlands; an agricultural primary stream riparian forest. In *Fresh Water Wetlands and Wildlife* ed. R. R. Sharitz & J. W. Gibbons, pp. 9–23, Aiken, SC, USA: Savannah River Ecology Laboratory.

Gambrell, R.P., Gilliam, J.W. & Weed, S.B. (1975). Nitrogen losses from soils of the North Carolina coastal plain. *J. Environ. Qual.*, **4**, 317–323.

Gilliam, J.W., Daniels, R.B. & Lutz, J.F. (1974). Nitrogen content of shall ground water in North Carolina coastal plain. *J. Environ. Qual.*, **3**, 147–151.

Labroue, L. & Pinay, G. (1986). Epuration naturelle des nitrates des eaux souterraines: possibilités d'application au réaménagement des lacs de gravières. *Annals Limnol.*, **22**, 83–88.

Lowrance, R. R., Todd, R. L., Fail, Jr., Hendrickson, Jr.O., Leonard, R. L. & Asmussen, L. (1984a). Riparian forests as nutrient filters in agricultural watersheds. *Bioscience*, **34**, 374–377.

Lowrance, R. R., Todd, R. L. & Asmussen, L. E. (1984b). Nutrient cycling in an agricultural watershed. I. phreatic movement. *J. Environ. Qual.*, **13**, 22–27.

Peterjohn, W.T. & Correll, D.L. (1984). Nutrient dynamics in an agricultural watershed: observations of the role of a riparian forest. *Ecology*, **65**, 1466–1475.

Peterjohn, W.T. & Correll, D.L. (1986). The effect of riparian forest on the volume and chemical composition of base flow in an agricultural watershed. In *Watershed Research Perspectives*, ed. D.L. Correll, pp. 244–262. Washington: Smithsonian Institute Press.

Pinay, G. & Décamps, H. (1988). The role of riparian woods in regulating nitrogen fluxes between the alluvial aquifer and surface water: a conceptual model. *Regulated Rivers: Research & Management*, **2**, 507–516.

Pinay, G. & Labroue, L. (1986). Une station d'épuration naturelle en nitrates transportés par les nappes alluviales, l'aulnaie glutineuse. *C.R. Acad. Sci. Paris*, **302**, 629–632.

Sanchez-Perez, J.M., Trémolières, M., Schnitzler, A. & Carbiener, R. (1991). Evolution de la qualité physico-chimique des eaux de la frange superficielle de la nappe phréatique en fonction du cycle saisonnier et des stades de succession des forêts alluviales rhénanes (Querco-Ulmetum). *Acta Oecol.*, **12**, 581–601.

Sanchez-Perez, J. M., Trémolières, M., Schnitzler, A. & Carbiener, R. (1993). Nutrient content in alluvial soils submitted to flooding in the Rhine alluvial deciduous forest. *Acta Oecol.*, **14**, 371–387.

Schnabel, R.R. (1986). Nitrate concentrations in a small stream as affected by chemical and hydrologic interactions in the riparian zone. In *Watershed Research Perspectives*, ed. D.L. Correll, pp. 263–282. Washington: Smithsonian Institute Press.

Welsh, D. J. (1991). *Riparian forest buffers, function and design for protection and enhancement of water resources*. United States Dept. of Agriculture Report NA-PR-07–91, Radnor, PA, USA.

Round Table 2 Biodiversity in groundwater/surface water ecotones: central questions

P. MARMONIER*, J.V. WARD** & D.L. DANIELOPOL***

* *Université de Savoie, GRETI, 73376 Le Bourget du Lac, France*

** *Department of Biology, Colorado State University, Fort Collins, Colorado 80523, USA*

*** *Limnological Institute, Austrian Academy of Sciences, A-5310 Mondsee, Austria*

Biodiversity is the object of a large international programme of the IUBS-SCOPE-UNESCO and is important for many scientific, economic and ethic reasons (Solbrig, 1992). In groundwater, the study of biodiversity can be considered as a promising research field (Marmonier *et al.*, 1993). Five major topics were discussed:

1. the importance of groundwater/surface water ecotone fauna in the estimation of global diversity;
2. the local and regional biodiversity;
3. factors promoting biodiversity;
4. biodiversity in gradients and impact of disturbances;
5. how to preserve biodiversity.

IMPORTANCE OF GROUNDWATER/SURFACE WATER ECOTONE FAUNA IN THE ESTIMATION OF GLOBAL DIVERSITY

Groundwater fauna is largely ignored in the calculation of global biodiversity. The number of species of tropical rain forests is estimated to be 10 or 20 times higher than those known to science (Cairns, 1988). Groundwater diversity is less well known than the diversity of surficial species in tropical forests globally: the total biodiversity of groundwater fauna is certainly higher than current estimations (in the Stygofauna Mundi, for example – Botosaneanu, 1986) and these organisms may represent an important part of the global biodiversity.

This is greatly important if groundwater fauna play an active role in groundwater system functioning (Fig. 1). It is especially true for microbes (microfauna, bacteria, fungi) which are still more or less unknown and may play an important role in the bank filtration of large rivers (the River Rhine for example is rather well studied from this point of view).

Macrophytes and microphytes which occur in ground-water/surface water systems (both aquatic and terrestrial) should also be included in groundwater/surface water ecotone biodiversity. For example, in the Netherlands, wetland macrophyte biodiversity decreased during the past centuries mostly because of the exploitation of groundwater. There is a strong need for such studies which can be grouped as 'Ecohydrology'.

In the same way, the state of the knowledge is not equal for all countries. The groundwater organisms of Africa, Latin America, Eastern Asia and Australia are still poorly or not at all known. There is also a need for methodological progress: for instance, standardized quantitative methods which can be exported from one region to another.

Two other types of biodiversity are still poorly known and need research:

– the great 'phylogenetic biodiversity' in groundwater; many rare taxa are represented in the underground environment, they are sometimes representatives of very old phylogenetic lineages (considered as 'living fossils'; Fig. 2);
– the phenotypic biodiversity between connected populations or between isolated ones is also still poorly known.

THE LOCAL AND REGIONAL BIODIVERSITY

Groundwater fauna is characterized by the high number of species with a high degree of endemism. Most of them are limited to one catchment area or one karstic system. *Phreatocandona motasi*, for example, is only known from a single well in Romania (Danielopol, 1978) and *Antrocamptus catherinae* was described from a single population of a karstic stream in Southern France (Rouch, 1988).

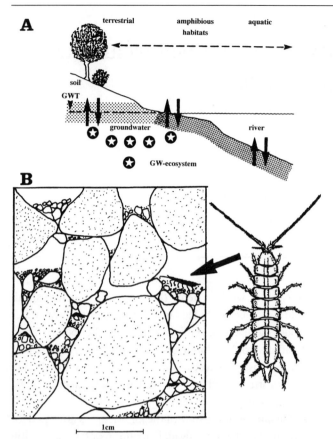

Fig. 1 The groundwater systems and their ecotones. Groundwater (GW) ecosystems of floodplain rivers connect through a variety of exchange zones to the surrounding soil and running water systems (A). The exchange of matter and organisms between the different systems (arrows) occurs in these ecotone/transition zones (stippled areas) between aquatic, amphibious and terrestrial biotopes (GWT: groundwater table below which the subsoil system is saturated with water). The stars symbolize the presence of various animal assemblages, dominated by exclusively hypogean dwelling species (in the subsurface waters). A diagrammatic view of a groundwater system (B) in the alluvial sediments of a floodplain shows an interstitial dwelling animal, the isopod *Proasellus sp.*, which through its feeding activity on fine sediments, bacteria and plant detritus, contributes to the ecological function of the interstitial ecosystem to which it belongs.

Many factors may influence the biodiversity in alluvial aquifers. Ward & Palmer (in press) proposed a list of factors that influence spatial distribution patterns of interstitial animals in alluvial aquifers (Table 1). Although not directly addressing biodiversity, the factors in Table 1 are responsible for structuring local and regional faunal assemblages which ultimately determine biodiversity patterns. The protection of groundwater fauna should consider these and other factors.

After all, there is a great need for articles on local measurements of biodiversity: data have to be published to be used and compared. These publications may also have great

Table 1. *Structure of interstitial faunal assemblages. Some of the variables are listed in the table below. The abiotic variables reflect geomorphic and hydrogeologic processes. Definitive data for assessing the importance of these factors in structuring biodiversity patterns are sparse, especially for the biotic variables (modified and simplified from Ward & Palmer, 1994)*

1 – Characteristics of the alluvium – particle size – particle size heterogeneity – pore size	3 – Exchange characteristics – hydraulic conductivity – clogging – water movement – oxygen concentration
2 – Food resource patterns – FPOM – bio-film – prey species	4 – Competition 5 – Predation

importance for local managers who may directly use them for the protection of groundwater fauna.

PROMOTING BIODIVERSITY STUDIES

Because of this high degree of endemism, the need for specialists on the systematics of each group of animals is increasing as the number of studies increases. This need is not supported by universities which have more and more problems finding students who want to invest their efforts in such an unpromising topic from the employment point of view. In a few years, who will identify the animals collected by stream and groundwater ecologists? There is also a need for taxonomic institutes, because this type of permanent structure may formalize training and pass that knowledge on to successive generations.

In the same way, there is also a great need for taxonomic books and keys to identify groundwater animals. These books may detail taxa at the species level (those intended for ecologists) or just consider the major groundwater high-level taxa (those to be used by managers).

The need for specialists and taxonomic books is not the only way of promoting studies of groundwater faunal biodiversity; lay people have to be made aware of the problem. For this, two main strategies are required: 1) development of academic education of students, by providing groundwater ecology courses in the university, and 2) development of the media awareness of groundwater fauna (expositions and T.V. films are very powerful tools).

Another important point is that environmental managers do not need large amounts of data. They need simple but efficient concepts to make decisions.

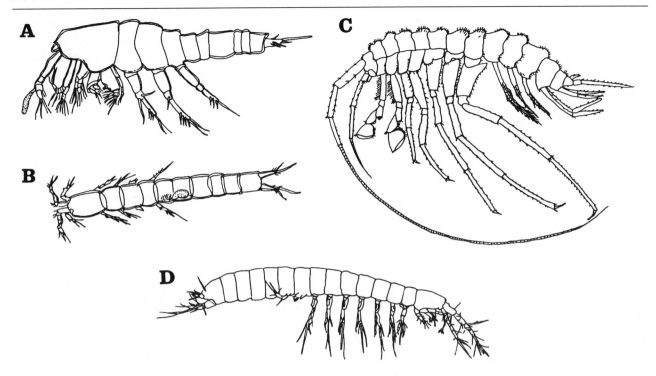

Fig. 2 Four examples of typical groundwater dwelling animals. Many groundwater dwelling species are crustaceans. For example, *Graeteriella droguei* Rouch & Lescher-Moutoue (A) is a minute copepod (0.35 mm length) discovered in a submerged karst system in Southern France. This species is a 'living fossil' of marine origin, a representative of a new order of Crustacea, the Gelyelloida, which is related to one of the dominant groups of copepods (Cyclopoida) currently living in surface waters. *Parastenocaris dianae* Chappuis (B) is also a minute copepod (0.45 mm length) that belongs to a highly diversified group (the Parastenocaridae, Harpacticoida), which inhabits all types of groundwaters around the world (a third of all the known exclusively dwelling hypogean harpacticoids, 441 species, belongs to this group). *Niphargus balcanicus* (C; 25 mm body length) is a remarkable blind crustacean, representative of the highly diverse, aquatic subterranean fauna of the Balkan Peninsula, in Europe. This amphipod occurs in the Vjternica Cave (Dinaric Karst). *Bathynella sp.* (D; about 1 mm body length) belongs to the most famous crustacean group commonly considered as 'living fossils': the Syncarida Bathynellacea. The present-day species of this group live in groundwater. They are related to fossil species that were common in surface waters during the Upper Palaeozoic (about 300 million years B. P.; Schminke, 1986). Redrawn, with permission, after *Stygofauna mundi* (Botosaneanu, Ed., 1986).

BIODIVERSITY IN GRADIENTS AND IMPACT OF DISTURBANCES

Western (1992) noted that the strategies one shall adopt for safeguarding the biodiversity depend on our ability to understand how communities are assembled and maintained. A lot of important information on the groundwater biodiversity can be obtained by studying gradients, because they offer theoretical templets to test general ideas on biodiversity patterns (Fig. 3).

For example, the role of disturbance on community structure can be efficiently tested in physical gradients. The 'intermediate disturbance hypothesis' (Connell, 1978), already verified in surface environments, could be easily tested on transversal gradients in large river floodplains (Dole & Chessel, 1986) or on a gradient of polluted wells (Notenboom, pers. comm.).

But the basic problem with this disturbance-biodiversity link is that environment managers and wider society do not want disturbed systems (such as rivers with natural flood regimes); they want static systems which are considered as less dangerous systems.

HOW TO PRESERVE GROUNDWATER BIODIVERSITY

There is, nowadays, a great danger of decreasing biodiversity resulting from destruction of groundwater animals by toxic chemicals. The ecological tolerance of groundwater species has to be studied. The choice of widely distributed groups of species seems to be more logical than utilizing endemic species. This allow facilitation of export of knowledge from one area to another. Representatives of the harpacticoid *Parastenocaris,* which are distributed world wide, form a good example.

Because groundwater animals often have long life cycles and low reproduction rates, there is also a need for 'long term

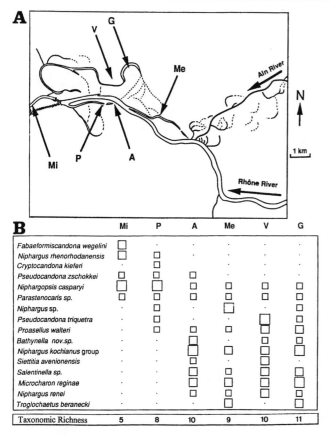

B	Mi	P	A	Me	V	G
Fabaeformiscandona wegelini	□	·	·	·	·	·
Niphargus rhenorhodanensis		□	·	·	·	·
Cryptocandona kieferi	·	□		·	·	·
Pseudocandona zschokkei	□	□	□	·	·	·
Niphargopsis casparyi	□	□	□	□	□	□
Parastenocaris sp.	□	□	□	□	·	□
Niphargus sp.	·	□		□	·	□
Pseudocandona triquetra	·	□	·	·	□	□
Proasellus walteri	·	□	□	□	□	□
Bathynella nov.sp.	·	·	□		□	□
Niphargus kochianus group	·	·	□	□		□
Siettitia avenionensis	·	·	□	·	□	
Salentinella sp.	·	·	□	□	□	□
Microcharon reginae	·	·	□	□	□	□
Niphargus renei	·	·	□	□	□	
Troglochaetus beranecki	·	·		□	·	□
Taxonomic Richness	**5**	**8**	**10**	**9**	**10**	**11**

Fig. 3 Spatial heterogeneity of the biodiversity. Interstitial milieu were previously considered as homogeneous compared to epigean biotopes, but are now recognized as heterogeneous, mainly because of the influence hydrology exerts on physical, chemical, and biological structure of the system. In the Rhone floodplain (A), the different hypogean species are distributed along a stability gradient (B): the interstitial fauna of areas close to the main channel consist mainly of ubiquitous stygobites (e.g. *Niphargus rhenorhodanensis, Niphargopsis casparyi*) whereas those of the floodplain's margin are dominated by phreatobites (e.g. *Microcharon reginae, Niphargus renei*). The resulting taxonomic richness of hypogean species assemblages increases with distance from the main channel – modified from Dole-Olivier *et al.* (1993).

toxicity tests', to assess population survival of groundwater animals during long periods in an environment with low pollution levels. These tests are especially important because of the low rate of recovery of groundwater communities.

River regulation has also a great impact on groundwater/surface water ecotone assemblages by modifying the dynamics of water exchange between the river and its aquifer. River regulation is not a marginal disturbance; most of the European rivers are strongly regulated and, in contrast to chemical pollution, there are no remediation techniques for river regulation.

But where does preservation of groundwater biodiversity have to be focused? 'Hot spot' areas for groundwater fauna biodiversity have to be defined at both local and global

scales. Some of them are already well known; at the local scale, interface zones located near the margin of alluvial plains can be an example; at the global scale, some karstic areas such as those of Slovania can be another example.

The protection of the groundwater/surface water ecotone fauna should be considered also by those planning natural reserves. See, for example, the proposals of Danielopol (1989) and Pospisil & Danielopol (1990). Finally, one should be remember that the conservation of the ecotonal biodiversities represents a multifacious topic where the utilitarian aspects can or should not be separated from cultural aspects. Similar arguments were addressed recently by one of the leading contemporary biologists (Stebbins, 1992).

CONCLUSION: GAPS IN KNOWLEDGE AND GUIDELINES FOR FUTURE RESEARCH

Seven main gaps in knowledge can be considered as recommendations for future studies.

1 The total number of groundwater species is far from being known, especially in poorly studied continents (Africa, Australia).

2 There is a need for international standardized sampling methods which could facilitate valuable inter-site comparisons

3 Available data on groundwater biodiversity have to be published.

4 There is also a need for taxonomists and books for identification of groundwater animals.

5 Education at both academic and popular levels has to be developed.

6 Ecological and ecotoxicological tolerance of groundwater fauna, acute and long-term, have to be studied for widely distributed taxa.

7 Finally, it would be efficient to define 'biodiversity hot spots' for groundwater/surface water ecotones, at both local and global scales.

UNESCO can contribute to the protection of groundwater/surface water ecotone biodiversity by supporting the publications of books (Points 3 and 4) and promoting the integration of groundwater ecology into education (Point 5).

REFERENCES

Botosaneanu, L. (1986). *Stygofauna Mundi, A faunistic distributional and ecological synthesis of the world fauna inhabiting subterranean waters*. Leiden: E.J. Brill.

Cairns, J. Jr. (1988). Can the global loss of species be stopped? *Speculations in Science and Technology*, **11**, 189–196.

Connell, J. H. (1978). Diversity in tropical rain forests and coral reefs. High diversity of trees and corals is maintained only a non-equilibrium state. *Science*, **119**, 1302–1310.

Danielopol, D. L. (1978). Uber Herkunft und Morphologie der Süsswasser-hypogäischen Candoninae (Crustacea. Ostracoda). *Ber. österr. Akad. Wiss. Math. Nat. Kl. Abt.*, **1**, 187, 1–162.

Danielopol, D.L. (1989). Groundwater fauna associated with riverine aquifers. *J. N. Am. Benthol. Soc.*, **8**, 18–35.

Dole, M.J. & Chessel, D. (1986). Stabilité physique et biologique des milieux interstitiels: cas de deux stations du Haut-Rhône. *Annls Limnol.*, **22**, 69–81.

Dole-Olivier, M. J., Creuzé des Châtelliers, M. & Marmonier, P. (1993). Repeated gradients in subterranean landscape – Example of the stygofauna in the alluvial floodplain of the Rhône River (France). *Arch. Hydrobiol.*, **127**, 451–471.

Marmonier, P., Vervier, Ph., Gibert, J. & Dole-Olivier, M.-J. (1993). Biodiversity in groundwaters: a research field in progress. *Trends Ecol. Evol.*, **8**, 392–395.

Pospisil, P. & Danielopol, D.L. (1990). Vorschläge für den Schutz der Groundwasserfauna im geplanten Nationalpark 'Donauauen' östlich von Wien, Österreich. *Stygologia*, **5**, 75–85.

Rouch, R. (1988). Sur la répartition spatiale des crustacés dans le sous-écoulement d'un ruisseau des Pyrénées. *Annls. Limnol.*, **24**, 213–234.

Schminke, H. K. (1986). Syncarida. In *Stygofauna Mundi. A faunistic distributional and ecological synthesis of the world fauna inhabiting subterranean waters*, ed. L. Botosaneanu, pp. 389–404. Leiden, E. J. Brill.

Solbrig, O.T. (1992). The IUBS-SCOPE-UNESCO Program of Research in Biodiversity. *Ecological Applications*, **2**, 131–138.

Stebbins, G.L. (1992). Why should we conserve species and wildlands? In: *Conservation Biology. The theory and practice of natural conservation, preservation and management*. ed. P.L. Fiedler & S.K. Jain, pp. 454–470, Chapman & Hall.

Ward, J.V. & Palmer, M.A. (1994). Distribution patterns of interstitial freshwater meiofauna over a range of spatial scales, with emphasis on alluvial river-aquifer systems. *Hydrobiologia*, **287**, 147–156.

Western, D. (1992). The biodiversity crisis: a challenge for biology. *Oikos*, **63**, 29–38.

Round Table 3 Modelling of flows at the interface

V. VANEK*, C. THIRRIOT**, A.M.J. MEIJERINK*** & G. JACKS****

* Université Lyon 1, URA CNRS 1974, Ecologie des Eaux Douces et des Grands Fleuves, Hydrobiologie et Ecologie

Souterraines, 43 Bd du 11 novembre 1918, 69622 Villeurbanne cedex, France

Present address: VBB VIAK Consulting Engineers, Geijersgatan 8, S-216 18 Malmö, Sweden

** Institut de Mécanique des Fluides, URA 0005 CNRS, Allée du Professeur Camille Soula, 31400 Toulouse, France

*** International Institute for, Aerospace Surveyard Earth Sciences (ITC), Enschede, 7500 AA, The Netherlands

**** Land and Water Resources, Royal Institute of Technology, S-10044 Stockholm, Sweden

Modelling the flows across the groundwater/surface water ecotone often concentrates around the following questions: (1) why should we model, (2) what kind of fluxes and scales should be considered in our models, (3) possibilities and difficulties when using models developed for other purposes, (4) communication gaps and future research needs.

(1) The modelling of ecotone-related fluxes may (and should) be used in the following cases:

- to test assumptions and hypotheses
- to improve our understanding and insight into the processes
- to design new experiments and to optimize sampling strategies
- to generalize and synthesize the results
- to develop scenarios and make predictions.

The development of scenarios is felt as one of most important tasks. The ecotones are subjected to the increasing pressure of various human activities (river flow regulation, groundwater extraction, irrigation, non-point pollution etc.), and we need more efficient tools to predict possible impacts of these activities and to improve the management practices. The situation is particularly critical in the third world where hundreds of millions of people live on the surface/groundwater ecotones, depending on them and affecting their function.

(2) The groundwater/surface water ecotones convey numerous fluxes – of water, heat, solutes, particulate matter, genetical information etc. (Gibert et al., 1990). All these fluxes occur on different spatial and temporal scales (Fränzle & Kluge, in prep.). With regard to the temporal variability, three main levels may be distinguished: (a) short-term scale (from diurnal or tidal cycles, snow melt or storm runoff generation to seasonal fluctuations), (b) 'steady state' (usually considered as a mean of several years' period), (c) long-term trends. Most important spatial scales are (a) vertical gradients sediment-water interface, vertical gradients driven by an equilibrium between advection, diffusion, dispersion and biogeochemical changes), (b) sampling-site scale (seepage patterns and similar gradients in a horizontal direction, large-scale heterogeneity structures), (c) catchment scale, (d) regional scale.

Usually, the different kinds of fluxes and the different scales will necessitate the use of different models or submodels. For example, modelling the diurnal variation in oxygen evapotranspiration or the response of an ecotone to a hydrological event will be quite different from modelling 'steady state' conditions or predicting long-term trends in ecotone permeability or buffering capacity.

(3) Most processes occurring within the surface water/groundwater ecotones may also be found in other ecotones or ecosystems, and some models developed there might readily be applicable on the ecotone-related fluxes. In general, however, the surface water/groundwater ecotones are characterized by higher dynamics and lower predictability than other similar systems. For example, the direction of flow or the ecotone boundaries may change almost instantaneously in response to sudden changes in surface water elevation, erosion or other processes.

In general, large surface water and groundwater systems can be modelled by what concerns the hydraulics, the heat flow and the biogeochemistry. Even if these systems usually are not homogeneous there is wide experience with the processing and averaging the data. The problems turn up when we try to apply these models on the ecotones which combine the characteristics of both adjacent systems and, in addition, often have a number of their own, ecotone-related characteristics. The downscaling of the large-scale models and the upscaling of the point measurements is further complicated by the fact that the field data usually are not homogeneous – even if they may be sampled at the exactly same point and time, some of the values such as hydraulic head or flow velocity are instantaneous while others such as soil chemistry or a number of organisms are integrated over different scales of time or space.

Recently, several models oriented on the running water/groundwater ecotones have been presented (Hunt *et al.*, 1993; O'Brien & Hendershot, 1993; Runkel & Broshears, 1991; Schälchli, 1992; Schenk & Poeter, 1990). Modelling interrelationship between lakes and groundwater seems less diversified (Winter, 1984; Lijklema, 1993; Fränzle & Kluge, in prep.).

When choosing the optimal size and complexity of the model, the following two factors should be considered.

(a) The number of in-data needed to model various processes within the ecotone is directly proportional to the ecotone heterogeneity in space and time. Therefore, the heterogeneity has to be considered in an early stage when choosing sampling sites and developing the models (Vanek, in prep.). Sensitivity tests of models should also be performed regularly to focus the efforts on the critical 'hot spots' in time or space.

(b) The number of processes which we want to model, and our knowledge about them. For example, the fluxes of reactive solutes, particulate matter or ecological fluxes are closely related or depend directly on the fluxes of water, non-reactive solutes or heat. The relationships between the various fluxes, however, often are difficult to describe or poorly known.

The increasing complexity of a model usually means an exponentially increasing need of high quality in-data, and an increasing number of assumptions with regard to initial and boundary conditions which seldom can be fulfilled. Therefore, it is sometimes wise to use simple, tailor-made empirical models which may only be valid within a narrow experimental range (Lijklema, 1993).

(4) There is a continuous need for closer co-operation between different ecotone-interested researchers such as ecologists, hydrologists and hydrogeologists. It is believed that multi-disciplinary meetings, workshops and projects will improve our understanding of ecotone functioning. Another communication gap which should be considered is that between the modellers and the scientists making observations. The modellers have difficulties in getting the right parameters and may sometimes not know what parameters are really important. The scientists observing the natural

phenomena may on the other hand not be focused on the key processes, the knowledge of which is needed for modelling. A statement which might be helpful in closing the gap is that we all have models in our mind when trying to treat a problem – simple conceptual ones or more complicated mathematical ones. Education in ecology for modellers and in modelling for ecologists (or, perhaps, ecotonologists) may be very helpful in closing the gap.

Some of the future research needs within this field are:

– new field methods and evaluation tools to describe in a realistic way the spatial and temporal heterogeneity of the ecotones, particularly within the initial stages of a project to design measurement and sampling strategies;
– the improvement of theoretical models of upscaling and averaging point data, and testing the possibilities of generalizing the results obtained in other environments or environmental conditions. For example, it is believed that the studies of strongly human-influenced ecotones with very steep hydraulic or other gradients may contribute significantly to our understanding of the rate limits and buffering capacity of both natural and human-influenced ecotones;
– increased effort is needed to develop long-term, predictive models which might serve as management tools.

REFERENCES

Fränzle, O. &. Kluge, W. (1996). Typology of water transport and chemical reactions in groundwater/lake ecotone. *Proc. Int. Conf. on Groundwater/Surface Water Ecotones*, Cambridge University Press, This volume.

Gibert, J., Dole-Olivier, M-J., Marmonier, P. & Vervier, P. (1990). Surface water- groundwater ecotones. In *The ecology and management of aquatic-terrestrial ecotones*, ed. R.J. Naiman & H. Décamps, pp. 199–225, Parthenon Publ.

Hun, J.R., Hwang, B-H. & McDowell-Boyer, L.M. (1993). Solids accumulation during deep bed filtration. *Environ. Sci. Technol.*, **27**, 1099–1107.

Lijklema, L. (1993). Considerations in modelling the sediment-water exchange of phosphorus. *Hydrobiologia*, **253**, 219–231.

O'Brien, C. & Hendershot, W.H. (1993). Separating streamflow into groundwater, solum and upwelling flow and its implications for hydrochemical modelling. *J. Hydrol.*, **146**, 1–12.

Runkel, R.L. & Broshears, R.E. (1991). *One-dimensional transport with inflow and storage (OTIS): A solute transport model for small streams.* Univ. of Colorado, CADSWES Technical Report 91–01, 85 p.

Schälchli, U. (1992). The clogging of coarse gravel river by fine sediments. *Hydrobiologia*, **235/236**, 189–197.

Schenk, J. & Poeter, E. (1990). *RIVINT – An improved code for simulating surface/groundwater interactions with MODFLOW.* Colorado Water Resources Research Institute, Completion Report 155.

Vanek, V. (1996). Heterogeneity of groundwater/surface water ecotones. *Proc. Int. Conf. on Groundwater/Surface Water Ecotones*, Cambridge University Press, This volume.

Winter, T. (1984). Modelling the interrelationship of groundwater and surface water. In *Modelling of Total Acid Precipitation Impact*, ed. J.L. Schnoor, pp. 89–119, Boston, Acid Precipitation Series Vol. 9, Butterworth Publ.

Round Table 4 Contribution of the groundwater/surface water ecotone concept to our knowledge of river ecosystem functioning

PH. VERVIER*, M. H. VALETT**, C. C. HAKENKAMP*** & M.-J. DOLE-OLIVIER****

* CERR – CNRS, 29 rue Jeanne Marvig, 31055 Toulouse, France

** Department of Biology, University of New Mexico, Albuquerque, New Mexico 97131, USA

*** Department of Zoology, University of Maryland at College Park, 1200 Zoology-Psychology Building, College Park, MA 20742–4415, USA

**** HBES, URA 1974, UCB Lyon 1, 43 Bd du 11 novembre 1918, 69622 Villeurbanne cedex, France

The aim of this work was to examine the contribution of the groundwater/surface water (GW/SW) ecotone concept towards a better understanding of ecosystem functioning in running water systems. It became clear that a consideration of the differences and similarities between the GW/SW ecotone and the hyporheic zone (HZ) concepts would be helpful. Four major parts could be pointed out. The first part focuses on a comparison between the GW/SW ecotone concepts. The second part emphasizes how the interaction zone between GW and SW influences ecosystem functioning in the two adjacent systems. The third part suggests important topics for future research, and finally recommendations for UNESCO and for managers of river ecosystems are made in the fourth part.

ECOTONE VERSUS HYPORHEIC ZONE

Origin of terms

The hyporheic zone was described in Germany 30 years ago (Orghidan, 1959; Schwoerbel, 1964) but has only recently flourished in the vocabulary of North American stream ecologists (e.g. Williams, 1984; Stanford & Ward, 1988; Triska et al., 1989). Though the term originally associated with a strong biological bias, it is now used in a more general manner to describe the deep sediments of stream beds where GW/SW exchange (see Valett et al., 1993; Hakenkamp et al., 1993).

The ecotone concept is much older since Clements (1904) used this term to describe contact zones between adjacent communities. The ecotone concept also used by Leopold (1933) and Odum (1971), had an initial organismal or com-

munity emphasis. However in 1988, Holland provided a functional interpretation by emphasizing all exchanges (i.e. water flow, biotic and abiotic fluxes) between adjacent systems. The GW/SW ecotone concept, initially presented by Janine Gibert in May of 1988 at the First International Workshop of Land/Water Ecotones in Sopron, Hungary, has been developed with this later perspective that emphasizes exchanges.

These exchanges between adjacent ecosystems create a zone of interactions which defines the ecotone. To describe the interactions between adjacent ecosystems, it is necessary to define the interface that separates these ecosystems. This has been accomplished by determining the main features distinguishing underground and surface systems, the presence of permanent darkness, and the fact that sediments, or fractures within karstic systems, control water flow (Gibert et al., 1990; Vervier et al., 1992). Having described the interface between light and darkness or between water and sediment, exchanges which occur through this interface create interactions that define a zone of interactions, i.e. the ecotone (Fig. 1).

Comparison of the concepts

The main difference between the ecotonal and hyporheic concepts is linked to the way that each is studied. The hyporheic zone (HZ) is first located, then processes and exchanges which occur within this zone are studied. Thus processes in the HZ are often studied alone without consideration of how these processes relate to or affect adjacent systems. On the other hand, ecotones due to their very definition are identified by where GW/SW systems interact and are always intrinsically connected to these adjacent systems. These

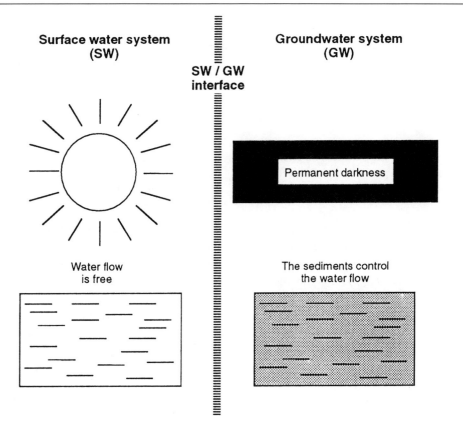

| Surface water system (SW) | SW / GW interface | Groundwater system (GW) |

Fig.1 To describe the interactions between adjacent ecosystems, it is necessary to define the interface that separates these ecosystems. This has been accomplished by determining the main features distinguishing underground and surface systems, the presence of permanent darkness, and the fact that sediments, or fractures within karstic systems, control water flow.

conceptual approaches entail self-limitations, both conceptual and practical (Table 1).

– Concerning the HZ, since it is first located by definition, some important interactions may be excluded. For example, upwelling of true groundwater at the tail of gravel bars is not, by definition, part of the hyporheic zone, but its implications for river functioning are potentially great.

– Concerning the ecotone, since its definition is based upon interactions, it is problematic to study the permeability of the ecotone to certain entities involved in the definition (i.e., the argument becomes circular). For example, if we define the ecotone by the simultaneous presence of epigean and hypogean macroinvertebrates, it is difficult to measure ecotone permeability to epigean species.

It can be concluded that despite their differing histories and applications, these two concepts are complementary and could be used with equal effectiveness, but with some preference according to context. Indeed, to specify the location of many studies, it is easier to use the HZ term since it is more

spatially intuitive. To emphasize the interactions between these two very contrasted systems (GW +SW), it is, perhaps, better to use the ecotone concept.

Value of the ecotone concept

Considering the differences and the similarities between the two concepts, it became clear that in addition to the fact that this zone of interactions has great influence on the functioning of surface water systems, and it is also important to the functioning of groundwater systems. Historically, the HZ concept has focused primarily on interactions with SW systems. Thus, one of the most important points of the ecotone concept is that it strongly emphasizes links between the adjacent SW and GW ecosystems. Consequently, it seems appropriate to answer an opening query by stating that an important contribution of the GW/SW ecotone concept is the recognition of the bi-directional, coupled functioning of river ecosystems and groundwater ecosystems.

ROLE OF GW/SW INTERACTIONS TO THE FUNCTIONING OF ADJACENT SURFACE WATER AND GROUNDWATER ECOSYSTEMS

Exchanges that occur in the zone of GW/SW interactions (Fig. 2) are strongly influenced by hydrological processes

Table 1. *Concepts and limitations of the hyporheic zone and ecotonal perspectives to groundwater/surface water contact zones*

GW/SW Ecotone	Hyporheic zone
Concept	*Concept*
1 – Interactions are studied around the GW/SW interface	1 – The HZ is defined in space
2 – The ecotone can be delineated	2 – Interactions are studied within this zone
Limitations	*Limitations*
Since the 'Ecotone' is defined by the presence of a given flux, then it is impossible to study the permeability of the ecotone to this particular flux	More static
	underestimation of exchanges between GW and SW
	Unidirectional perspective (SW to GW) often taken

(Gibert *et al.*, 1990; Hakenhamp *et al.*, 1993). Indeed, the hydrology of both the adjacent systems determines the direction and flux of water flow which strongly controls the functioning of the ecotone (Vervier *et al.*, 1992). The role of the ecotone will change according to the direction of water flow. For example, when groundwater is discharging into a river, the ecotone which takes place under riparian forest may protect the river against overloads of nitrate by increasing denitrification processes (Haycock *et al.*, 1993). When infiltration of surface water within the deep sediments occurs, the ecotone (i.e., HZ in this case) will influence metabolism and nutrient cycling of the river (e.g., Grimm & Fischer, 1984; Triska *et al.*, 1989).

When functioning of rivers is addressed, it is now recognized that interactions between groundwater and surface water have to be considered. Concerning groundwater systems, the consequences of interactions between hydrology and ecotone processes for the functioning of these systems are more known in karstic systems (Vervier & Gibert, 1991) than in alluvial river valleys. In both cases, i.e., for the functioning of groundwater and surface water systems, it is most important to determine hydrologic exchanges first, then the implications of associated fluxes on biotic and biogeochemical physicochemical variables can be addressed.

recognized as discrete entities that need to be quantified and understood.

Future research should include two types of comparative efforts. First, researchers should focus on 'natural' variation in exchange. Studies could be designed to assess differences or similarities in the structure and functioning of biotic communities in geomorphic settings with differing patterns of GW/SW exchange, such as studies undertaken on the Rhône River (e.g., Creuzé des Châtelliers & Reygrobellet, 1990) or proposed longitudinally by Stanford & Ward (1993). Further, studies should focus on GW/SW exchange and its influence on fundamental processes of stream metabolism (i.e., decomposition, primary production). The second type of comparative study should focus on the effects on anthropogenic impacts on exchange between groundwater and surface water systems. It should be assumed that management practices such as impoundment, channelization and organic enrichment due to sewage disposal decrease the extent and functional efficiency (e.g., purification, filtration . . . processes) of GW/SW ecotones. The natural abundance or extent of GW/SW ecotones is just now being assessed and the consequences of loss or alteration of ecotones may be best envisioned by comparative studies of impacted and less altered rivers.

FUTURE RESEARCH

Due to the complexity of biotic and abiotic exchanges that occur in the GW/SW interaction zone, it is important that future studies have an interdisciplinary nature (i.e., must include hydrobiology, hydrogeology, geomorphology, biogeochemistry, microbiology). Future efforts will require coordination among these various researchers to resolve the complexity of GW/SW interactions.

The role of GW/SW interactions could be better understood by comparative studies among, and within, streams or rivers. These studies should be conducted to emphasize the differences but also the similarities observed between zones of interactions. With this perspective, interaction zones are

RECOMMENDATIONS FOR UNESCO

Education

The importance of integrating the GW/SW ecotone concept into education concerning the management of aquatic systems should be underlined. Incorporation of the GW/SW ecotone concept into traditional classes of limnology will result in increased appreciation of groundwater as a representative and important freshwater ecosystem. Indeed, it should be emphasized that these interactions are important both for rivers and associated groundwaters environments, and thus also important to the long-term quality of freshwater resources.

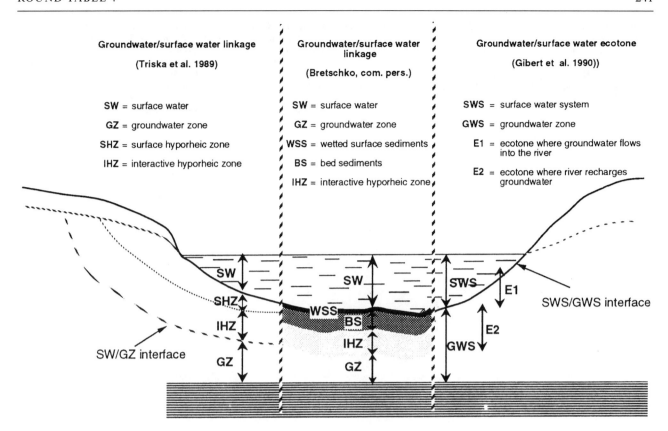

Fig. 2 Ecotonal and hyporheic concepts are used to describe
linkages between groundwater and surface water systems.

Management issues

The role of GW/SW ecotones in the self-purification potential of rivers should be emphasized to management per-
sonnel. This will be more efficiently accomplished if running
water ecologists recognize and begin to research this potentially crucial aspect of GW/SW ecotones. Many running-
water systems in Europe and the USA are the subject of
aquatic restoration programs and SW/GW interactions
should be integrated into such long-term studies (c.f.
Osborne *et al.,* 1993).

ACKNOWLEDGEMENTS

We would like to thank the congress coordinators, especially
Janine Gibert, for the opportunity to hold the discussion that
we have summarized above. The contents of the summary
reflect the ideas and suggestions of a number of interested
and insightful participants.

REFERENCES

Clements, F.C. (1904). *Research methods in ecology.* University
Publishing Co, Lincoln, Nebraska, USA.

Creuzé des Châtelliers, M. & Reygrobellet, J.L. (1990). Interactions
between geomorphological processes, benthic and hyporheic communities: first results on a by-passed canal of the French Upper Rhône
River. *Regulated Rivers,* **5,** 139–158.

Gibert, J., Dole-Olivier, M-J., Marmonier, P. & Vervier, P. (1990).
Surface water- groundwater ecotones. In *The ecology and management
of aquatic-terrestrial ecotones*, ed. R.J. Naiman & H. Décamps,pp.
199–225, Parthenon Publ.

Grimm, N.B. & Fisher, S.G. (1984). Exchange between surface and
interstitial water: implications for stream metabolism and nutrient
cycling. *Hydrobiologia,* **111,** 219–228.

Hakenkamp, C.C., Valett, H.M. & Boulton, A.J. (1993). Perspectives on
the hyporheic zone: integrating hydrology and biology. Concluding
remarks. *J. N. Am. Benthol. Soc.,* **12,** 1, 94–99.

Haycock, N.E., Pinay, G. & Walker, C. (1993). Nitrogen retention in
river corridors: European perspective. *Ambio,* **22,** 340–346.

Holland, M.M. (1988). SCOPE/MAB technical consultations on landscape boundaries: report of a SCOPE/MAB workshop on ecotones.
Biology International, Special Issue, **17,** 47–106.

Leopold, A. (1933). *Game management.* Scriber, New York, USA.

Odum, E.P. (1971). *Fundamentals of ecology.* third edition. W.B.
Saunders Company, Philadelphia, PA.

Orghidan, T. (1959). Ein neuer Lebensraum des unterirdischen Wassers,
das hyporheische Biotop. *Arch. Hydrobiol.,* **55,** 15–33.

Osborne, L.L., Bayley, P.B. & Higler, L.W. (1993). Lowland stream restoration: theory and practice. *Freshwater Biology, special issue,* **29,** 2, 342 p.

Schwoerbel, J. (1964). Die beduntung des Hyporheals für die
Lebensgemeinschaft der Fliessgewässer. *Verhandlungen der
Internationalen Vereinigung für Theoretische und Angewandte
Limnologie,* **15,** 215–226.

Stanford, J.A. & Ward, J.V. (1988). The hyporheic habitat of river ecosystems. *Nature,* **335,** 64–66.

Stanford, J.A. & Ward, J.V. (1993). An ecosystem perspective of alluvial
rivers: connectivity and the hyporheic corridor. *J. N. Am. Benthol.
Soc.,* **12,** 1, 48–60.

Triska, F.J., Kennedy, V.C., Avanzino, R.J., Zellweger, G.W. & Bencala, K.E. (1989). Retention and transport of nutrients in a third-order stream in northwestern California: hyporheic processes. *Ecology,* **70**, 1893–1905.

Valett, H.M., Hakenkamp, C.C. & Boulton, A.J. (1993). Perspectives on the hyporheic zone: integrating hydrology and biology. Introduction. *J. N. Am. Benthol. Soc.*, **12**, 1, 40–43.

Vervier, Ph. & Gibert, J. (1991). Dynamics of surface water/groundwater ecotones in a karstic aquifer. *Freshwater Biology*, **26**, 241–250.

Vervier, Ph., Gibert, J., Marmonier, P. & Dole-Olivier, M.J. (1992). A perspective on the permeability of the surface freshwater-groundwater ecotone. *J. N. Am. Benthol. Soc.*, **11**, 1, 93–102.

Williams, D.D. (1984). The hyporheic as a habitat for aquatic insects and associated arthropods. In *The ecology of aquatic insects*, ed.Resh V.H. & Rosenberg D.M., pp. 430–455, Praeger Publisher, New York.

Round Table 5 Groundwater/surface water interface and effective resource management

J. SIMONS* & J. NOTENBOOM**

*Environmental Protection Agency, 401 M Street SW, Washington DC 20460, USA

**RIVM Laboratory of Ecotoxicology, P.O. Box 1, Bilthoven 3720 BA, The Netherelands*

The scientific community is increasingly interested in becoming involved in management of groundwater resources and of related surface water and riparian ecosystems impacted by groundwater/surface water interactions. Researchers want more control over how the results of their work are applied and they want to be more involved in the important task of environmental protection. They also want to have more influence on research funding. In fact, many scientists feel their involvement is necessary for their work to continue receiving support and funding. This new involvement of researchers is welcomed by enlightened water resources managers. Managers are under increasing pressure to balance protection policies with an array of other needs. They realize that the most effective management must be based on the best scientific information. Because of these factors, scientists are more frequently devoting a portion of their programmes to management issues (Stanford & Simons, 1992). However, groundwater management is complex, dealing with scientific as well as social, economic and political issues. There are many problems that make solutions difficult, and progress is often slow. In spite of this, scientists and managers must continue to push for a more co-ordinated approach because protection of our vital groundwater resources needs the involvement of everyone who can help.

GOAL STATEMENT

To base the management of groundwater resources on the best available scientific information in order to achieve maximum ecological benefits. By doing this, groundwater will also be protection as a safe source of drinking water.

A list of actions is needed in order to achieve this goal, and how these actions could be implemented. These actions are really the basis for developing a strategic plan of action. Such a plan should address the many issues involved in co-ordination of science and management and provide practical guidance to achieve objectives. The development of a plan, which would also require wide scientific and management review, was beyond the scope of this discussion.

EXAMPLES OF PROBLEMS THAT PREVENT ACHIEVEMENT OF THE OVERALL GOAL

• Lack of appropriate laws to allow an ecosystem approach. This lack was a recurring theme. While much is often achieved without statutory authority, there are situations where enforcement is needed. Where laws do exist, they need to form an integrated, national approach for effective management of resources.

• Lack of political will. There may be no political will to promote wise use of resources and protect the environment, even though laws may or may not be present.

• Lack of funding. This is a major problem that is expected to continue. A comprehensive strategy could provide guidance and ideas on how scientists can better work within funding constraints to influence management. A secondary benefit is that scientists could be more aware of priority research needed by management and this knowledge could help scientist to formulate and describe their research to compete more successfully for limited funds.

• Lack of scientific information. At the present time, there is a lack of empirical data on which to base management policies. For example, predictive models linking hydrological and biological information would be a useful tool. A strategy should help direct research to fill knowledge gaps.

243

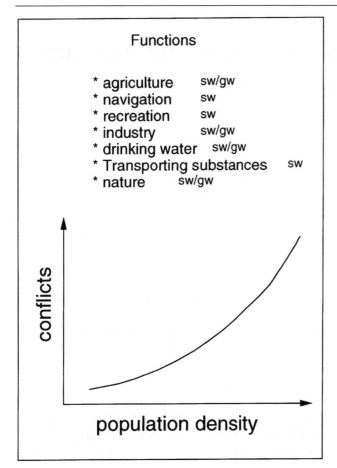

Fig. 1 The different functions that can be attributed to surface water and groundwater; and a schematic presentation of the relation between conflicting uses and population density within a certain area.

• Difficulty of communication between scientists and water managers. Efforts to provide a forum for scientists and managers to talk are ways this problem can be mitigated.

EXAMPLES OF HOW SCIENTISTS CAN BETTER SOLVE THESE PROBLEMS

• Promote significance of their findings in a policy or regulatory framework.
• Explain to non-scientists how parts fit together, using non-technical words when possible.
• Look at their research to explore ways that it forms part of a larger system or can be expanded to do this.
• Join with other organizations to gain influence and gain maximum benefits from limited resources.
• Promote understanding and acceptance at all political levels, especially at the local level. This is especially useful if laws are not adequate.

• Consolidate local level organizations to bring changes at higher levels of government.
• Because groundwater is out of sight, it can more easily be ignored in favour of more obvious problems. Scientists need to make themselves visible and heard so that the political structure is adequately aware of the need to protect groundwater resources.
• Use news media to advantage; videos are especially effective for use on television.
• Promote an integrated approach to water protection, viewing efforts in terms of the entire hydrological cycle, including groundwater, The watershed approach is probably the best way of doing this, but groundwater must be included.

METHODS AND TOOLS

Groundwater and surface waters are used for a variety of purposes (e.g. drinking water, industrial and agricultural). In more densely populated areas, the intensity of conflicting uses attached to a water system will increase as the number and scope of these uses increase (Fig. 1). It is the job of management to resolve conflicts over competing uses and, especially in densely populated areas, they should base management practices on scientific information regarding the interrelationships among quantitative, qualitative and ecological aspects of water systems. Management strategies can be grouped according to their bases in these relationships (Fig. 2). The most conservative strategy is preservation management with limited objectives to prevent further pollution of water resources. This type of management demands minimal integration between qualitative and quantitative aspects and generally ignores ecology as a primary influence. More sophisticated management tools are required if the aim is to restore ecological functions while preserving other uses.

It is clear that many scientists recommend that groundwater management use a drainage basin or hydro-system approach. This implies an integration of water, land use, and groundwater management. A legislative context, in the sense of a set of coherent laws forming the umbrella for such a strategy, is lacking in most countries. The legislative context needs to be designed so that differentiation at lower political levels is possible and that optimal safeguards for specific local situations are guaranteed. Scientists need to provide managers with appropriate information and scientifically based tools to assist them in coping with the complexity of the ecosystem in its socio-economic context. Examples of widely applicable tools are:

1) Hydrological Facies Analysis (Styfzand, 1989) and similar approaches coupling hydrological and ecological models on national and regional scales;

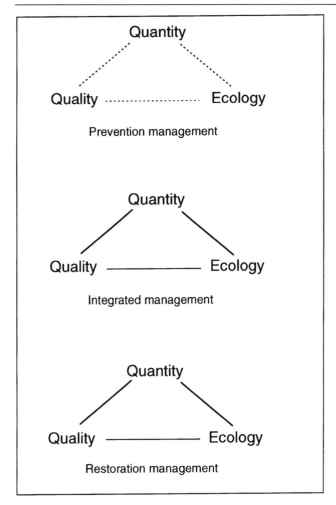

Fig. 2 Three different approaches in water management and the strength of interaction among quantitative, qualitative, and ecological knowledge on which these strategies are based. SW = surface water; GW = groundwater.

2)   the use of ecological risk assessments (EPA, 1992; Suter, 1993). This general approach emphasizes protection of the ecological integrity of a system and is also applicable to other groundwater/surface water problems.

Applied environmental researchers can support environmental managers and policy makers in three main ways:

a)   by considering how all phases of their work results may be incorporated in improved management;

b)   by being aware how results from a particular study can be extrapolated to other areas, and

c)   by realizing that management tools are by definition based on interdisciplinary research.

A point of concern in developing scientifically sound management tools for groundwater is the limited number of scientists involved in these types of studies. Therefore, it is recommended that UNESCO and other international organizations take concerted actions to stimulate more

research in groundwater ecology for the purpose of developing tools for management.

In management strategies related to the protection of groundwater, it was evident that the surface water/groundwater ecotones must be viewed in a wider perspective that crosses boundaries into surface water and terrestrial ecosystems.

The following two questions should be asked.

How can UNESCO help achieve the best possible science?

The science of groundwater/surface water ecology is a relatively new area of study that is growing rapidly, partly because it strives to gain knowledge on one of the most vital and threatened elements of human existence – a safe source of water. We feel that it would be highly advantageous to the management of this resource and to the development of a comprehensive body of scientific knowledge if there were an overall strategic plan. This plan would provide a number of benefits, including:

- provide scientists with guidance to deal with the problems that exist;
- help find solutions to these problems;
- better answer the needs of managers;
- help make the importance of this resource known to the average citizen;
- help gain support and funding for research;
- help scientists compete for research funds;
- assist communication between scientists and non-scientists;
- promote this new field of science;
- provide better protection of groundwater resources.

UNESCO should be urged to support development of an overall strategic plan. This plan must be more than a paper exercise. It must be developed in co-operation with leaders of the scientific community and water resource managers in order to have wide acceptance. The plan should contain specific implementation steps in order to ensure the recommendations are followed so that the objectives are actually achieved. We feel such a plan could make a very positive contribution to achieving the best possible science in this new field.

How can science make the quality of life better? How can the results of science be applied to improving living conditions around the world?

A lack of a safe source of water for drinking and basic hygiene is a major cause of human misery around the world. The need to protect groundwater resources grows each day in proportion to the growth in world population and the loss of aquifers through misuse. It is becoming apparent that to protect groundwater resources as a sustainable source of drinking water requires the protection of natural ecosystems that support it. These systems are best protected by integrated management practices based on the best possible science. This, in turn, is best done by having an overall strategic plan as described above. The advancement of this science and the application of its principles are too important to be left to issues sorting themselves out. Again we call for UNESCO to support the development of a strategic plan that is endorsed by both parties, the scientists and the managers.

REFERENCES

EPA (US Environmental Protection Agency). (1992). *Framework for Ecological Risk Assessment*. Risk Assessment Forum. EPA/630/R92/001. US Environmental Protection Agency, Washington, DC. 20460.

Stanford, J.A. & Simons, J.J. (Eds.). (1992). *Proceedings of the First International Conference on Groundwater Ecology*. American Water Sources Association, Bethseda, ML, U.S.A.

Stuyfzand, P.J. (1989). A new hydrochemical classification of water types. *IAHS-Publ.*, **182**, 89–98.

Suter, G.W. (1993). *Ecological Risk Assessment*. Lewis Publ., Chelsea, MI, U.S.A.